Society on the Line

Information Politics in the

Information and Communication Technologies

Visions and Realities

Edited by
William H. Dutton

A companion volume to *Society on the Line*, this edited collection illuminates the social and economic implications of advances in information and communication technologies. Into its second printing, reviewers have judged it as:

'. . . a coming of age for the social-science study of communications and information technology.'

Colin Scott, the *Times Higher Education Supplement*

'. . . the most important collection in print on the topic of communication technologies.'

Donald Case, *Information, Communication and Society*

'. . . the sort of book that will appeal to anyone seriously interested in ICTs and their pervasive influence in all advanced societies. Who can resist the considered views of leading researchers in such an appealing form . . . ?'

Frank Webster, *Oxford Brookes University*

'. . . a pleasure to read and wonderfully presented, . . . social science research can "inform" public decisions . . . This work succeeds fully in doing so . . .'

Thierry Vedel, *Communication & Strategies*

ISBN 0-19-877459-1
ISBN 0-19-877496-6 (pbk)

Society on the Line

Information Politics in the Digital Age

Written and Edited by William H. Dutton

With the assistance of Malcolm Peltu and with essays by Margaret Bruce, Martin Cave, James Cornford, Rod Coombs, Nicholas Garnham, Andrew Gillespie, Leslie Haddon, David Knights, Dale Littler, Donald MacKenzie, Robin Mansell, Ian Miles, Vincent Porter, Charles Raab, Ranald Richardson, Roger Silverstone, David Stout, John Taylor, Juliet Webster, Robin Williams, Hugh Willmott, and Steve Woolgar

A Synthesis of Research Based on Britain's Economic and Social Research Council Programme on Information and Communication Technologies (PICT)

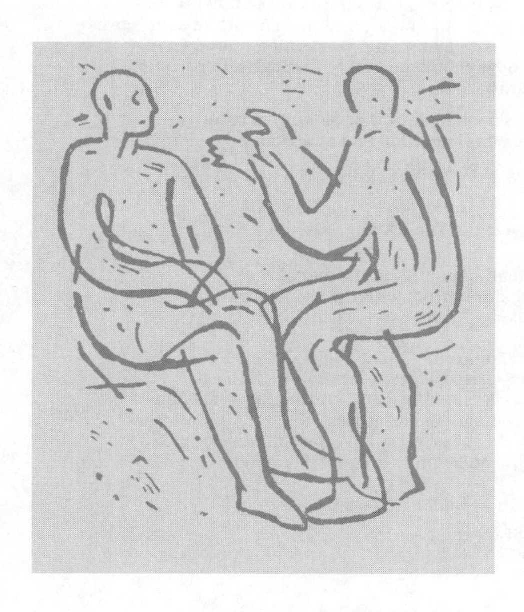

OXFORD
UNIVERSITY PRESS

Great Clarendon Street, Oxford OX2 6DP

Oxford University Press is a department of the University of Oxford.
It furthers the University's objective of excellence in research, scholarship,
and education by publishing worldwide in

Oxford New York

Athens Auckland Bangkok Bogotá Buenos Aires Calcutta
Cape Town Chennai Dar es Salaam Delhi Florence Hong Kong Istanbul
Karachi Kuala Lumpur Madrid Melbourne Mexico City Mumbai
Nairobi Paris São Paulo Singapore Taipei Tokyo Toronto Warsaw

with associated companies in Berlin Ibadan

Oxford is a registered trade mark of Oxford University Press
in the UK and in certain other countries

Published in the United States
by Oxford University Press Inc., New York

The opinions expressed in this book are those of the authors and do not necessarily reflect those of the UK Economic and Social Research Council or the Programme on Information and Communication Technologies.

First published 1999

British Library Cataloguing in Publication Data

Data available

Library of Congress Cataloging in Publication Data

Dutton, William H., 1947–
Society on the line: information politics in the digital age/
written and edited by William Dutton; with the assistance of
Malcolm Peltu and with essays by Margaret Bruce . . . [et al.].
p. cm.
Includes bibliographical references.
1. Information technology—Social aspects. 2. Information
technology—Political aspects. 3. Information technology—Economic
aspects. 4. Telecommunication—Social aspects.
5. Telecommunication—Political aspects. 6. Telecommunication—
Economic aspects. I. Peltu, Malcolm. II. Bruce, Margaret.
III. Title.
HM221.D88 1998 303.48'33—dc21 98–29516

ISBN 0–19–877461–3 (Hbk)
ISBN 0–19–877460–5 (Pbk)

10 9 8 7 6 5 4 3 2

Typeset in 11/13 pt Dante by Graphicraft Limited, Hong Kong
Printed in Great Britain on acid-free paper by
Bookcraft (Bath) Ltd, Midsomer Norton, Somerset

Preface

This book presents a new way of thinking about the social and economic implications of the revolution in information and communication technologies (ICTs). It offers a critical perspective on information politics in the digital age—the way social and technical choices about ICTs influence access to information, people, services, and technologies themselves. I have called this process 'the shaping of tele-access' and this book shows how this concept challenges prevailing theoretical perspectives on the information and communication revolution. It also proposes a framework for policy and practice, for the shaping of tele-access is directly connected to choices in everyday life, to strategies in the workplace, to the application of ICTs in government, education, and the household, and to issues of public policy.

My focus on the social shaping of tele-access evolved from an effort to synthesize a decade of research undertaken by the UK's Programme on Information and Communication Technologies (PICT). Supported by grants from the Economic and Social Research Council (ESRC) of Britain, PICT was one of the most ambitious social research initiatives in Europe to be focused on the role of ICTs in social and economic development (see Appendix).

PICT involved numerous social scientists from different disciplines—geographers, management scientists, political economists, and sociologists—at six university research centres across the UK. They published dozens of books and hundreds of articles based on their PICT research, contributing many new concepts and themes to the literature, and influencing contemporary understanding of the social shaping and impacts of ICTs.

The wide range and significance of PICT research led to a programme-wide synthesis during the final years of the programme. I joined PICT as its national director at the launch of this final-synthesis phase. One of my major responsibilities was to determine if the programme as a whole added up to more than the sum of its parts. This goal led me on a search for an integrative concept or theme that could encapsulate the central contributions across the full range of PICT research.

I needed an all-embracing concept that would achieve three related objectives. First, it was important to synthesize and extend all the most important themes of PICT research, and not focus on one particular stream of work. Secondly, I wanted a concept that applied well to the full range of ICTs, from business networks to the mass media, as well as the many different arenas of ICT adoption and use. These arenas, most of which were covered by PICT, include such different social contexts as the business and management use of ICTs, and the ways ICTs are used in government, education, and the household. Finally, it was critical that this concept should signify the opportunities as well as the dangers—the benefits as well as the risks—of the development and diffusion of ICTs.

I found this all-embracing theme in the shaping of tele-access. Having worked with this idea, I believe it can help integrate the findings of research on ICTs across the social sciences generally. Many social and economic issues—ranging from issues of information inequality, privacy, and censorship, to the role of the Internet and information superhighways in economic development—can be better understood if viewed as products of a process that is quite literally reshaping social and economic access in this digital age of new ICTs.

The main narrative of this book is my personal synthesis of research and focuses on the process of shaping tele-access. The meaning of 'tele-access' is explained in Chapter 1. The remaining chapters elaborate this perspective on the politics of information, distinguish it from other concepts guiding research on ICTs, and discuss the factors shaping tele-access and its significance in a variety of arenas, including the production of ICTs and their use in business and industry, in households and communities, and in public policy.

At the end of each chapter you will find a set of original essays contributed by other PICT researchers. Their analytical approach to challenging commonly accepted assumptions about the information society generated many creative ideas, and I have included these essays to expand on the contributions of my colleagues, and also to show how my understanding of the technical and social processes shaping tele-access is built on their work.

This book is one of two companion volumes. The first, entitled *Information and Communication Technologies—Visions and Realities* (Dutton 1996*a*), pools the key themes of major research projects, lectures, and forums of PICT. *Society on the Line* also draws extensively on PICT as well as on my own research, but goes beyond the first volume to offer a new perspective on social and policy choices of the coming digital age.

Acknowledgements

I wish first to thank all of those who contributed essays to this book. I appreciate their willingness to devote their own work to the PICT-wide synthesis effort, but I also wish to acknowledge my debt to them and their colleagues more broadly. This synthesis owes much to my study of PICT research, which allowed me to look at decades of social-science research on the information revolution in a new light.

In this regard, I am indebted also to the many participants in a series of workshops organized to further the synthesis of PICT research. In addition to the contributors of essays to this volume, they included: Brian Allison, Victoria Bellotti, Brian Bloomfield, Jay Blumler, Jean Claude Burgelman, Chris Caswill, Martin Elton, James Fleck, Moyra Forrest, Helen Foster, Chris Fowler, Andy Freedman, Michael Gannaway, John Goddard, Richard Hawkins, Christine Hine, Richard Hull, James Katz, Kenneth Kraemer, Richard Kramer, Alfonso Molina, Nick Moore, Thomas O'Malley, Brian Oakley, Charles Portman, Paul Quintas, James Rule, Harry Scarbrough, Stuart Shapiro, Betty Skolnick, Puay Tang, Graham Thomas, Thierry Vedel, Frank Webster, Howard Williams, and Joyce Wood. I participated in these workshops, listened to audiotapes of the discussions, and benefited greatly from these exchanges.

The PICT Management Group (Rod Coombs, Nicholas Garnham, John Goddard, Robin Mansell, Robin Williams, and Steve Woolgar) advised me on all aspects of the programme, including the synthesis. Steve Woolgar, in particular, made it possible for me to direct PICT as a Visiting Professor at Brunel University.

Likewise, the PICT Committee, which included David Stout, Brian Allison, Jay G. Blumler, Arthur Francis, David Kaye (until 1994), and Brian Oakley, invited me to take on the role of directing PICT and provided valuable feedback throughout the course of my work. They gave top priority to the synthesis effort, of which this book is a central part.

I owe special thanks to Malcolm Peltu, an editorial consultant and London-based journalist, who helped me throughout the course of this book to address a broader and more international audience than I could have done on my own. He reacted to early outlines, provided detailed comments on every chapter, and made many creative contributions to the argument of this book. Antony Mason, who has edited a number of PICT publications, helped me to make this book more accessible to readers outside the social sciences and less enamoured with the world of information technology.

Steve Russell's drawings, which are incorporated in the design of the book's cover and part titles, show how an artist can help convey the social dimensions of ICTs. David Musson and his editorial team at Oxford University Press provided valuable support and advice in shaping this book for a wider audience, and gave me the time and moral support to complete it. My students at Brunel University and the University of Southern California provided constructive feedback on my work.

Many other colleagues provided critical comments on particular ideas, outlines, chapters, or essays, for which I am most grateful. They include Brian Allison, Walter Baer, Jay Blumler, Chris Caswill, Manuel Castells, Donald MacKenzie, Robin Mansell, Michael Noll, Walt Fisher, and David Stout. They and several anonymous referees for Oxford University Press have improved this book, and I thank them for their help.

Society on the Line and the research on which it is based was supported by a series of grants from the ESRC. All of us involved with PICT are grateful to the ESRC for its foresight in initiating PICT and providing long-term support of research on ICTs. This carried the programme through not only the heady days of enthusiasm over the launch of a new cable policy, and the wave of public fascination with the Internet, but also in the years between, when ICTs were off the public agenda, and not fashionable in academia.

Finally, I wish to thank my family for interrupting their lives to join me in Britain during my tenure as director of PICT. Diana, and my daughters, Eva and Sophia, make my involvement with PICT one of the highlights of my career.

I am indebted to all who have contributed to this manuscript, but I accept full responsibility for any errors or shortcomings that remain. I have learned a great deal from my colleagues in PICT and I hope that the reader does not judge PICT, and its contribution, only by my interpretation of what is a larger, more diverse, and more fascinating body of work than one book, and one general theme, can convey. I hope this synthesis and the accompanying essays will lead you to read well beyond this volume in exploring one of the most important subjects of enquiry in the social sciences.

W.D.

Contents

Contributors

Margaret Bruce is a Professor of Design Management and Marketing at the University of Manchester Institute of Science and Technology (UMIST). Her PICT research focused on the strategic marketing of new technology.

Martin Cave is Professor of Economics and Vice-Principal at Brunel University and a consultant to the UK's Office of Telecommunications. Professor Cave directed several PICT projects.

Rod Coombs is a Professor in the Manchester School of Management at UMIST. He was Director of the UMIST–PICT centre from 1987 to 1995.

James Cornford is a Senior Research Associate at Newcastle University's Centre for Urban and Regional Development Studies (CURDS).

William Dutton is a Professor at the Annenberg School for Communication, University of Southern California. He was the Director of PICT and a visiting professor at Brunel University during PICT's last phase.

Nicholas Garnham is Professor of Media Studies and formerly Director of the Centre for Communication and Information Studies (CCIS) at the University of Westminster, from where he coordinated Westminster's PICT research.

Andrew Gillespie is Professor of Communications Geography and Executive Director of the Centre for Urban and Regional Development Studies (CURDS) at the University of Newcastle upon Tyne.

Leslie G. Haddon is Senior Research Fellow in the Graduate Research Centre in Culture and Communication at the University of Sussex.

David Knights is Midland Bank's Professor of Finance at the School of Management and Finance at the University of Nottingham, and was Deputy Director of UMIST–PICT research.

Dale Littler is Professor of Marketing and Head of the Manchester School of Management at UMIST and a principal investigator on UMIST–PICT projects.

Donald MacKenzie is a Professor in the Sociology Department at Edinburgh University and was a principal investigator on projects at the Edinburgh PICT centre.

Contributors

Robin Mansell is Director of the Information, Networks & Knowledge (INK) Research Centre within Sussex University's Science Policy Research Unit (SPRU), where she is Professor of Information and Communication Technology Policy, and was Director of the SPRU–PICT centre.

Ian Miles is Director of the Programme of Policy Research in Engineering, Science and Technology (PREST) at the University of Manchester. He was formerly a Senior Fellow at the Sussex University PICT centre.

Malcolm Peltu is an editorial consultant and IT journalist who has edited many research-based publications aimed at a non-academic audience.

Vincent Porter is Professor of Mass Communications and Director of the Doctoral Programme in the Centre for Communication and Information Studies at the University of Westminster.

Charles Raab is Reader in Politics at the University of Edinburgh, where he has conducted research on privacy and data protection.

Ranald Richardson is a Research Associate at Newcastle University's Centre for Urban and Regional Development Studies (CURDS).

Roger Silverstone is Professor of Media and Communications at the London School of Economics and Political Science. He was the first Director of the Brunel–PICT centre.

David Stout directed the Centre for Business Strategy at the London Business School and chaired the PICT Committee.

John Taylor is a Professor and Director of Research in the Business Faculty, Department of Management, Glasgow Caledonian University. He was a Research Fellow at the Newcastle PICT centre.

Juliet Webster is conducting research for The Tavistock Institute and was a Senior Research Fellow at the Edinburgh PICT centre.

Robin Williams is a Reader and Director of the Research Centre for Social Sciences (RCSS) and the Science and Technology Studies research network at Edinburgh University. He was coordinator of the Edinburgh PICT centre.

Hugh Willmott is Professor of Organizational Analysis in the Manchester School of Management at UMIST and was a joint principle investigator for PICT.

Steve Woolgar is Professor of Sociology and Head of the Department of Human Sciences at Brunel University, where he directs the Centre for Research into Innovation, Culture and Technology (CRICT).

Essays

Essays

Figures

Tables

Boxes

Boxes

Abbreviations and Acronyms

AI	artificial intelligence
ALN	Asynchronous Learning Network
AOL	America Online
ARPA	Advanced Research Projects Agency (USA)
ATM	asynchronous transfer mode
ATM	automatic teller machine
ATV	advanced television
BA	British Airways (UK)
BBC	British Broadcasting Corporation
BBS	bulletin board system
BPR	business process re-engineering
bps	bits per second
BST	Big Sky Telegraph
BT	British Telecommunications
CAD	computer-aided design
caller ID	call line identification
Cal–SPAN	California–Satellite Public Affairs Network (USA)
CAM	computer-aided manufacture
CAN	Community Area Network (Center for Global Communications, Japan)
CAPTAIN	Character and Pattern Telephone Access Information Network
CASE	computer-aided software engineering
CATV	community antenna television
CCIS	Coaxial Cable Information System (Tama New Town, Japan)
CCIS	Centre for Communication and Information Studies
CCITT	Consultative Committee on International Telegraphy and Telephony
CCTA	Government Centre for Information Systems, formerly Central Computing and Telecommunications Agency (UK)
CD	compact disc
CD-ROM	compact disc read only memory
CDA	Communications Decency Act (USA)
CENIC	Consortium for Education Network Initiatives

CEO	chief executive officer
CEPT	Conference of European Postal and Telecommunications Administrations
CERN	European Laboratory for Particle Physics (Switzerland)
CICT	Centre for Information and Communication Technologies
CIM	computer-integrated manufacture
CNC	computer numerically controlled
CPU	central processing unit
CRICT	Centre for Research into Innovation, Culture and Technology
CROMTEC	Centre for Research on Organizations, Management and Technical Change
CSCW	computer supported cooperative work
C–SPAN	Cable–Satellite Public Affairs Network (USA)
CURDS	Centre for Urban and Regional Development Studies
DBS	direct broadcast satellite
DG	Directorate-General (EU)
DNet	Democracy Network (USA)
DP	data processing
DTI	Department of Trade and Industry (UK)
DVD	digital video disc
EBU	European Broadcasting Union
EC	European Community (earlier name for the EU)
ECU	European currency unit
EDI	electronic data interchange
EDP	electronic data processing
EFTPoS	electronic fund transfer at point of sale
e-mail	electronic mail
EPOS	electronic point of sale
ESD	electronic service delivery
ESPRIT	European Strategic Programme for R&D in Information Technology
ESRC	Economic and Social Research Council
ETSI	European Telecommunication Standards Institute
EU	European Union
EURICOM	European Institute for Communication and Culture
EVH	electronic village hall
FBCS	Fujitsu Business Communication Systems
FCC	Federal Communications Commission (USA)
FM	facilities management

FM	frequency modulation
FTC	Federal Trade Commission
FTP	file transfer protocol
FTTC	fibre to the curb
FTTH	fibre to the home
GAO	General Accounting Office
GATS	General Agreement on Trade and Services
GATT	General Agreement on Trade and Tariffs
GDP	gross domestic product
GII	Global Information Infrastructure (USA)
GIS	The Government's Information Service (UK)
GNA	Global Network Academy
GNP	gross national product
GUI	graphical user interface
GVU	Graphic, Visualization & Usability Center (USA)
HCI	human–computer interaction (or interface)
HDTV	high-definition television
Hi–OVIS	Highly Interactive–Optical Visual Information System
HLEG	Higher Level Expert Group (EU)
HMSO	Her Majesty's Stationery Office (UK)
HPCC	High Performance Computing and Communication initiative (USA)
HTML	hypertext markup language
IBA	Independent Broadcasting Authority (UK)
ICBM	inter-continental ballistic missile
ICT	information and communication technology
IITF	Information Infrastructure Task Force (USA)
IP	intellectual property
IPR	intellectual property rights
IRS	Internal Revenue Service (USA)
ISDN	Integrated Services Digital Network
ISH	information superhighway
ISI	Information Sciences Institute (USA)
IT	information technology
ITAP	Information Technology Advisory Panel (UK)
ITC	Independent Television Commission (UK)
ITEC	IT, Electronics, and Communications (UK)
ITN	Instructional Television Network
ITU	International Telecommunications Union
JAMIS	Japan American Institute for Management Sciences
JANET	Joint Academic Network (UK)
JIT	just in time

JPL	Jet Propulsion Laboratory (USA)
K	1,000 in general terms; 1,024 for measurements specific to computing
killer app	killer application
KITE	Kinawley Integrated Teleworking Enterprise Ltd.
LAN	local area network
MFJ	Modified Final Judgment (USA)
MFP	Multifunciton Polis (Australia)
MIS	management information system
MIT	Massachusetts Institute of Technology
MPEG	Motion Pictures Expert Group
MPT	Ministry of Posts and Telecommunications (Japan)
MSO	multiple system operator
NC	network computer
NC	numerical control
NCB	National Computer Board (Singapore)
NCES	National Center for Education Statistics (USA)
NHS	National Health Service (UK)
NIC	newly industrializing country
NIE	newly industrializing economy
NII	National Information Infrastructure (USA)
NLM	National Library of Medicine (USA)
NLS	Englebart's oN-Line System
NPM	new public management
NREN	National Research and Education Network (USA)
NSF	National Science Foundation (USA)
NSFNET	National Science Foundation Network (USA)
NTIA	National Telecommunications and Information Administration (USA)
NTSC	National Television Systems Committee (TV standard)
OECD	Organization of Economic Cooperation and Development
Oftel	Office of Telecommunications (UK)
OMB	Office of Management and Budget (USA)
ONA	open network architecture
ONS	Office of National Statistics (UK)
OPAC	Online Public Access Catalogue (British Library, UK)
OPCS	Office of Population Censuses and Surveys (UK)
ORL	Olivetti & Oracle Research Laboratory (Cambridge, UK)
OSI	Open Systems Interconnection
OST	Office of Science and Technology (UK)

OTA	former Office of Technology Assessment (USA)
OU	Open University (UK)
PABX	private automatic branch exchange
PAL	Phase Alternate Line
PAL	phase alternate line (TV)
PARC	Palo Alto Research Center (USA)
PBX	private branch exchange
PC	personal computer
PEN	Public Electronic Network (USA)
PICS	platforms for internet content selection
PICT	Programme on Information and Communication Technologies (UK)
PIE	Public Information Exchange (RP)
PITCOM	Parliamentary Information Technology Committee (UK)
PMS	Picturephone Meeting Service (AT&T, USA)
PoS	point of sale
POST	Parliamentary Office of Science and Technology (UK)
POTS	plain old telephone service
PPP	point-to-point protocol
PREST	Policy Research in Engineering, Science and Technology
PRP	Policy Research Paper (PICT)
PSTN	Public Switched Telephone Network
PTO	Public Telecommunications Operator
PTT	Postal, Telephone, and Telegraph
RBOC	Regional Bell Operating Company
R&D	research and development
RCSS	Research Centre for Social Sciences
RP	ReferencePoint (USA)
RTD	research, training, and development
SECAM	Sequential Colour and Memory (TV standard)
SPRU	Science Policy Research Unit
SRI	Stanford Research Institute
SST	social shaping of technology
STAR	Special Telecommunications Action for Regional development programme (EU)
SuperJANET	broadband Joint Academic Network (UK)
Telco	telecommunications company
TCP/IP	Transmission Control Protocol/Internet Protocol
TIIAP	Telecommunications and Information Infrastructure Assistance Program

TRS	telecommunications relay services
TTD	Telecommunication Devices for the Deaf
UMIST	University of Manchester Institute of Science and Technology
V-chip	violence-chip
VAN	value-added network
VCR	video cassette recorder
VLSI	very large-scale integration
VOD	video on demand
VR	virtual reality
WAIS	wide area information servers
WAN	wide area network
Web	World Wide Web
WELL	Whole Earth 'Lectronic Link
WEN	Whittle Educational Network
WINGS	Web Interactive Network of Government Services (USA)
WTO	World Trade Organization
WWW	World Wide Web

Part I

A New Perspective on the Information Revolution

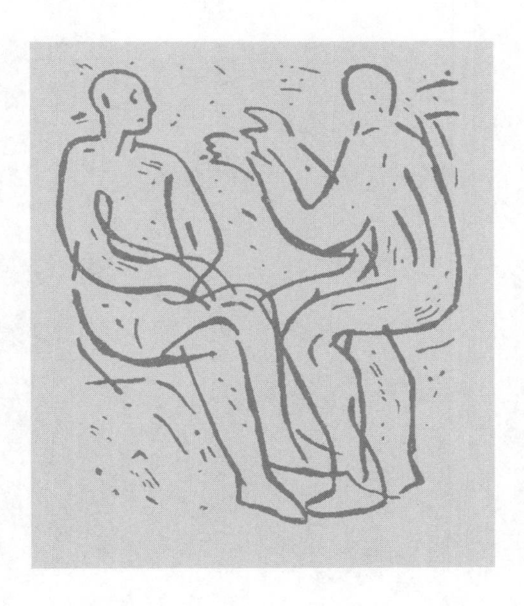

1 Introduction: Tele-Access—The Outcome of an Ecology of Games

Advances in information and communication technologies (ICTs) are among the defining technological transformations of the late twentieth century (Castells 1996; Dyson 1997). The widespread fascination with personal computers (PCs), video games, interactive TV, cell phones, the Internet, electronic payment systems, and a multitude of other ICTs often focuses on the technical ingenuity of their designs and their potential capabilities. These innovatory dimensions can be fascinating. But a narrow focus on technical advances overlooks the ways in which ICTs are also redefining your social world.

Popular conceptions of the 'information society'—like the virtual society, or the cyberculture—capture the social significance of the ICT revolution. While these concepts emphasize the increasing centrality of ICTs to society, they fail to provide insights about the role of ICTs in social change. Moreover, if taken literally, many prevailing conceptions of the information society are misleading. They suggest that information is a new economic and social resource, when there is nothing new about the importance of information (Robins 1992; Castells 1996). What is new is how you get access to information, and to much more.

Tele-Access

ICTs are social in that they define how people do things, such as how they get information, work, communicate, and are educated. A child may now find information via the World Wide Web rather than through a book. A family can keep in touch with one another by e-mail as well as by telephone and travel. A secretary's technology is no longer limited to a typewriter, telephone, and copier, but includes office equipment and document-management systems which might be globally networked. A manager could send an e-mail instead of distributing a memo and get news over the Internet rather than from the newspaper or TV. Students of all ages, in all parts of the world, could ask questions of major world figures in any field—whether educators, celebrities, or experts—via multimedia information superhighways.

Dimensions of Tele-Access

ICTs not only shape how people do things; they also shape and reshape what, when, and where things are done and what you know, who you know, and what you consume. Most importantly, ICTs shape access to:

1. **Information.** ICTs not only change the way you and others get information, but also alter the whole corpus of what you know and the information available to you and others at any given time and place. ICTs play a role in making some people information rich and others comparatively information poor.
2. **People.** ICTs not only provide new ways to communicate with others, but also influence whom you meet, talk to, stay in touch with, work with, and get to know. ICTs can connect or isolate people.
3. **Services.** ICTs do more than simply change the way we consume products and services. They also influence what products and services we consume and whom we purchase them from. ICTs can render obsolete a local business or an entire industry, but also create a new business or industry.

4. **Technologies.** Access to particular technologies—equipment, know-how, and techniques—shapes access to other technologies as ICTs interconnect and depend on one another in many ways. For instance, the Internet can provide access to vast numbers of computers around the world, yet you need a computer and other ICTs (such as a telephone line or cable connection) to access the Internet.

The activities involved in these four dimensions (see Table 1.1) are all concerned with what I call 'tele-access', a concept which highlights how ICTs involve much more than just the information processing or telecommunications transmission with which they are more usually associated. Tele-access covers the multifaceted interactions available through ICTs and how they shape access to information, people, services, and technology.

Table 1.1. Four Dimensions of Tele-Access: Information, People, Services, and Technologies

Access to	Kind of activities	Examples
Information (ICTs affect what you read, hear, see; what you know)	Storing, retrieving, analysing, printing, and transmitting facts, statistics, images, video, data, sounds, etc.	Watching TV news; reading information at Web site; viewing a CD-ROM; accessing a database
People (ICTs shape who you know; with whom you communicate)	Communications—with individuals, groups, multitudes; one to one, one to many, many to many, one to millions	Publishing; broadcasting; talking on a (video)phone; faxing; mailing a letter; e-mailing; teleconferencing
Services (ICTs influence what you consume; who pays what to whom)	Conducting electronic transactions and obtaining electronic services from remote or nearby locations	Pay-per-view TV; electronic banking and shopping; renewing a licence at a multimedia kiosk; booking a flight by phone
Technologies (ICTs shape access to other ICTs)	Producing, using, and consuming ICT equipment, techniques, and know-how shape access to other ICTs	Modems to link a computer to a network; the Internet and other information infrastructures; ICT service providers to support e-mail and Internet access

Just consider what routinely happens when a person looks for information about a topic on the Web. Abstracts are more likely to be found than full articles, and recent material rather than historical sources. Results of searches for a topic will be presented in a priority order, but the work of leading world authorities might be found side by side with the unsubstantiated views of a crank. As the number of search 'hits' could run into hundreds or thousands, the user is most likely to have access choices identified by the search tool used on the Web, or limited by a filter employed by an Internet service provider.

Similar processes will affect the information you locate, the people you meet, the services you choose, and the other ICTs you use. All are features of tele-access, and serve to shape our access—both electronically mediated and unmediated—to social and economic resouces.

There are many other ways in which ICTs can reduce, screen, and change the content and flow of information—by accident or design (Danziger *et al.* 1982; Garnham 1994*a*). ICTs do not just provide access to more people, many of whom who you would not be in touch with otherwise: they change patterns of interaction between people, communities, and organizations. As a substitute for face-to-face communication, for example, ICTs can provide benefits such as reducing travel, saving time, and extending the geography of human community. They may replace valuable human contact with a much less rewarding form of communication, or permit communication among people who might never have an opportunity to meet face to face. Tele-access encompasses all these substitutions, enhancements, and much more, by highlighting how technological, economic, social, and political factors are reshuffling society, influencing who's in and who's left out.

Social Choices Shaping Tele-Access

Technological change is just one factor that enables or constrains tele-access. Global electronic networks may make it easier to communicate around the world, but new technologies in themselves spell neither greater freedom nor stricter control. Instead, tele-access is shaped most directly by a history of separate but interdependent social choices made by a multitude of actors.

ICTs will be of most significance in social and economic developments in their role of opening, closing, regulating, distorting, and otherwise filtering tele-access. The consequences of such changes will be as profound and far-reaching as the way in which the development of the printing press in Europe in the fifteenth century allowed control of the written word to move away from scribes and priests to a much wider group of people involved in printing, translating, and distributing texts (Eisenstadt 1980). Understanding tele-access processes and their policy implications is, therefore, vital to each person, community, and private and public organization. Everyone in society has a vested interest in helping to shape tele-access.

Social, Economic, and Policy Implications

The 'ICT revolution' has been perceived and described in a variety of ways since it was first identified in the 1960s. For instance, in the early 1970s journalist Ralph Lee Smith (1972) saw innovations in areas such as interactive cable television as a way of providing an 'electronic communications highway' for a 'wired nation' in which all kinds of services might be supplied to businesses and households. Others have seen interactive TV as an 'electronic nightmare' in which there will be increased surveillance and invasions of personal privacy (Weicklin 1979). Many more have dismissed the social and economic significance of ICTs (e.g. Traber 1986; Winston 1989).

By the 1990s, many influential protagonists argued that new information superhighways would help realize Marshall McLuhan's vision of a 'global village' (Gore 1991), as well as providing the tools for managers to reinvent government and re-engineer business for the twenty-first century (Hammer and Champy 1993). Others saw the new networks and services on the Internet and the Web as creating a virtual society that would undermine the cohesion of real communities, destroy jobs, and fail to achieve the promises of the most zealous promoters of ICTs (Slouka 1995; Stoll 1995)

Such views of technology as either an 'unalloyed blessing', or an 'unmitigated curse', or 'not worthy of special notice' oversimplify its role and fail to provide an understanding of the 'actual mechanism by which technology leads to social change' (Mesthene 1969). The continuing debate between utopian and dystopian perspectives illuminates the social and economic issues at stake, but it does not inform policy and practice. Understanding the forces shaping tele-access addresses broad social, economic, and political issues, and also assists in the effective application of ICTs in a variety of contexts.

Factors Affecting Social and Technical Choices

The outcomes of ICT innovations are not random or unstructured. For example, patterns of tele-access, such as the distribution of information haves and have-nots, are enabled and constrained by the social and economic contexts within which relevant actors at all levels make choices (Robins 1992; Dutton 1996*a*). PICT research highlighted five major sets of factors that shape the design and use of ICTs and tele-access (see Box 1.1):

Box 1.1. Factors Shaping Tele-Access

- Economic resources and constraints
- ICT paradigms and practices
- Conceptions and responses of users
- Geography of space and place
- Institutional arrangements and public policy

1. **Economic Resources and Constraints.** The size, wealth, and vitality of nations, companies, and other actors place major constraints on the development and use of ICTs in all arenas of activity (Dutton *et al.* 1987*a*). For example, the options open to a 'high achiever, high-tech teleworker' differ from those of the 'low-income, low-tech lone parent' (Silverstone 1996: 225–8). Likewise, the strategies of the dominant telephone companies will differ from those of new entrants to the communication industry (Mansell 1993). Tele-access can improve or undermine the economic vitality of a nation or household, as well as enhance or exacerbate socio-economic disparities.
2. **ICT Paradigms and Practices.** Ideas have great power (Derthick and Quirk 1985; Robins 1992; Dutton 1996*a*). They can become the foundation of powerful belief systems or 'paradigms' which create a way of interpreting reality that is very different from that perceived by people whose thinking is embedded in another paradigm. The very idea that we work in a virtual organization or live in an information society can influence public policy and the behaviour of individuals, such as the career a student chooses to pursue. At the same time, experience and knowledge about ICTs can influence or even create a paradigm shift (Freeman 1996*a*, *b*). Concepts such as the 'value of competition', the emergence of a 'virtual society', and the building of 'information superhighways', for instance, have been important factors in shaping social and political choices in the design and implementation of ICTs, irrespective of their descriptive validity (Bloomfield *et al.* 1997). Technologies themselves make a difference, even if they do not determine social outcomes. ICT designs can bias social choices by making some avenues more economically, culturally, or socially rational than others (Woolgar 1996).

3. **Conceptions and Responses of Users.** ICTs are designed with a more or less well-founded conception of the user in mind. Users—whether workers, consumers, managers, citizens, or audiences—can also play an active role in shaping the implications of ICTs in ways that would not be expected by simply extrapolating from the perceived potential of the technology. Silverstone (see Essay 9.1) illustrates this in terms of the 'domestication' of ICTs—showing how households shape ICTs to fit their own needs and interests. Many innovative technological and market failures can also be understood as a consequence of having a weak conception of the user (Woolgar 1996).
4. **Geography of Space and Place.** One of the most prominent attributes of ICTs is the relative ease with which the new electronic media can overcome constraints of time and distance. Yet, rather than undermining the importance of space and place as had been previously claimed, ICTs might make geography matter more (Goddard and Richardson 1996). As ICTs make location decisions more flexible, other criteria, such as facilitating access to skilled or low-cost labour, can take an even higher priority in how a firm chooses to locate particular jobs and functions (see Chapter 6).
5. **Institutional Arrangements and Public Policy.** Technical, social, and organizational innovation are interdependent (Molina 1989; Fincham *et al.* 1994; Freeman 1996*a*; Bloomfield *et al.* 1997). The design of an organization influences the use of ICTs, since managers often implement systems that promise to strengthen existing structures and processes. However, ICTs also create a variety of new options for radically redesigning organizations and interactions between organizations, such as the 'virtual organization' composed of separate private firms or public agencies that employ ICTs to enable them to act as if they were part of the same real unit. At a broader societal level, patterns of tele-access, such as the gaps between the information rich and poor, are strongly affected by institutional arrangements and policies in areas such as telecommunications regulation, standards, copyright, public-service broadcasting, and education. Nevertheless, public policy at local, national, and international levels can be responsive to technological change, as evidenced in a variety of policy initiatives aimed at supporting advanced information infrastructures—notably the information superhighway.

By showing how these and other social factors shape and are shaped by tele-access, this book seeks to inform the social, personal, business, technical, and other choices that can have both immediate and cumulative long-range consequences on most areas of modern life.

The Roots of Tele-Access

Most popular descriptions of the ICT revolution have celebrated the abundance of information or new communication channels it brings. They have generally ignored how ICTs affect control over access to a much broader array of activities and resources. However, the value of focusing on tele-access was recognized decades ago by Marshall McLuhan (1964) in his seminal book *Understanding Media*, where he coined the phrase 'the medium is the message' to encapsulate the argument that a communication medium, like television, shapes the way in which you come into contact with information.[1]

The importance of ICTs in shaping tele-access to information, people, services, and technology is analogous to the impact on physical access created by transportation innovations such as railroads, freeways, and air flight. In the USA, for example, it is often easier and cheaper to fly from one coast to another than to much closer locations. In a city, mass transit makes it easier to travel to some areas than to others. Likewise, access is affected by physical structures, such as a building with many stairways that makes it inaccessible by wheelchair.

Physical shapers of access will remain important. But they are changing at a much slower rate than the electronic technologies affecting tele-access, such as mobile cellular phones, multimedia computers, cable TV, satellite for personal communications as well as broadcasting, and the continuously expanding Web. Furthermore, ICTs are providing more opportunities—for better or worse—to substitute electronic for physical access by moving 'bits rather than atoms' (Negroponte 1995: 11–20). As a UK government panel on the future of transportation has argued, ICTs offer increasingly attractive alternatives to physical mobility as a means of undertaking 'work, shopping, leisure, or commerce' (OST 1995: 9).

Moving Beyond Computers

When computers first began to emerge as a significant technology in the 1960s, they were treated primarily as calculators. Only a few individuals understood them to be far more general-purpose information-processing systems. For example, the convergence of computing and telecommunications was defined as information technology (IT) as early as the 1950s (Leavitt and Whisler 1958), even though the significance of this concept was not generally understood until the 1980s. As McLuhan (1964: 9) noted: 'When IBM discovered that it was not in the business of making office equipment or business machines, but that it was in the business of processing information, then it began to navigate with a clear vision.'

As the changing business fortunes of IBM have shown, computers and other ICTs are concerned with much more than just information. More recently, the focus has switched to communications, with many IT suppliers beginning to claim that they are really 'in the communication business' (Gates 1995). A character in the novel *Disclosure* by Michael Crichton (1994: 224–5) gets close to the truth when he says of a modem company that grew into a multimedia firm: 'Your business is not hardware. Your business is communications. Your business is access to information.' The founders of PICT showed much foresight in the early 1980s by talking about 'ICT' rather than 'IT'. But even adding the C for Communication is not enough.

In fact, neither information nor communication tells the whole tele-access story. They are but two dimensions to a much larger and more profound picture.

Lost in Information

Information has always been a key factor in social and industrial processes. What is new is the way in which we come into the presence of information. This helps to explain why the so-called information revolution that began in the 1960s has not actually led to a more 'informed' society. British social scientist John Bessant (1984: 176) has captured this irony well in his observation: 'It is interesting to speculate as to how much information can actually be gleaned from the mass of data collected in research; it seems to be a characteristic of the "information society" that we actually know relatively little about it.' We can become 'lost in information', as the poet T. S. Eliot once said.[2] Growing concerns over the lack of real information, the prevalence of misinformation, and increasing problems with information overload should therefore not be viewed as aberrations within an information society. These failures are actually caused by inadequate regulation of access to information—the incorrect treatment of all information as being equal and benign. Decades ago, this was recognized in organizations, where computer-based information systems were first applied (see Box 1.2). The Internet and the World Wide Web have made the need to manage information more apparent to the public at large.

Information is not like air. You might view access to it as a public good or an inalienable right. But the power of ICTs lies not just in creating greater access to information, but in creating the opportunity for you to have more control over access, and over the terms of access. You might

Box 1.2. Misinformation Systems: Poor Assumptions about Tele-Access

Russell Ackoff (1969) identified a number of incorrect assumptions which led to the development of management 'mis'information systems in the 1960s. All these have direct parallels in the assumptions guiding discussions of ICTs like the Internet in households, business, and industry of the 1990s:

- *Users should get more information.* In many cases, users suffer from an overabundance of information.
- *Users need the information they want.* However, most users fail to consider, filter, and prioritize what they already have.
- *If users get the information they want, they will make better decisions.* Yet many people do not know how to understand or correctly to apply complex or specialized information.
- *More communication will result in improved performance.* In practice, however, it can create conflicts, waste time, and displace more important work.
- *Users do not need to know how an information system works, only how to use it.* This limits the ability of users to invent new applications of the system, evaluate the information they are provided with, and solve problems encountered in its use.

wish to block an obscene phone call, choose your own TV programmes, and not be forced to use particular service providers. But tele-access is also a political process of conflict and negotiation over who gets access to whom, how, and when.

Policy and practice that promote the notion of an 'accessible society' in which anyone could get access to you at any place, any time, or any instant hold out the prospect of many exciting new ways of working, playing, and living. But they could also lead to George Orwell's (1949) nightmare scenario described in *Nineteen Eighty-Four*—a surveillance society in which Big Brother would always be watching you (see Box 12.2). Tele-access is a double-edged sword.

Tele-Access: The Big Picture

The concept of tele-access helps to illuminate the role of ICTs in a wide range of contexts across many disciplines, such as social relations between:

- insiders and outsiders;
- senders and receivers, for example the mass media and their audiences;

- the geographical centre and periphery;
- élites and masses; or the government and the public;
- producers and consumers; and employees and supervisors.

It also helps to define the meaning of a minimum level of 'universal service' for those in telecommunications. Universal service is of particular relevance to ICT infrastructures. It has long been recognized as a vital principle (Anderson *et al.* 1995; Thomas 1995) which was relatively easy to define when telecommunications involved just the plain old telephone service (POTS). But the vast variety of new ICT capabilities had led to much contemporary debate about defining what universal should mean in an age of multimedia, multi-service ICT capabilities. Focusing too narrowly on the question of universal access to ICT network infrastructures promotes the false assumption that access to technologies *per se* is good, thus marginalizing other complex issues of tele-access, such as the need to filter information.

Universal access, however, is just one aspect of tele-access. As can be seen from Table 1.2, the issues of tele-access encompass subjects

Table 1.2. ICT Issues from the Perspective of Tele-Access

Arenas	Contemporary issues	Concerns access to
Technical choices	telephone v. cable v. satellite v. the Internet	households, consumers
	caller ID (call-line identification)	phone numbers
	violence-chip (v-chip)	TV content
Business, management and work	Information overload	unscreened, unprioritized information
	telework	the workplace, customers
	unemployment	ICTs (infrastructures, know-how)
Politics, governance, education	teledemocracy	citizens, voters
	virtual universities	knowledge, information producers
	security, vandalism	proprietary information
Households, cities, and electronic communities	information rich/poor	information
	mediated reality	real, not simulated, experience
	privacy: disclosure	personal information
	privacy: intrusion	unwanted information
Industrial and public policies	freedom of speech, press	uncensored content
	cultural sovereignty	one's own language and culture
	intellectual property rights	an author's work
	regional development	infrastructures, services
	decline of the public sphere	public information, discussion
	universal service	telephone service

that range well beyond the inequalities of the information have/have-not divide to encompass such issues as individuals coping with 'infoglut', new forms of democratic participation, radical innovation in work processes and organizational structures, and far-reaching challenges to national and international economic policy-making.

The Outcome of an Ecology of Games

Analyses of tele-access, and the forces shaping it, provide a uniquely coherent framework to connect the many apparently disparate issues that have been associated with the ICT revolution (see Table 1.2). The decisions of individuals and organizations about these issues are interrelated facets of a single, broad 'ecology of choices' that ultimately shapes tele-access.

This concept represents an important step towards enhancing our understanding of why ICTs are not on a predetermined technical path that will redefine tele-access in predictable ways. ICTs could be deployed to make society as a whole, and the individuals and groups within it, more accessible or more isolated. They will enable people to get access or to close it off. For example, the answering machine can be used to keep in better touch with one's friends, or to screen unwanted calls and avoid contact with some people. How these outcomes unfold will be products of countless numbers of both strategic and everyday decisions made by a multitude of actors in many separate arenas: small businesses, shops, households, large organizations, complete industries, schools and universities, hospitals and libraries, international trade conferences—in fact, in almost every area of social and economic activity. Each of these arenas could be seen as involving a set of players following an established set of traditions, rules, and disciplines—defined by the individuals and their unique contexts—that comprise a 'game'. It is the outcome of this 'ecology of games' (Dutton 1992*a*, 1995) which ultimately shapes tele-access (see Box 1.3).[3]

The metaphor of a game is a useful way of viewing actors as purposive players in a variety of activities defined by their own rules and assumptions in trying to achieve particular goals (Dutton 1992*a*). Every actor is involved in one or more 'games' within the broader 'ecology' that shapes the design and development of ICTs and public policy. The metaphor does not imply that all actors are simply self-interested. The goals of many actors could be to further their conception of the public interest or a corporation's profits. But in order to understand the development of a large

Box 1.3. The Concept of an Ecology of Games

A game is an arena of competition and cooperation structured by a set of rules and assumptions about how to act to achieve a particular set of objectives. An ecology of games is a larger system of action composed of two or more separate but interdependent games. Aspects of an ecology of games —games, rules, strategies, and players—offer a grammar for describing the system of action shaping the development of households, organizations, and public policy (Dutton 1992*a*, 1995*b*). Norton Long (1958) first used the idea of an ecology of games to describe the governance of communities. He argued that influential people in local communities are rarely focused on governing the community. That assumption would oversimplify the system of action governing the course of public affairs. Instead, they are more focused on such matters as selling real estate, being elected to office, developing land, creating a general plan, and finding a home. To understand the behaviour of these players, it was therefore more useful to think of them as real-estate agents, candidates for the council, land developers, planners, and homeowners or renters, rather than as élites seeking to govern their community. In this way, the development of a community could be understood as the outcome of an unfolding history of events driven by the often unplanned and unanticipated interactions among individuals playing relatively separate but interdependent games. Individuals make decisions as the occupant of a particular role within a specific game. The evolution of tele-access in an organization or nation—like the development of a local community—can also be viewed as the outcome of an ecology of games.

system, such as an organization's ICT arrangements, the broad ecology of games governing the actions of key players must be understood.

In some arenas, the choices made by powerful individuals, groups, or companies can determine the tele-access opportunities available to other players. For example, a growing number of companies use electronic data interchange (EDI) networks to link themselves directly with suppliers and retailers for automatic ordering and just-in-time (JIT) delivery. If suppliers are not 'on line', even if they are across the street, they are virtually invisible. More generally, the costs of keeping pace with state-of-the-practice in business communications can be a barrier for the many small businesses who cannot afford advanced services or equipment and do not have the expertise to support the development and maintenance of sophisticated ICT-based applications.

The perspective of an ecology of games admits that a variety of motives —other than improving tele-access—will drive choices about ICTs. For instance, you might buy a computer to make it cheaper and easier to work at home. But this decision could also affect your access to new entertainment opportunities and electronic discussion groups. If tele-access becomes a more explicit objective in ICT developments, then this ecology might evolve in more desirable ways. In challenging many previous assumptions about the coming information society and offering alternative perspectives, the tele-access concept also helps us to move towards a more coherent view of the social, political, cultural, geographical, and technical factors that decisively shape the outcomes of the revolution in ICTs.

About This Book

The remainder of this book explores the ways in which tele-access unfolds as a consequence of the specific decisions by individuals and groups acting in a variety of arenas through an ecology of games. This provides a new way of getting to grips with the theoretical and practical specifics of the social choices shaping tele-access, in addition to the way they are pieced together to build a bigger picture of the social and economic implications of ICTs. The book synthesizes the most valuable findings of PICT, as well as my own and other research on the myriad of topics it encompasses, using examples from every sector of the economy.

This chapter has defined the nature and significance of tele-access. Chapter 2 explains how this concept differs from alternative perspectives on ICTs that have underpinned public debate and social-science research. It provides an overview and background to the different ways in which social scientists think about advances in ICTs and shows how the ecology of games provides a framework that is more suited to the task of understanding the interdependent processes shaping tele-access.

Part II (Chapters 3 and 4) introduces the interactions of technical and social processes that shape tele-access. Chapter 3 looks at how ICTs are designed and produced in ways that can structure tele-access, while Chapter 4 focuses on the many ways in which social, cultural, and economic factors shape not only the design of technologies, but also the ways individuals and organizations decide to reject them or embed them in their everyday lives.

Part III (Chapters 5 and 6) looks at ways ICTs have been adopted and used in management and business and also in the workplace. These chapters show how tele-access is being used to redesign organizations and the workplace, but they also demonstrate how tele-access is becoming more critical to the competitiveness of the firm and the employment prospects of an individual. Chapter 5 focuses on business and industry, particularly the manner in which the private sector has used ICTs in strategic ways to restructure access to skilled personnel, labour, and markets. Chapter 6 shifts to a discussion of how these tele-access strategies are reshaping the workplace, including the geography of the firm.

Part IV turns to issues of public access, focusing on government and education. Chapter 7 looks at technological change in governments that could have major implications on relations between governments and the public at large, which are particularly acute with respect to political participation, personal privacy, and freedom of expression. Chapter 8 deals with knowledge gatekeepers because it looks specifically at ICTs in education, research, and scientific communication, where traditional linkages between the producers and users of knowledge are being challenged.

Part V moves to the local community and the household, the topic of Chapter 9. This chapter moves to the more widespread use and consumption of ICT products and services by the public at large, focusing on the important role that consumers play in 'domesticating' ICTs—fitting them into their everyday routines and family values. Chapter 10 extends this analysis by looking at how various information and communication industries seek to wire the household and business. The outcomes of growing competition to wire communities has major consequences for industry, but also for the public at large. Debates over the technical advantages of cable versus satellite versus telephone systems can mask the social implications of ICTs.

Part VI (Chapters 11 and 12) extends the analysis by addressing the major policy issues and processes shaping tele-access. Chapter 11 moves to industrial and economic development policies that have dramatically changed the broad ecology of information and communication policy. This chapter shows how many key information and communication policy issues, such as privacy, freedom of expression, and standards, can all be understood as tele-access policies. Viewed in this way, ICT policies could be less fragmented, contradictory, and hamstrung by conflicting values and interests.

Finally, Chapter 12 concludes with a discussion of the politics of tele-access. Personal and societal outcomes of the continued advance and diffusion of ICTs can be qualitatively improved by people who recognize the processes shaping tele-access and the central importance that access will play in building a more open and inclusive society. How people choose to shape tele-access through the design and use of ICTs will have far-reaching implications on who is brought into the centre, and who is left out at the margins of society.

Notes

1. One of McLuhan's most imaginative examples of a communication medium is the light bulb. The importance of the light bulb is that it provides access to the contents of a room. The social consequences of this illumination are far more fundamental that the contents of any given room.
2. In choruses from *The Rock*, a 1934 play, Eliot's (1969: 147) words were:

   Where is the wisdom we have lost in knowledge?
   Where is the knowledge we have lost in information?

3. The 'ecology of games'—like the concept of 'tele-access'—is a sensitizing concept, developed within a qualitative, participant-observer mode of enquiry. It provides an abstract representation of concrete empirical observations. It gives 'a sense of reference, a general orientation, rather than a precise definition, to a phenomenon under study' (Bruyn 1966: 32).

2 Information Politics, Technology, and Society

Advances in ICTs have generated many theoretical perspectives on the 'information revolution' and what it will mean for society. The most dominant ideas, such as the concept of an information society, have influenced policy and practice around the world. This chapter provides a brief background to some of the main ways in which social scientists have thought about ICTs. I have grouped them into five broad and overlapping

Table 2.1. Perspectives on Information and Communication Technology

Focus	Key problem	ICT focus
Influence	Effects of mass media on public opinion and political behaviour	Interactivity of new media; expansion of channels; changing role of gatekeepers
Technology	The role of technology in society	Effects of technological change on the locus of control of social and technical systems
Impacts	Psychological, social, economic impacts; planned and unanticipated	Impacts of technical features of ICTs, such as processing capability
Strategy	The strategic use of computers and telecommunications in organizations	Information system and network design as a management strategy
Information	Role of ICTs in advancing stages of economic development	Information as a new economic resource
Tele-access	Social, economic, geographical consequences of tele-access—to information, people, services, and technology	Ecology of technical, organizational, social, and policy choices shaping tele-access, and the factors constraining choice

categories: influence, technology, impacts, strategy, and information, which I contrast with this book's perspective on the shaping of tele-access (see Table 2.1).[1]

A common denominator among the best social-science research has been a willingness to challenge commonly accepted views and assumptions (see Essay 12.2). This has helped to identify blind spots in many viewpoints guiding debate over ICTs and to assist the development of new perspectives, such as the 'tele-access' concept explored in this book. The significance of tele-access can be usefully explained against the background of earlier perspectives that failed to appreciate its centrality to the revolution in ICTs.

Influence

One traditional perspective on ICTs has evolved out of the study of the mass media: newspapers, radio, television, and film. In the aftermath of the Second World War, concern over propaganda and the political implications of radio and television underpinned studies of the effects of the media on public opinion and voting behaviour. Initial survey research focused on political campaigns and elections. It found that the mass media had much more limited effects than were anticipated by the simplistic models of a relatively passive audience being influenced by their exposure to messages (Berelson *et al.* 1954; Campbell *et al.* 1964).

Media Effects and the Impact of New Media

Reformulations of theory since these early studies have led to a variety of more sophisticated models of media effects (Schramm and Roberts 1971; McQail 1994). For example, some argued that the mass media can influence the opinion leaders most actively seeking and using the media to gain information, who in turn influence others (Katz and Lazersfeld 1955).

Later studies of media effects shifted attention to the media's critical role in agenda-setting—for example, in deciding which issues to cover. In this way, 'gatekeepers' within the mass media were seen to be influencing what people think about, rather than determining how people feel about the issues and candidates (Shaw and McCombs 1977). As the distinguished American journalist Walter Lippmann (1922: 3) put it, the media can shape 'the pictures in our heads'. Many scholars raised major concerns about the conservative influence television and other mass media can have on the long-term agenda of the public—determining what people ignore and what they take for granted (Bachrach and Baratz 1970; Chomsky 1984; Schiller 1989).

This 'influence' tradition in media studies continues to drive a great deal of media research in communications, sociology, and political science. It is anchored in traditional models of communication, such as 'information theory' (see Box 2.1), and extends into more recent work on new electronic media, such as electronic networks, PCs, and virtual reality (Rice and Associates 1984; Rogers 1986; Heeter 1995). The tradition focuses on the contents of messages conveyed through the media and the messages' influence on those exposed directly or indirectly to them. It has led many social scientists to ask whether the interactive character of emerging media will make them more engaging and, therefore, more powerful in shaping attitudes, beliefs, and values. Others have focused on the way the profusion of media and channels can segment audiences in ways that might erode the quality and integrative effect of the mass media, and no longer provide the common experiences or shared text of a community (Blumler 1992).

Box 2.1. Information Theory in the Study of Communication

The communication field adopted many of its fundamental concepts from 'information theory' (Shannon and Weaver 1949). This generic model of communication underpins the study of influence. Information theory breaks down all communications into a process that entails a sender transmitting information to a receiver over a channel. The information must be encoded to be sent over a communication channel, which is subject to varying amounts of 'noise' interference. The message must then be decoded for the receiver.

This model was of much value to telecommunications engineering. However, the application of this model to the social sciences has been more contentious; here it was primarily used to reinforce the study of one-way communication between a sender (the candidates, the press, broadcasters) and the receiver (the audience). The theory's concept of information as a transmittable entity, like electricity, which is also uniquely reusable, fostered this relatively narrow research focus on the nature of messages (information) transmitted from senders to receivers.

The Bias of Media: The Significance of the Medium of Communication

Concern over audience segmentation provides an example of how technological change has challenged research in the influence tradition, which is built on a presumption of access to a mass audience. Marshall McLuhan anticipated this problem in the early 1960s when he argued that there had been too much emphasis on the content of the message rather than the more significant effect of the medium itself. In claiming 'it is the medium that shapes and controls the scale and form of human association and action', McLuhan (1964: 9) indicated his belief that television's ability to convey sights and sounds from round the world instantaneously to create a 'global village' was more significant than whatever message was conveyed.

Although McLuhan's perspective was of broad value, it was generally ignored by those working within the major paradigm governing communication research, who focused on the influence of messages in mass media such as newspapers, radio, and television and did not look at other ICTs until the late 1970s.[2] A focus on tele-access builds on the insights of McLuhan and others who argued that different media can provide access to different kinds of information, sets of people, assortments of services, and technologies. McLuhan also pointed out that the processes shaping tele-access can be even more important than the particular messages, people, services, and technologies involved in the interactions.

Media researchers continue to focus on the influence of messages, but a major rift divides those who assume the audience is relatively passive from those who assume the audience is actively and selectively processing and reinterpreting the meaning of messages (Blumler and Katz 1974; Silverstone and Hirsch 1992). The concept of an active audience is important to the study of all ICTs (see Chapter 9).

However, media research has tended to take access for granted, even though it has implicitly accepted the power of access by focusing on the gatekeeper function and the effects of messages on people who have access to them in the mass media (Rosengren 1997). But technological change has made assumptions of media access to a mass audience more problematic (EURICOM 1995). As a result, gaining access to audiences has become a central preoccupation of the communication industry (see Chapter 10).

McLuhan had something to say about access. But his deterministic assumption that broadcasters would have global access has been undermined to some extent by ICT innovations which are being used not only to expand mass-audience broadcasting, but also to fragment audiences into specialist 'narrowcasting' streams. The influence perspective becomes even less applicable as one moves outside media studies, to assess the implications of ICTs for the economy, for example.

Technology

The Role of Technology in Society

Outside the communications field, social scientists have generally looked at ICTs as just one case in a more general concern with the impact of technology on society. Treatments of technology and society have flourished since the 1960s. For example, the French sociologist, Jacques Ellul (1964), and others have provided broad philosophical treatments of how society is affected by the growing role of technology—and how technology, as a means, has become an end in itself. Researchers in this field have also analysed the impacts of specific ICTs, such as computers (Mesthene 1969) and the video phone (Dickson 1974). Such technology assessments bring to the surface a wide range of multidisciplinary issues that go far beyond concerns with the mass media; they show, for example, how ICTs have become an integral component of a 'high-tech society' that pervades all aspects of our lives (Forester 1987).

Another major contribution to this literature has focused on control as one impact of technological change (Winner 1977; Thompson 1991). Many dystopian novels and films, such as *2001: A Space Odyssey* and *Jurassic Park*, capture the fear that technology is out of control and has created a new, unelected, high-tech élite which increasingly dominates decision-making (Ellul 1964; Weizenbaum 1976).

Technology, Expertise, and Social Control

Early discussion of computers and telecommunications reflected this concern that modern technology is becoming an unstoppable force which would support increasingly centralized control in organizations and society. Many observers of the social impacts of more recent advances in ICTs argue the opposite—that the microelectronics revolution in personal computing and networking has decentralized control over the technology and placed more control in the hands of users (de Sola Pool 1983*a*, 1990; Gilder 1994). Nevertheless, many others continue to argue that ICT innovations are empowering a new élite of 'cybercrats' (Ronfeldt 1992) who have the skills and knowledge to control networks in organizations and society.

Technology concerns more than just equipment. It also encompasses the knowledge which is essential to its use (MacKenzie and Wajcman 1985: 3). The control of technology is, therefore, bound up with issues of who has access to the skills, equipment, and know-how essential to design, implement, and employ technology. Changes in technology can, therefore, restrict access to all these resources. For instance, early mainframe computers depended on programmers with knowledge of very specialized programming languages. But technological change can equally expand access—for example, by the way simple graphical user interfaces with the Web have made access to information stored on millions of computers available to users round the world at the click of a button. Likewise, social choices, such as the decision to learn a new human or computer language, affect access to technology, jobs, and people. The ways in which technical and social choices shape access to ICTs have been central to the technological perspective, but have been lost in a body of work covering myriad concerns at all levels of analysis and across all technology.

Impacts

The classic dystopian novel of the twentieth century, George Orwell's *Nineteen Eighty-Four*, pivoted around the emergence of television and the logic of electronic surveillance (see Box 12.2). Half a century later, the growing centrality of the Internet has generated updated Orwellian visions, as portrayed in *The Net* (Columbia Pictures 1995), where the heroine's life and identity are transformed, deleted, and restored on-line.

These scenarios frighten because they are often based on the actual capabilities of ICTs. The avoidance of such futures, however, is less easy to devise because they are determined primarily by social and political choices. The same can be said of more immediate concerns over the implications of ICTs on privacy, freedom of speech, and employment. Some of the earliest research on the social impact of ICTs was led by specialists involved directly in the development of computing and telecommunications who sensed that the public were failing to understand the full potential of the computer and the risks it posed to society. They formed interest groups within their respective professional associations and attempted to explain how the technology works and how it could affect individuals and society in terms of major issues, such as privacy and surveillance, that still remain central to policy debates (Gotlieb and Borodin 1973; Wessel 1974; Berleur *et al.* 1990).

The Limits of Determinism: Technology as a Product of Social Choice

The social-impact approach was usually based on rational forecasts of the social opportunities and risks created by particular features of the technology. For instance, the enormous storage capacity of the computer raised the potential for organizations to create huge databanks of information about individuals. However, empirical research usually found that rational expectations based on such capabilities of the technology were seldom realized (Danziger *et al.* 1982).

For example, a landmark study of the impact of computers on privacy found that computerized organizations did not change their record-keeping practices in the ways anticipated by those who were alarmed about databanks (Westin and Baker 1972). Instead, they tended to use computers to make their existing information practices more efficient.

Likewise, forecasts of the paperless office were generally based on a rational expectation that people would want to save the time and costs associated with paper—but ignored the reader's habits and values and the countervailing technical advantages of paper, such as the relative difficulty of reading an electronic screen.

This has led to fundamental criticisms of theories of social change based on deterministic assumptions that particular real or potential technical features will have predictable effects (see Essay 2.1). The public is often told that technology is on a predetermined trajectory, but different scientists forecast different trajectories. By contrast, much social research and analysis on the design and impacts of technologies have emphasized the decisive role played by social choices in determining ultimate outcomes (MacKenzie and Wajcman 1985; Mansell and Silverstone 1996).

Studies of the social bases of choices shaping technological design have, therefore, led to a valuable shift in the focus of social-science enquiry. Social scientists no longer look only at the social implications of technological change, but also consider the psychological, social, political, geographical, and economic factors that comprise the 'social shaping of technology' (see Essay 2.2). When I mentioned this approach to a multimedia development specialist, he assumed that I was questioning the motives of engineers. 'Many of us are not in this for the money. We simply love technology!' he exclaimed. His argument that passion matters and that technology can be pursued for technology's sake helps to illustrate the important role that social factors can play in technological change.

Research on the 'social shaping' of technological change has been highly critical of the simplified, linear models of cause and effect used by technological determinists. This has led to a greater emphasis on exploring the underlying 'processes' of technical and social change, rather than predicting their long-term impacts (Mansell and Silverstone 1996: 39–41). As American political scientist Lester Milbrath (1996: 104) has argued:

> **Our culture generally teaches us to think linearly: to connect cause to effect, to break down an object of inquiry into its constituent parts, to take orders from above and give orders to those below. Systems thinkers focus on wholes rather than parts. Within wholes they concern themselves with relationships more than objects, with process more than with structures, with networks more than hierarchies. In a system, a given effect not only radiates through the system, it also generates feedbacks which change the factor that caused it.**

The Processes Shaping Tele-Access: Technology Still Matters

The study of technological change has also suggested that tele-access—which includes the key social impacts attributed to ICTs (see Chapter 1)—should be viewed as the outcome of an indeterminate social and political process rather than being set on a predetermined technological trajectory. For example, the much promoted view that technological advances are democratizing access to ICTs is empirically questionable (Danziger *et al.* 1982; Mulgan 1991). It is also dangerous because it paints a deterministic future that might well undermine the political will to make hard choices to ensure there is wide democratic access to ICTs.

The design and development of ICTs is not as deterministic of social outcomes as assumed by early theorists. Neither are they neutral. Advances in ICTs affect access, but they do not determine access. Technology is like policy, because it tells us how we are supposed to do things (Laufer 1990: 2). However, the biases designed into technological artefacts and systems can be more enduring than legislation (Winner 1986: 19–39).

Strategy

A focus on the goals and strategies of actors has been most fully developed within the management field, where some of the earliest empirical studies into the design and use of ICTs were conducted. Management researchers began to study computers in the 1950s (Kraemer 1969). From early on, leading management science thinkers, such as Leavitt and Whisler (1958), also highlighted the significance of the convergence into information technology (IT) of computing, telecommunications, and related techniques, such as modelling and computer simulation.

Managing Information System Design

Through the early decades of its use, the computer was seen by management theorists as a strategic tool for enabling executives to boost centralized control by giving them access to information about the organization's resources and everyday operations (Whisler 1970; Danziger *et al.* 1982). Centralization was expected not as an inevitable consequence of computerization, but as an extension of the prevailing management paradigm. It was a technical fix for problems that seemed immune to structural reforms (Laudon 1974; Kraemer *et al.* 1981: 1–27).

From this strategic perspective, the major implications of ICTs extend from the strategies of management rather than the characteristics of the technology (Child 1987). This view remains central to contemporary management research, but with an ICT management paradigm that places less emphasis on top management control and more on the virtues of innovation and networking (see Chapter 5; also Drucker 1995; Handy 1995; Bloomfield *et al.* 1997).

Instead of finding that computers inevitably increased centralization, early empirical research on computers (e.g. Laudon 1974; Danziger *et al.* 1982) discovered that those who control decision-making tend to adopt and use ICTs in ways that follow and reinforce existing patterns of control within the organization, whether they be highly centralized or decentralized (Danziger *et al.* 1982).

Top managers are only one set of actors in a more complex and interdependent set of decisions about the design and use of IT in organizations. The organizational and social implications of ICTs are the uncertain outcome of a struggle between conflicting viewpoints and actions—what some have characterized as a dialectic (Mansell and Silverstone 1996: 213). However, this struggle among actors over the design and use of ICTs is not taking place on a level playing field. Instead, it is marked by great inequalities in existing institutions, cultures, and social and technical systems that favour some viewpoints and actions over others (Mansell and Silverstone 1996).

The Interaction of Strategies within a Broader Ecology of Actors

I would go a step further and argue that this struggle generally takes place in a variety of different fields at the same time. This means the idea of a dialectic is misleading because of the degree to which it suggests that all actors are involved in the same debate. Instead, individuals and groups are pursuing different goals within their own domains in an ecology of games (see Box 1.3). For instance, a technical expert might be pursuing a technically elegant solution to a network's design, while a top manager is primarily seeking cost reductions. This places major constraints on the predictability of outcomes based on an assessment of strategic goals, unless the broader ecology is well understood and orchestrated.

Global advances in ICTs, the inertia of technologies already in place, and the widespread application of ICTs throughout society place limits on the ability of any individual, organization, or nation to control the design and implementation of ICTs in predetermined ways. It is more realistic to view tele-access, and its consequences, as the outcome of a process of social and technical choices by many different actors within a variety of separate but interrelated technical, organizational, social, and policy arenas.

Information

A fresh and enduring perspective was introduced in the 1970s by discussion of the information society as a new stage in economic development (Bell 1974, 1980; Masuda 1980; Nora and Minc 1981; Berleur *et al.* 1990; Drucker 1993). The information society theorists cite many of the social threats identified by others, such as to personal privacy and information inequality, but they have focused more attention on the role of ICTs in the economy.

The American sociologist Daniel Bell wrote the seminal work on the information society, which he first called the 'post-industrial society' (Bell 1974, 1980). He said information would be the key economic resource in this new era—not raw materials, or financial capital as in earlier agricultural and industrial societies. Citing the US economy as an exemplary case, Bell identified three major trends in the development of an information society (see Box 2.2).

Box 2.2. Trends in the Development of an Information Society

1. The growth of employment in information-related work.
2. The rise of business and industry tied to the production, transmission, and analysis of information.
3. The increasing centrality to decision-making of technologists—managers and professionals skilled in the use of information for planning and analysis.

Source: Bell (1974, 1980).

The most significant trend towards an information society is the shift in the majority of the labour force from agriculture (the primary sector) and manufacturing (the secondary sector) to services (the tertiary sector). Since the Second World War, the growth in information work has contributed to substantial growth in the service sectors of the advanced industrial economies (Machlup 1962; Porat 1976, 1977). Information work has been defined to include a broad array of jobs related to the creation, transmission, and processing of information, ranging from programmers, and software engineers, to teachers and researchers.

Another trend has been the increasing importance of knowledge and its codification to the management of social and economic institutions. Theoretical knowledge and techniques and methods such as systems theory, operations research, modelling, and simulation are viewed as vital to what Bell defined as central problems of the post-industrial era: forecasting, planning, and managing complex organizations and systems. According to Bell, the complexity and scale of emerging social and economic systems require systematic forecasting and foresight to replace a previously trusted reliance on common sense or reasoning based on surveys and experiments. One of Bell's critics (Beniger 1986) has argued that this need for control structures has been the central economic and social factor driving the revolution in ICTs.

A third set of information-society trends involves power shifts, particularly the growing prominence of a professional and managerial class who understand and know how to work with data, knowledge, information systems, simulation, and related analytical techniques. These 'knowledge workers' will become increasingly vital to decision-making processes in situations of growing complexity (Downs 1967; Bell 1974: 12–33). Thus, reflecting other early studies of technology and society, the relative power of experts—a knowledge élite—is expected to rise with the emergence of an information society.

The information-society thesis has provoked much criticism (see e.g. H. I. Schiller 1981; Webster and Robins 1986; Garnham 1994*a*). Changes in work, technology, and power posited by this thesis have all been challenged. Yet its basic thrust has increasingly defined the public's understanding of the information age. Public and corporate policies and academic institutions and journals have been built on the concept of information as a new strategic resource. However, this focus can mislead policy and practice, particularly by the way it has substantially shifted attention to the information sector and away from other parts of the economy.

This is a problem in at least two respects. First, the revolution in ICTs does not encompass the degree to which employment in services is itself subject to displacement by manufactured consumer goods, such as the PC or the washing machine (Miles and Robins 1992: 4–7). Advances in ICTs can also be used to reduce employment among information workers, as in the telecommunications industry in the 1990s.

Secondly, the revolution in ICTs has not just stimulated new information industries, such as the providers of on-line data and multimedia services.

It has also affected every other sector of the economy (Miles 1990: 7–25), including agriculture and manufacturing. In this respect, the occupational shifts associated with the information society do not necessarily imply a decline in the relevance of primary or secondary sectors to national or global economies. Instead, they suggest a diminishing need for labour within these sectors as computing, telecommunications, and management-science techniques are used to redesign how work is accomplished.

Most importantly, the identification of information as the key central resource of the economy has also been widely questioned (D. Schiller 1988). For a start, information has a highly variable currency: it is not always wanted or valued. As a resource, information can lose its value, such as in the case of a day-old newspaper, or retain its value after repeated use, as with a literary classic. If defined narrowly in the sense of being about 'facts', the term 'information' becomes far too limiting as a depiction of the social role of ICTs. But, if defined very broadly—for instance, as anything that 'reduces uncertainty'—then information seems to be so all encompassing that it becomes virtually meaningless. The definition of information offered by Marc Porat (1977: 2), and accepted by Manuel Castells (1996: 17 n. 27), is bounded and consistent with its use throughout this book: 'Information is data that have been organized and communicated.'

Moreover, information can create, rather than reduce, uncertainty. Miguel de Unamuno's (1913) wise observation that 'True science teaches, above all, to doubt and be ignorant' is well accepted. Donald MacKenzie's research underscored the degree to which those more informed about a topic are less certain about its properties than many who are less informed. There is no linear relationship between information and certainty, but rather what MacKenzie has called a 'certainty trough' (see Essay 2.3).[3]

Some take such criticisms as a call for developing a new economics of information (Oettinger *et al.* 1977). However, this problem is more fundamental. As all know, and many have pointed out, information is not a new resource (Miles 1990; Robins 1992; Castells 1996). It has been important in every sector of the economy throughout history. As I argued in Chapter 1, what is new is how people get access to information.

The Politics of Information

Given that the value of seeing information as the pivotal concept in the study of ICTs has been shown to be limited, the central role of processes that shape access to information has become more evident. The cliché that 'information is power' is seldom examined critically. It suggests that experts will gain power over politicians, managers, and the lay public. Yet this thesis ignores the degree to which 'access to knowledge (and to the means of producing and disseminating knowledge) is itself facilitated by the ownership of material capital' (Robins 1992: 5). It is not information *per se*, but the ability to control access to information that seems key. The structures and properties of social systems, such as ownership, can shape access, as well as technologies.

The role of ICTs in shaping access to information might be called 'information politics' (Danziger *et al.* 1982: 133–5). This is also the concern of Nicholas Garnham (1994*a*: 49), who has argued for a research agenda focused on the sociology of information production, dissemination, and consumption, based on the assumption that 'informational power is differentially and hierarchically distributed and involves social specialization. It would not assume that the power of information definers went unchallenged or that dissemination was not problematic. But nor would it accept that consumers of information are free to make of it what they will.' Garnham directs us to study the processes involving producers, distributors, and users in negotiations over how information is created, disseminated, and consumed (see Essay 3.3).

ICTs are only one element in what can be viewed as a far more general politics of information. As I argued in Chapter 1, ICTs also shape access to people, services, and other ICTs. So it is critical to define information politics broadly to include the choices shaping tele-access and the factors that constrain and facilitate them.

The Evolution of an Information Society: Confusing Carriers and Content

The idea of a staged progression of an economy from an agricultural, to an industrial, to an information society (or information age), and beyond has led to many forecasts as to what will come next, such as a 'knowledge society', or 'age of creation' (Murakami, Nishiwaki, *et al.* 1991).

The understanding of the computer as a creator of information or a source of creativity fits what personal computing pioneer Alan Kay (1991: 100) has called a 'confusion between carriers and content'. Making an analogy with a piano, Kay (1991: 100) argued:

> Pianists know that music is not in the piano. It begins inside human beings as special urges to communicate feelings. . . . The piano at its best can only be an amplifier of existing feelings, bring

> forth multiple notes in harmony and polyphony that the unaided voice cannot produce.
>
> The computer is the greatest 'piano' ever invented, for it is the master carrier of representations of every kind.

Similarly, computers and telecommunications are not the source of information. They have instead revolutionized the way in which individuals access information.

Another pattern of social and economic evolution is the notion of the existence of long economic development cycles caused by successive technological revolutions.[4] This approach argues that the invention of a new technology, like the steam engine or the computer, can have applications across the economy that affect many facets of our lives—for instance, ICTs having effects well beyond what might be labelled as information work or information-processing.

The concepts of staged and cyclical theories of economic and technological change are both attractive because they promise some level of predictability. However, there is no consensus on the identification and validation of these stages or cycles of economic development (Robins 1992). Also, there is a great lag between the invention of a technology and its impact on the way things are done. It takes time for people to change their habits and beliefs and to accept a new paradigm for how they do things. This led to a new focus of attention on the role of ICT paradigms in shaping the impacts of technology in organizations and society (Perez 1983; Freeman 1996*a*, *b*). From this perspective, ideas such as the information society and economic cycles are important in part because they shape our views about the way the world works and, thereby, influence our decisions. Similarly, the idea of tele-access could influence decisions about ICTs in ways that would have implications for organizations and society as a whole.

Tele-Access: The Ecology of Choices Shaping Access

There have been significant flaws—as well as important and enduring insights—in previous theories and analyses of the social implications of ICTs (see Table 2.2). I have given more attention to notions of the information society, because these have become such a dominant perspective on ICTs. However, all these perspectives provide support for discussions of tele-access, even though all tend to marginalize the importance of access, or take it as given.

Table 2.2. Strengths and Weaknesses of Competing Perspectives

Perspective	Insights tied to tele-access	Flaws from vantage point of tele-access
Influence	Identified the key role of gatekeepers in shaping the impacts of media	Assumed access to audiences; emphasized content to the neglect of media–ICTs
Technology	Broadened range of consequences tied to technologies, identified control of and access to technologies as key impacts	Diffused focus from ICTs and issues of access by grappling with all technologies at all levels of society
Impacts	Established the thesis that technology matters, and merits study by social scientists	Technology is not deterministic, and its design and impacts are socially shaped
Strategy	Shows that ICTs are malleable and that individuals seek to use ICTs to reinforce their interests	Individuals/managers often lack control over design, implementation, and use of ICTs
Information	Identified the increasing centrality of ICTs across sectors	Revolution in access mistaken for change in the role of information

The ICT revolution throws access up for grabs in many respects and therefore helps to recognize its importance to many perspectives on the social role of the information revolution (Rosengren 1997). New and more empirically anchored perspectives on tele-access need to be identified to guide policy and practice. Critics of the early influence, technology, impacts, strategy, and information perspectives summarized in this chapter have provided various elements of such a new perspective on tele-access.

A concern with tele-access—the ecology of games shaping access to information, people, services, and technology—will not replace other perspectives on ICTs. But it can complement, integrate, and extend research from every perspective in the field.

What you see is largely dependent on what you look for. Research is not altogether different. It is designed to discover new insights, but research is most often guided by theory which tells researchers what to look for and what to ignore. As Karl Weick (1987: 98–9) has argued, if we 'prefigure what we see' then we should intentionally choose theoretical frameworks to 'improve the quality of understanding': 'If believing affects seeing, and if theories are significant beliefs that affect what we see, then theories should be adopted more to maximize what we will see than to summarize what we have already seen.'

Notes

1. This extends a categorization that Elihu Katz (1996) has offered to distinguish mass-media studies of 'influence' from studies of new technologies focused on 'information'.
2. McLuhan's ideas developed out of a qualitative approach to research and were not amenable to quantitative approaches, which dominated research in the media-effects tradition through the 1970s. For example, the concept of a 'global village' is not easily quantifiable, and, if quantified, is likely to veer from the broad meaning McLuhan conveyed by this term.
3. An excellent extension of the 'certainty trough' to the study of innovation in general is provided by Woolgar (1997).
4. This has been called a neo-Schumpeterian approach because of its ties to the work of the economist Joseph Schumpeter, who called long cycles of economic growth and decline Kondratiev waves (Miles and Robins 1992: 8–10).

Essays

2.1. Technological Determinism

Donald MacKenzie

The common view of the relationship between technology and society stresses the role of technology as the key driver. However, Donald MacKenzie argues that this perspective—known as technological determinism—has major weaknesses and he indicates how the social impacts of technology can be addressed more adequately.

2.2. The Social Shaping of Technology

Robin Williams

The concept of 'social shaping' offers an important perspective on the interaction between technological developments and the socio-economic environment in which they are used. Looking particularly in the workplace, Robin Williams explains what is meant by social shaping in relation to ICT-based innovations.

2.3. The Certainty Trough

Donald MacKenzie

How does intimate knowledge of information and communication technologies affect confidence in them? Donald MacKenzie reports on the results of research which challenge the conventional expectation that those closest to a technology have the highest level of confidence in it.

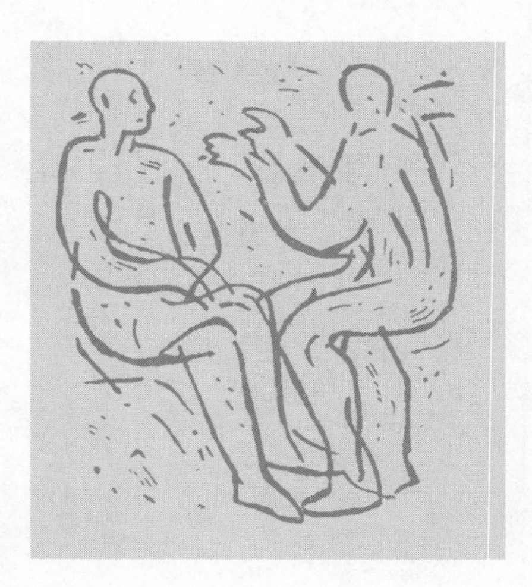

2.1. Technological Determinism

Donald MacKenzie

Two ideas are at the core of the multitude of forms in which the case for technological determinism is expressed (see e.g. Winner 1977; Smith and Marx 1994). First, there is the notion that technological change follows a logic of its own which, at least to some extent, is independent of human will. Secondly, there is the belief that changing technology brings with it social changes. Robin Williams's essay (Essay 2.2) in this chapter on the social shaping of technology addresses the first of the basic ideas of technological determinism. I shall focus on the second.

Clearly, this second idea can be held in weaker or stronger versions. The stronger forms of technological determinism, for example, assert not only that technological change causes social change, but that it is indeed the most important cause of social change. Notions of 'the information society' or 'the information technology revolution' often attribute some such causal role to technological change.

The kind of explanation typical of the idea that technological change brings social change can be clarified through the following example from a sort of technology quite different from ICTs. It is taken from a discussion by historian Lynn White (1978) on feudal society. I have simplified White's account in order to bring out key technologically determinist features.

White attributes the coming-about of a feudal society in Western Europe to the invention, and diffusion westwards, of the stirrup. Prior to the stirrup, fighting on horseback was limited by the risk of falling off. Swipe too vigorously with a sword, or lunge with a spear, and horse-borne warriors could find themselves lying ignominiously in the dust.

Because the stirrup offered riders a more secure position on their horses, it 'effectively welded horse and rider into a single fighting unit capable of violence without precedent' (White 1978: 2). But the mounted shock cavalry it made possible was an expensive, as well as an effective, way of doing battle. It required intensive training, armour, and war horses. White (1978: 38) suggests it could be sustained only by a reorganization of society designed specifically to support an élite of mounted warriors able and equipped to fight in this 'new and highly specialized way'.

Detailed historical research (such as Hilton and Sawyer 1963) usually tends to soften such technologically determinist stories that reduce the status of technological change to one factor amongst many. For example, another factor in the rise of feudalism was the decline in long-distance trade in the latter part of the first millennium AD, which made land the only real source of wealth. White (1978: 28) himself notes that 'a new device merely opens a door; it does not compel one to enter'. Perceived compulsion more often comes from economic or military competition than from technology itself.

Despite such provisos, the appeal of technological determinism remains. This is partly because most people experience technological change in their everyday lives as an external process, in which they have no involvement and over which they have no control. It is also intellectually very tempting to imagine that we can discover, and even predict, the social effects of innovations like ICTs. Indeed, that is often precisely what policy-makers expect from social-science research on these topics. Unfortunately, answers to questions about the 'social effects' of new ICTs are usually far from simple.

There are two reasons for this. One is general and the same reason that technological determinism largely fails as a theory of history. Social change is a complex, multifaceted process within which technological change is only one aspect—and one which is not independent, but is linked to other aspects of social change by complex feedback loops.

The other reason is specific to new ICTs and to the device at their core: the digital computer. In the words of computing pioneer Alan Turing (1950: 441–2):

> **This special property of digital computers, that they can mimic any discrete state machine, is described by saying that they are universal machines. The existence of machines with this property has the important consequence that, considerations of speed apart, it is unnecessary to design various new machines to do various computing processes. They can all be done with one digital computer, suitably programmed for each case.**

While, perhaps, no technology is entirely single purpose, the digital computer is, arguably, in this sense uniquely, if not entirely, flexible (MacKenzie 1991). This makes the software which takes advantage of its flexibility of enormous practical importance. It also means, however, that the social effects of the digital computer, or of the technologies based upon it, are unlikely to be straightforward.

PICT researcher James Fleck (1993) usefully distinguishes between technologies that are provided as 'systems' and those that are provided as 'configurations'. He characterizes those that come as systems as having strong constraints on how components must relate to one another, as well as clear overall development paths. Those that come as configurations are depicted as being more open to the contingencies of particular applications, with highly uncertain development paths.

Although there are systemic aspects of the technologies for handling both information and communication, the configurational aspects of current ICTs are at least as important as their systemic dimensions. For example, ICTs typically cannot be implemented successfully according to a pre-set, standard blueprint; they have to be tailored carefully to local circumstances, such as the specific features and tasks of the organizations employing them. Users need to be involved in shaping their development and implementation.

If the effects of even a simple technology like the stirrup turn out to be complex, then we cannot expect these flexible, diverse ICT configurations to have a straightforward, unidirectional, and universal social impact. That clearly complicates the task of the researcher interested in these effects.

More broadly, however, it is grounds for optimism rather than pessimism. It means that the 'information and communication technology revolution' is not taking a predetermined pathway to a definite outcome. Rather, it is a complex, contingent, and fundamentally open process. It is a process that we can potentially influence—as citizens and employees, not just as policy-makers and managers. That vision of the potential democratization of technology is one that has inspired many social researchers.

References

Fleck, J. (1993), 'Configurations: Crystallizing Contingency', *International Journal on Human Factors in Manufacturing*, 3: 15–36.

Hilton, R. H., and Sawyer, P. H. (1963), 'Technological Determinism: The Stirrup and the Plough', *Past and Present*, 24: 90–100.

MacKenzie, D. (1991), 'The Influence of the Los Alamos and Livermore National Laboratories on the Development of Supercomputing', *Annals of the History of Computing*, 13: 179–201.

Smith, M. R., and Marx, L. (1994) (eds.), *Does Technology Drive History? The Dilemma of Technological Determinism* (Cambridge, Mass.: MIT Press).

Turing, A. (1950), 'Computing Machinery and Intelligence', *Mind*, 59: 433–60.

White, L. T. (1978), *Medieval Technology and Social Change* (New York: Oxford University Press).

Winner, L. (1977), *Autonomous Technology: Technics-out-of-Control as a Theme in Political Thought* (Cambridge, Mass.: MIT Press).

2.2. The Social Shaping of Technology

Robin Williams

Policy-makers and the public have often taken the course of technological progress for granted—as if technology developed according to some predetermined technical rationality—and assumed that the content and direction of technological innovation were not amenable to social analysis and explanation. Such a view limits the scope of social-scientific enquiry to monitoring the 'impacts' of technological change upon economic and social life.

However, social scientists have increasingly recognized that technological change is shaped by social factors. This perspective, known as the Social Shaping of Technology (SST), emerged in part from critiques of technological determinism (see Essay 2.1). SST studies show that the generation and implementation of new technologies involve many choices between technical options. A range of social factors affect which of the technical options are selected. These choices influence the content of technologies and, thereby, their social implications. In this way, technology is a social product, patterned by the conditions of its creation and use (Williams and Edge 1996).

Studies of the generation of ICTs have identified social-shaping processes occurring at even the most detailed technical level. For example, technology design has been shown to be shaped by concerns of local significance to the R&D process. Thus Molina (1990) has shown how developers' desire for technological sophistication drove technical choices in the design of microprocessors in a way that, paradoxically, inhibited the commercial success of their products, as they required skills in a programming language that were not widespread among their prospective users. This is one illustration of why products emerging from R&D must often be subsequently transformed to meet the exigencies of commercial production and use.

This finding does not fit conventional 'linear' models of innovation, which see innovation only in initial technology supply and presume an essentially one-way flow of ideas and artefacts from R&D to production, diffusion, use, and consumption. Instead, much SST research highlights the degree to which the design of technologies is open to negotiation and the way many factors bear upon its development, application, and use.

The classic illustration of social shaping is the historical study by David Noble (1979) of the development of automatically controlled machine tools in post-war USA. Noble noted that two options existed: 'record playback', whereby the machine merely replicated the manual operations of a skilled machinist; and 'numerical control' (NC), in which tool movements were controlled by a mathematical programme produced by a technician. He showed how the machine-tool suppliers, technologists, and managers in the aerospace companies deliberately suppressed record playback in favour of NC in order to reduce their reliance on the, often unionized, craft workers.

This attempt to obviate manual skills was only partly successful as metal-cutting skills were still needed in preparing and 'proving' NC programmes. Subsequently machines were developed—notably in Japan and Germany—for programming on the shop floor. These were more attractive, especially to smaller engineering firms, as they allowed greater flexibility and a lower division of labour.

This case highlights the divergent requirements and presumptions of technology developers and users. This is one reason why technologies are not fixed at the design stage but evolve in their implementation and use. PICT researcher James Fleck (1994) has argued that the implementation arena is comparable to an industrial laboratory, where suppliers and users jointly learn about the utility of technological products and about user requirements. This is particularly important in integrated corporate information systems which support a wide array of complex activities. These must be closely adapted to the structure and methods of the user organization—a lack of fit may result in unwanted organizational adaptation, or even the failure of the system. On the other hand, the struggle to apply and use technologies may also generate valuable innovations that can be fed back into future technological supply. These innovations may become widely available as cheap, standard commodities, such as software packages, and incorporated as components of future technological systems.

This is one way in which pre-existing models of work organization, together with visions of how these might be transformed, become embedded in industrial ICT applications. It shows how social relations become crystallized as 'technology'.

Although SST researchers initially tended to conceive of 'the social' as an external force influencing 'the technical', subsequent studies have tended to blur boundaries between the technical and the social. Thus it is important to view workplace technologies not as an external force which transforms work. Instead, technology and work organization develop in tandem. The 'technical' is socially constituted. Conversely, social structures, such as organizations, cannot be analysed in isolation from their material underpinnings (Clausen and Williams 1997).

References

Clausen, C., and Williams, R. (1997), *The Social Shaping of Computer-Aided Production Management and Computer Integrated Manufacture* (European Commission/COST A4; Luxembourg: Office for Official Publications of the European Communities).

Fleck, J. (1994), 'Learning by Trying: The Implementation of Configurational Technology', *Research Policy*, 23: 637–52.

Molina, A. (1990), 'Transputers and Transputer-Based Parallel Computers: Socio-Technical Constituencies and the Build-Up of British–European Capabilities in Information Technologies', *Research Policy*, 19: 309–33.

Noble, D. F. (1979), 'Social Choice in Machine Design: The Case of Automatically Controlled Machine Tools', in Zimbalist (1979), 18–50.

Williams, R., and Edge, D. (1996), 'The Social Shaping of Technology', *Research Policy*, 25: 865–99.

Zimbalist, A. (1979) (ed.), *Case Studies on the Labour Process* (New York: Monthly Review Press).

2.3. The Certainty Trough

Donald MacKenzie

The relationship between close proximity to knowledge and confidence in that knowledge is not a simple linear one. The more complicated relationship illustrated in Fig. 2.1 is likely to be a more realistic pattern, according to several pieces of research (e.g. MacKenzie 1990, 1994).

The horizontal axis in Fig. 2.1 shows 'social distance' from the technology in question. On the left are the 'insiders' directly involved in producing knowledge about the technology—for example, its designers or those who conduct and analyse its testing. Next along the horizontal axis come those who are not insiders in this sense, but who nevertheless have a commitment to the technology in question. Many users of a technology would fall into this category, as, typically, would the senior managers of firms producing or using it.

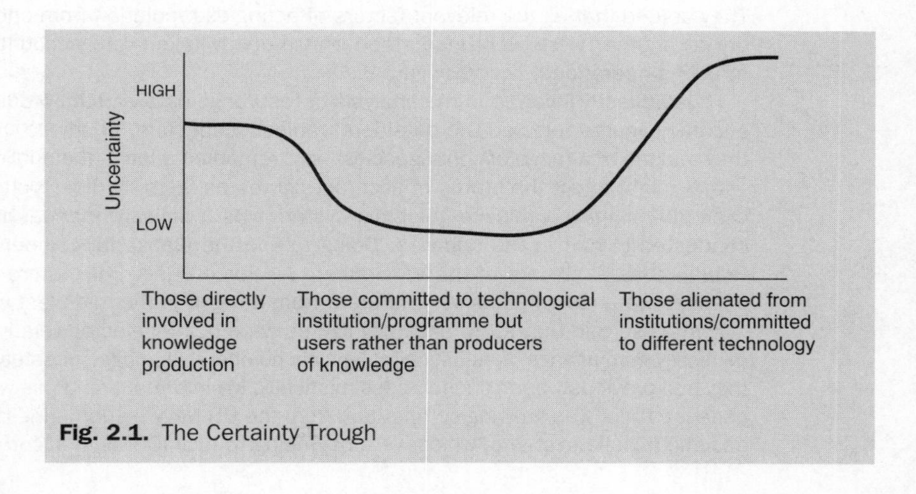

Fig. 2.1. The Certainty Trough

Moving to the right of the horizontal axis, we encounter first the uncommitted and then those who are hostile to the technology in question. These latter 'outsiders' might include people who are committed to a competing technology or people who distrust the institutions responsible for the technology.

The vertical axis in Fig. 2.1 represents perceived uncertainty about matters such as the reliability, safety, or predictability of the technology in question. Unsurprisingly, such uncertainty can be expected to be high amongst outsiders. Equally unsurprisingly, it is much lower amongst those who are committed to the technology, but are not directly involved in the production of knowledge about it, the inhabitants of what I call the 'certainty trough'. More surprisingly, though, uncertainty seems often to rise again amongst insiders.

I first found this pattern while studying the high-precision guidance systems of modern American inter-continental ballistic missiles (ICBMs). These have inertial guidance systems, in which an on-board computer receives data from electromechanical motion sensors, such as gyroscopes, and adjusts the direction of rocket thrust to bring the missile warheads onto a trajectory to their targets (MacKenzie 1990). The uncertainty that interested me had to do with whether the figures summarizing the accuracy of such missiles were actually reliable facts about them.

The possibility for doubt arose because ICBMs could not be tested over their war-time trajectories—going from their US silos north over the Arctic to their intended targets in the Soviet Union. Test ranges were used instead.

Outsider critics of ICBMs were often either opponents of the entire nuclear arms race—or proponents of the manned bomber, an alternative technology to the missile. The term 'manned' is not an accidentally sexist expression in this context, as many in the US Air Force saw sitting in a hole in the ground waiting to push a button as an unmanly and deskilled activity, compared to flying a military aircraft (see MacKenzie 1990 for more on the battles in the USA between proponents of the missile and the bomber).

Outsider critics were deeply sceptical about extrapolating from test-range missile accuracy figures to predict operational performance. At their most extreme, they believed missile accuracies to be both unknown and unknowable. In contrast, those heading ICBM programmes defended missile accuracy figures as true facts. They argued that all the relevant factors affecting extrapolation from one trajectory to another were well understood and properly taken into account in constructing operational accuracy figures.

Those directly involved in the analysis of test-range data and the production of accuracy figures rejected the outsiders' radical scepticism. In private, however, they told me of a range of contingencies—not remarked upon in the public debate—that might affect the status of accuracy figures as facts. In older systems, for example, aligning a missile guidance system was a delicate manual operation conducted in part in the open air. Doing this in the pleasantly temperate Californian coastal climate of the Vandenberg Air Force Base, from where the test missiles were launched, is very different from doing it in the mid-West winter.

I was also told that ICBM accuracy figures were being selected relatively optimistically from a range of statistically plausible numbers, although, for safety's sake, they had previously been chosen pessimistically. My informant said this was done because there was growing competition from the US Navy's submarine-launched ballistic missile systems, which were beginning to rival ICBMs in accuracy.

More recent research suggests the certainty trough may also help to make sense of different responses to less esoteric ICTs. In researching computer safety, for example, I found a vigorous insider culture of distrust of specific aspects of safety-critical computer systems (MacKenzie 1994). This was expressed in the long-standing Risks news group on the Internet and in the corresponding section of *Software Engineering Notes*, which is published by the US professional computer society, the Association for Computing Machinery. This insider perspective existed alongside both user and management confidence (indeed sometimes overconfidence), as well as the popular fears about out-of-control computer systems encapsulated in films such as *2001* and *War Games*.

Ethnographic studies of the development and use of ICTs have found similar patterns (Woolgar 1994). In a quite different context, research on computer modelling of the global climate has also found certainty-trough phenomena (see e.g. Shackley and Wynne 1996). The greatest degree of certainty is evinced by those who, for example, use the results of models for studies of the impact of global warming, without themselves being involved in their construction. Those directly involved in building the relevant computer models point to uncertainties and approximations in this modelling, as well as to marked differences between the results of different models.

Although I have portrayed the certainty trough in the form of the graph, it is a qualitative, not a quantitative, phenomenon: the axes in Fig. 2.1 have no unit of measurement and I know of no quantitative research which directly addresses this issue. Conventional attitude surveys would not be expected to pick up the insider uncertainty discussed here. This is partly because there are very few insiders in this sense relative to the population at large, and partly because questionnaires with pre-set questions are not good instruments for elucidating highly specific, nuanced, esoteric uncertainties.

There are two main social-science roots of the certainty-trough idea. One is work by sociologist of science Harry Collins (1987) on the uncertainties evinced by those performing 'frontier science', especially in areas of controversy. The other is studies by anthropologist Mary Douglas on how outsider status tends to lead to distrust of the central institutions of modern industrial society—and thus to a perception of great risk and uncertainty associated with the technologies they manage (Douglas and Wildavsky 1982).

The certainty trough matters for policy in two ways. The first is that technological decision-makers and senior managers should be aware that their situation may generate undue confidence, as it demands deep commitment to the technologies they manage, although it often prohibits intimate knowledge of them.

One reason for my original research on the topic was a concern that an international crisis might lead to a pre-emptive nuclear strike being launched by politicians and top generals who took the accuracy figures for their missiles unproblematically to be facts. More mundanely, there is evidence that overconfidence by management or users has been at the heart of several ICT disasters which have been expensive in both lives and money (MacKenzie 1994; Peltu *et al.* 1996).

The second way the certainty trough matters concerns the consequences when insider uncertainty surfaces in public. If we expect science and technology to yield certain knowledge and utterly reliable systems, the surfacing of insider uncertainty can lead to shock and overreaction, as happens in many controversies relating to health implications of eating certain types of food. Instead, we have to learn to accept that a degree of uncertainty in science and technology—as in all other human activities—is a normal, rather than a pathological, state of affairs (Collins and Pinch 1993).

References

Collins, H. (1987), 'Certainty and the Public Understanding of Science: Science on Television', *Social Studies of Science*, 17: 689–713.

—— and Pinch, T. (1993), *The Golem: What Everyone Should Know about Science* (Cambridge: Cambridge University Press).

Douglas, M., and Wildavsky, A. (1982), *Risk and Culture: An Essay on the Selection of Technological and Environmental Dangers* (Berkeley and Los Angeles: University of California Press).

Dutton, W. H. (1996) (ed.), with Malcolm Peltu, *Information and Communication Technologies: Visions and Realities* (Oxford: Oxford University Press).

MacKenzie, D. (1990), *Inventing Accuracy: A Historical Sociology of Nuclear Missile Guidance* (Cambridge, Mass.: MIT Press).

—— (1994), 'Computer-Related Accidental Death: An Empirical Exploration', *Science and Public Policy*, 21: 233–48.

Peltu, M., MacKenzie, D., Shapiro, S., and Dutton, W. H. (1996), 'Computer Power and Human Limits', in Dutton (1996), 177–96.

Shackley, S., and Wynne, B. (1996), 'Representing Uncertainty in Global Climate Change Science and Policy: Boundary-Ordering Devices and Authority', *Science, Technology, and Human Values*, 21: 275–302.

Woolgar, S. (1994), *The User Talks Back* (CRICT Discussion Paper, No. 40; Uxbridge, UK: Brunel University).

Part II

Social Dimensions of the Technical: Social, Cultural, and Political Processes Shaping Tele-Access

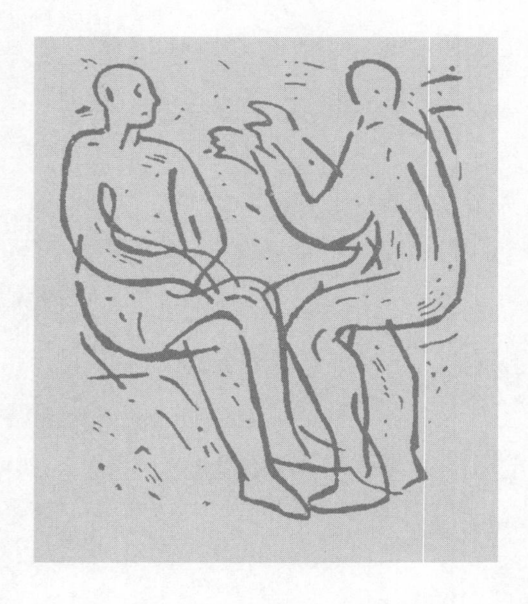

3 The Technological Shaping of Tele-Access: A Force for Social Change

Two Faces of Determinism

ICTs encompass the old as well as the new, the physical as well as the electronic. Choices among ICTs, as well as changes in their design, such as a move from print to electronic media, or from analogue to digital, can change features such as their cost, complexity, and capacity in ways that shape tele-access.

That major social and economic implications of ICTs revolve around changes in design as well as use could be seen by some as technologically deterministic, but it is not that simple. Technical choices are among the key factors influencing tele-access. However, this is not the same as saying, as the determinist might, that technology is the 'motor of history'; to do so would be to ignore the 'whole series of social choices that have been exercised in the production of the technology' (Webster and Robins 1986: 21). On the other hand, such valid critiques of technological determinism, particularly by scholars from the 'social shaping of technology' perspective (see Essay 2.2), have too often encouraged a form of social determinism that has led to the erroneous conclusion that technological change is of no particular significance. Reality lies somewhere between these two extremes.

Technical choices in the design and selection of ICTs are tied to what Nicholas Garnham has called 'communicative power' (see Essay 3.3). The results of particular design choices can bring you into the presence of some information and blind you to other information, and can make some social relations easier, others more challenging. Moreover, technologies can take 'a particular fixed version of social relations as the basis of action' and '[freeze] it in material form' (Woolgar 1996: 90).

Technologies are like public policy in that they are a product of social choices and can influence social behaviour long after the choices are made. Unlike public policy, however, they are made by individuals who are not accountable to an electorate, nor often to the market place. Too frequently, technical choices are delegated to others as if they had no social significance. Nevertheless, choices about the design and use of ICTs have consequences for communicative power, the politics of information, or—most broadly—tele-access.

The Technological Revolutions

So much has been said about the revolution in ICTs that many now see it as a foregone conclusion. However, ICTs are not inherently driving the freedom, control, or surveillance of society. The explosion resulting from the fusion of ICTs is shaping tele-access—for the better and the worse—in more complex and multidimensional ways.

To understand this uncertainty, it is useful to distinguish between:

1. the core innovations;
2. the ways in which these innovations are tied to the convergence and fusion of ICTs;
3. the shape of the ICT industry; and
4. the ways in which ICTs are redesigned and applied within specific settings.

These four factors are often lumped together in discussion of the information revolution. But, although they are interrelated, they should be viewed as separate developments.

Rapid Innovation at the Core of the ICT Revolution

Box 3.1 summarizes the six elements at the core of the revolution in ICTs: enabling technologies, equipment, content, user interface systems, telecommunications infrastructures, and people.

Box 3.1. Core Elements of Information and Communication Technologies

- *Enabling technologies*: microelectronics, optoelectronics, and successive generations of operating systems, programming languages, and application programmes that control the operations of a computer.
- *Equipment*: telephones, televisions, personal computers, scanners, game machines, compact disks, digital cameras, telecommunications switches, books, and other hardware used to receive, store, process, or display information.
- *Content*: the words, text, pictures, voices, services, or other contents that are stored, used, or transmitted.
- *User interface systems*: combinations of hardware (such as a keyboard, a mouse, or virtual-reality 'eyephones and datagloves') and software (such as graphical user interfaces, or voice recognition systems) for human interaction with computers.
- *Telecommunications infrastructures*: advances in narrow and broadband networks, including switching technologies, which are used in wired and wireless transmission, including terrestrial broadcasting, cable TV, fibre optic, cellular, and satellite communications.
- *People*: secretaries, authors, business planners, software developers, accountants, students, engineers, doctors, and others who use, create, manage, read, and support ICT-based systems, services, and equipment.

A critical and symbolic engine of the ICT revolution has been the advances in microelectronics, such as the integrated circuity which permits ever-increasing numbers of transistors to be arrayed on a single 'chip' of silicon. Microchips are at the heart of all ICT systems, from laptop PCs to supercomputers and from mobile phones and faxes to TV sets and virtual-reality games machines. The rapid and relentless progress made by microelectronics has been encapsulated in Moore's Law, which was first formulated in the 1960s by Gordon Moore, co-founder of the leading microelectronics company Intel. It has accurately predicted that the number of transistors that could be placed on a single chip would double every eighteen months (Miles *et al.* 1990: 16).

The escalating growth in the capacity of microprocessors represents a doubling of microprocessor capacity every eighteen months (Gates 1995: 31), resulting in greater speed, greater capacity, more functionality, and increased reliability for less cost (see Box 3.2). Fig. 3.1 shows the rapid escalation in the speed of processors and networks over two decades. That is creating real and perceived advantages for digital over competing analogue technologies in activities as diverse as print, film, recording, radio and television, telecommunications, and other computer-based media. This is the major economic driving force behind the eventual digitalization of all media.

Box 3.2. The Speed of Computers, 1971–1995

Capability of a microprocessor in additions per second

Year	Additions per second
1971	60,000
1974	290,000
1979	330,000
1982	900,000
1984	1,500,000
1985	5,500,000
1989	20,000,000
1993	100,000,000
1995	250,000,000

Note: Intel and other manufacturers no longer use 'additions per second' as a benchmark for the capability of microprocessors.

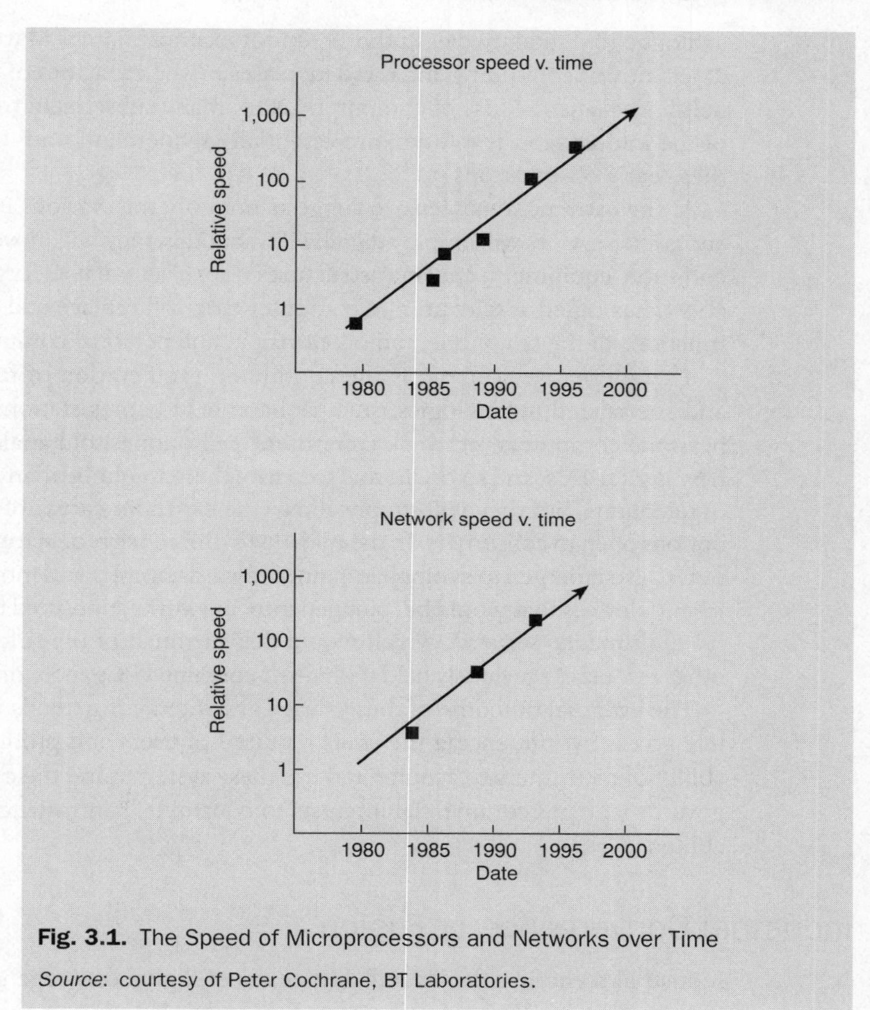

Fig. 3.1. The Speed of Microprocessors and Networks over Time

Source: courtesy of Peter Cochrane, BT Laboratories.

The Impact on ICTs: Digital Convergence or Multimedia Explosion?

Technological progress in such enabling ICTs as microelectronics and optoelectronics could result in both a convergence of media and an explosion of options, or various paths in between.

For decades, proponents of digital media have argued that advances in computer-based systems for encoding, transmitting, storing, processing, and displaying information should lead to the transformation from analogue to digital media. In the 1970s, for example, James Martin (1977: 27) identified continuing and rapid increases in 'the capacities of the channels in use' and the 'use of computers'. Very many subsequent treatments of the information revolution are essentially refinements and updates of these early observations.

At the extreme, this scenario suggests not only that various ICT appliances and services will employ digital ICTs, but that they will move towards common equipment and infrastructures—such as what George Gilder (1994) has called a 'teleputer', a computer that will replace and serve the functions of the telephone, radio, television, and personal computer.

The alternative scenario is the continuing proliferation of many new and specialized technologies, such as hand-held games stations, pagers, personal computers, network computers, cell phones, old analogue and new digital TVs, and so on. In this scenario, there might be many versions of teleputers, but they will simply add to the expanding array of technical options open to consumers. In the mid-1990s, the concept of 'network computer' also emerged to symbolize a move towards simpler and more diverse 'client' devices that would be 'plugged into' networks supported by powerful multimedia 'servers', which would deliver much of the software and other resources previously held within self-contained PCs and workstations.

The eventual outcome of these varied and interacting trends will affect tele-access by influencing the skills required of users and producers, the ability of consumers to acquire and use these systems, and the capacity of government and commercial interests to control ICT infrastructures and content.

Industrial Convergence or Fusion

Industrial scenarios often parallel forecasts of digital convergence. There have been many long-term forecasts that technological convergence will lead to the convergence of the information and communication industries. In the communications industry, for example, George Gilder (1994) has argued that the growing cost-effectiveness of ICTs has made the existing telephone, film, and television industries technologically obsolete because they have changed the cost and revenue structure of these industries in ways that are unsustainable.

The economic foundation of the telephone companies is based on people willing to pay a significant amount of money for the small amount of bandwidth required to make a telephone call. A typical telephone conversation lasts only a few minutes and requires a bandwidth of about 4,000 Hertz (4KHz) to convey the required information with enough fidelity for people to understand (see Pierce and Noll 1990: 49–50 for an excellent explanation of bandwidth requirements). In contrast, a normal analogue television signal requires 1,000 times as much bandwidth, about 4MHz. If a switched (like the telephone network) broadband (terrestrial, cable, and satellite broadcasting) network to the household was used to transmit television, then the cost of delivering a Hollywood movie would need to be competitive with alternatives, such as a video cassette rental. This would make telephone calls virtually free and undermine the economic foundation of the largest communication industry in every industrial nation.

However, there are many informed observers who are sceptical about convergence and its impact on the media industries because it has been long forecasted and very slow in arriving. In the 1960s, for example, the marriage of computers and coaxial cable TV created an unwarranted optimism about the convergence of all kinds of information and communication services (Dutton *et al.* 1987*a*). Computers might have made interactive communications over cable systems possible and coaxial technology might have made it feasible to transmit more channels and therefore make room for cultural, minority, and educational content. However, neither the computer nor coaxial systems made these uses more profitable or more popular with audiences. Many other factors, such as the culture, skills, and knowledge of the various industries, constrain the convergence of media (see Essay 3.1).

In the short term, at least, advances in ICTs have supported an explosion of new digital, analogue, and various hybrid media that provide new services for old industries. You can get your newspaper online. Microchips are in your TV. You can access the Internet through some cable TV companies. Telephone switches are digital. More and more computers are networked. The new media formed by the convergence of ICTs have also permitted existing companies to create new services and subsidiaries. Book publishers are distributing books on tape and CD-ROM, not just in print, and are creating whole new businesses around the provision of multimedia CD-ROMs and services on the Web. Cable companies offer Internet access. Phone companies provide wireless cable TV.

Although sometimes destructive of jobs, or old media, these advances have also been a creative force in unexpected areas. For example, the development of multimedia has brought some artists into the computing field and attracted some ICT experts into the world of computer-aided design and desktop publishing. Likewise, the film industry is bringing in ICT experts to create computer-generated special effects and creative artists are learning how to animate their work on the computer. Professor David Stout of the London Business School calls this creative cross-fertilization process one of 'fusion' (see Essay 12.1). The technological convergence around digital media could be leading to a fusion of individuals and industries that will generate many new alternatives, rather than a convergence around a single new medium of communications.

Applications of ICTs for Tele-Access

All the major media can be used to provide tele-access, which can affect how we come into the presence of information, people, services, and other ICTs (see Chapter 1). For example, if a library buys computers instead of books, or vice versa, a difference is made in what information individuals are likely to access. The expanding array of services that are available electronically make issues of tele-access increasingly central to all areas of everyday life and work (see Essay 3.2). The way that ICTs are being applied to shape tele-access is summarized in Table 3.1.

Table 3.1. How Different Media Shape Access to Information, People, Services, and Technologies

Media	Information	People	Services	Technology
Print	Newspapers, magazines, books	Mail, letters, memos	Mail-order catalogue, bank by mail	Print guides to TV and films, computer typesetting, satellite distribution of newspapers
Film	Photographs, documentaries, surveillance	Hollywood movies, cinema	Video cassette rental, training, instructional video	Digital film editing, animation
Recording	Audio tape transcription, books on tape	Popular music	Juke box, answering machine	Synthesizer, digital audio tape
Radio/TV	News, weather, sports	Entertainment, sitcoms, game shows, talk radio, talk shows	Pay-per-view, televised courses, video conferencing, business television, surveillance	Cable and satellite distribution of radio and TV, links to Web pages
Telecoms	Audiotext services, directory assistance	Telephone calls, facsimile, conference calls	Security alarm services, paging, ordering goods and services	Infrastructure for e-mail, fax, anwering machines
Computer, digital media	On-line news, magazines, World Wide Web, list servers	Electronic mail, video mail, electronic bulletin boards, computer conferences	On-line commerce, banking, airline and hotel booking, betting	Modems for networking over phone lines; telephone calls over the Internet

Applying ICTs to tele-access has multifaceted and interacting dimensions. For instance, a book is an ICT artefact, but of equal significance are the activities that enable people to produce and use this particular technology, like literacy and printing. In the same way, the Internet encompasses many of the ICT elements identified in Box 3.1, as well as know-how and skills, such as how to use a PC, create a Web page, or 'surf' the Internet to explore many areas of interest. Access to the Web can be limited by the lack of know-how as much as by a lack of equipment.

The Bias of ICTs: The Technology Perspective

A Canadian historian, Harold Innis, advanced a thesis in the 1950s to counter prevailing economically deterministic perspectives. He argued that a society's communication infrastructure was more important than the basic mode of production in shaping—he used the term 'biasing'—its social and political structures (Innis 1950). For example, he explained that oral communication is less adaptable than the written word for transmission over long distances and less durable over time—but more difficult to control centrally. His historical analysis illustrated the complex connections between communication technologies and the expanse, duration, and control structures of empires and civilizations. Based on this perspective, Innis and his followers claimed that the electronic media compressed time and space—the 'global village' of McLuhan (1964)—as well as creating the potential for more centralized control.

The idea that ICTs can enable a message to reach further, endure longer, and be more centrally controlled lies at the heart of the question of how ICTs shape tele-access. According to Innis, and others such as Langon Winner (1986: 19–39), technologies can be inherently more compatible with some structures of control than with others. In this respect, Innis suggested that print technology was more compatible with centralized control structures than was oral communication. It was also inherently more capable of being transmitted over a distance and retained over time.

Innis has been dismissed at times as a technological determinist. But he has a strong sense of indeterminacy, as underscored by his historical accounts which are packed with contradictions and exceptions to any general bias of technology. For example, he argued that any media, if carried to an extreme, can have a completely opposite bias. A written memorandum might have more endurance than an oral instruction, but a flood of memos in an organization might lead some to be totally ignored.

It is also possible that bias can be designed into a technology (Winner 1986). Traffic planners can design a street to speed or slow the flow of traffic, such as by using 'speed bumps' or 'sleeping policemen'. Landscape architects can guide college students through a campus with walkways. Engineers designed the Querty keyboard of the typewriter to avoid the jamming of mechanical parts by repositioning keys in ways that slowed down the typist. The design does not determine outcomes—look at the paths criss-crossing the lawns of college campuses or how a child can jam a manual typewriter—but it can make it easier and less costly to walk, or type, in particular ways.

Other ICTs, such as word-processing or spreadsheet-accounting programs, also make it easier to do things one way better than another. As Steve Woolgar (1996: 90) has argued in a discussion of computer software, the 'users of a technology confront and respond to the social relations embodied within it'. Users can adapt software for applications not intended by the developers, but this is likely to be more difficult, costly, or less socially acceptable because technologies establish a preferred course of behaviour (Woolgar 1996).

Technologies of Freedom, Control, or Surveillance

Much discussion of the social impact of ICTs has focused on control and has taken a deterministic view of its consequences. As discussed in Chapter 2, in the early decades of computing, technological advance was associated most often with more centralized control, such as the creation of huge databanks about individuals. The sheer complexity of computer technology, and the increasingly large-scale systems they support, was expected to distance technocratic élites from a public that would be increasingly uninvolved and uninformed about the decisions of public and private institutions (McDermott 1969).

Many social scientists continue to regard the ICT revolution as primarily a means for extending the control of large organizations (Beniger 1986; Carey 1989; H. I. Schiller 1989, 1996). For example, some have argued that ICTs are enabling the construction of a 'surveillance society' in which every citizen imagines their actions to be on camera (Flaherty 1985; Gandy 1989: 61; 1993).

However, a more dominant recent perspective has been that advances in ICTs—such as the telephone, fax machine, PC, and the Internet—create more democratic systems, undermining rather than extending centralized control. From this perspective, these technologies will inevitably spread information far and wide, thereby eroding hierarchies and monopolies of power across all sectors of society (de Sola Pool 1983*a*; Cleveland 1985; Gilder 1994; Dyson 1997).

Problems with Teatments of the Bias of Technologies

There are several critical weaknesses with these arguments about the biases of ICTs, which can be addressed by shifting our focus to tele-access.

First, much research in a variety of organizational and social settings runs counter to these predictions. As explained earlier in the chapter, the great malleability of ICTs tends to support their use as tools to reinforce prevailing patterns of control and to further key actors' values and interests. For instance, although the Internet might support inter-company communication, corporations often set up 'intranets' that facilitate intra-company networking. Likewise, new entrants in the entertainment industry can use the Web to market their work, but so can the established film and TV studios, who can also afford to advertise their Web addresses on TV.

Secondly, while actors attempt to shape ICTs to serve their interests, they are not equally capable of doing so. A top manager can exercise more control over technological change than an entry-level employee. A well-to-do high-tech professional can use ICTs to better advantage than a low-income household without a technical background. In addition, all actors pursue a variety of other objectives that might compromise rather than reinforce their values and interests. For example, consumers give personal information to companies in order to get credit, caring more about their convenience than privacy.

Actors are also affected by the choices of other actors in other arenas. If a competitor uses ICTs to streamline production, it becomes a greater economic imperative for the competition to emulate its move. Thus, the design and selection of ICTs are not controlled by any single actor, whether as a nation, company, or individual, nor any single motive. Instead, an ecology of social choices is shaping the design and selection of ICTs. The interaction of separate but interdependent social choices will have implications for tele-access that are not always in line with expectations based on the notion of reinforcement politics.

Finally, the technological perspective can often create an exaggerated set of expectations about the social and political implications of ICTs. Information might be an important resource, but there are many others, such as time, money, and status. Moreover, power and influence are historically and institutionally anchored in law, policy, social, and economic structures.

As Nicholas Garnham (Essay 3.3) argues, it is unrealistic to expect advances in ICTs to have a profound impact on, for example, the distribution of power or wealth in a nation. Discussion of the impacts of ICTs on the economic structure of societies or the expanse of an empire can mask the significance of more incremental implications that can be influenced by technical choices. Garnham also emphasizes the very real and meaningful implications that ICTs can have on 'information politics'—'the differences between individuals and groups in their ability to mobilize communication power in pursuit of their goals'. I would define the communicative power of actors by their relative ability to shape tele-access.

How Technological Advances Shape Tele-Access

Tele-access poses a number of conundrums. At a time of multiple phone lines, fax machines, and wireless communications, 'half the world's population lives two hours from a telephone'.[1] Many individuals with hearing or speech disabilities cannot communicate over the phone without special technological support, such as the Telecommunication Devices for the Deaf (TTD) terminals which the US Disabilities Act of 1990 requires all states to provide. Even in the late 1990s in California, which boasts one of the highest levels of on-line activity in the world, many home-computer users have experienced difficulties in getting access to the Internet.

But there are also many less obvious ways in which the problems of tele-access have consequences on power, control, privacy, community, and other long-range social outcomes.

Changing Cost Structures

The major social implications of ICTs stem from how technology is used. Advances in printing have made books affordable to more people. Advances in microelectronics have made computing affordable to the general public and permitted ICTs to become a component of an increasingly wide range of consumer products and services (see Essays 3.2 and 4.2). The Internet permits low-cost telephone calls for those with the necessary equipment. In such ways, ICT innovations can contribute substantially to lowering the cost of some consumer products and services and raising others, thereby, redistributing tele-access.

Even before the widespread introduction of competition in telecommunications, technological advances were bringing steady decreases in the costs of long-distance telephone calls (Noll 1994; Cochrane 1995: 22). It is difficult to separate the cost reductions driven by technological advances from the impact of competition, particularly since competition might spur technological innovation. However, there is evidence that technological change has been of comparable significance to competition in driving down costs (Baer 1996).

Not all costs are lowered by technological advances. For example, the costs of producing a major motion picture, launching satellites, developing complex applications software, or a large information database remain very high (Bortz 1985; Garnham 1990). R&D costs in some areas, such as that required for the development of optical telecommunications switching, have increased. This can advantage the largest companies and dominant firms, who have the financial muscle to invest larger sums. That combines with factors like the high cost of constructing telecommunications infrastructures to create barriers to entry for all but the biggest firms in certain markets and industries (Melody 1996).

Policies that support competition in telecommunications have therefore relied on regulations which force the existing and dominant carriers to open access to their networks so that new entrants can use the existing infrastructure to provide competing services. Nevertheless, as Robin Mansell argues (see Essay 10.1), those who own and operate the existing telecommunications infrastructure can design the technology in ways that advantage them in relation to their would-be competitors.

Expanding and Contracting the Geography of Access

The telegraph was a radical innovation in communications, arguably the first efficient system to permit access to information to be independent of transportation and to free 'communication from the constraints of geography' (Carey 1989: 204). By overcoming such constraints on tele-access, telecommunications can change one's psychological neighbourhood. As Suzanne Keller (1977: 282) noted in her work on the telephone: 'One of the most interesting questions is the meaning of 'near'—once human yardsticks are displaced by electronic ones.' This is ever more apparent with the Internet, which enables you to keep in regular, informal touch with people in distant locations and delivers files from around the world to your desktop as if they were stored on your own PC.

The dimensions of access identified by Innis—distance, time, and control—remain relevant to contemporary discussions of the social and economic role of ICTs. For example, the 'information superhighway' is an egalitarian vision of anyone getting access to any information, anywhere, at any time (see Box 4.3). Likewise, the Internet supports patterns of tele-access that are distinctly different from the printing press, the telegraph, or the telephone. Although it is possible to argue that many ICT-based innovations are essentially extensions of existing print and electronic media, rather than entirely new media (Eisenstadt 1980), advances in ICTs have created qualitative differences that change the kinds of equipment, techniques, and people involved in their design, production, and use.

More importantly, radical innovations in ICTs have altered patterns of tele-access across the economy, not just in communications. Changes in the cost and ease of gaining access to information or people, wherever they are located, can have dramatic implications for the structure, size, location, and competitiveness of business and industry throughout every sector of the economy (see Chapter 5 and Essay 12.1).

One potential consequence is that the largest players will be advantaged by virtue of ICTs expanding the geographical reach of firms. According to Bill Melody (1996: 308):

> **it is the largest national and transnational corporations and government agencies which have the greatest need for, and the ability to take full advantage of, these new opportunities. For them, the geographic boundaries of markets are extended globally—and their ability to administer and control markets efficiently and effectively from a central point is enhanced. These changes have been a significant factor in stimulating a wave of mergers and takeovers involving the largest transnational corporations throughout the 1990s.**

It is in these ways that the revolution in tele-access can support either greater competition and diversity in the provision of services or more oligopolistic control, depending on the strategies of users, providers, and regulators.

Restructuring the Architecture of Networks

A fascinating historical example is provided by the development and use of a line-of-sight telegraph system in France, as summarized in Box 3.3. The design of this 'visual-telegraph' network was so suited to the structure of the French state that, despite its many technical advantages, the electric telegraph was resisted by French authorities because 'it introduced two-way dialogue into a domain that . . . had been entirely conceived in terms of one-way monologue' (Attali and Stourdze 1977: 97).

Many scholars have viewed the telephone as a democratic medium of communication because of its simplicity, privacy, and support for person-to-person communication. By removing a dependence on a skilled telegrapher, voice telephony enabled electronic communications to become

Box 3.3. Development of the Visual Telegraph in France

Frenchman Claude Chappe invented the visual telegraph in the 1790s. It was constructed as a series of towers that radiated from Paris to cities within the various provinces of France. Initially built in response to military needs, each tower had gates, with high walls and guards to protect those who operated the movable arms used to relay coded messages by line of sight. The towers employed illiterate operators further to safeguard the security of messages that were encoded and decoded by specialists located at the end of each line.

The first line of telegraph towers was constructed between Paris and Lille, a financial centre north-east of Paris. By 1844 over 500 towers fanned out from Paris to reach twenty-seven cities. In 1856, roughly six years after the electronic telegraph was opened to the French public, the last visual telegraph tower was dismantled.

The longevity of the visual telegraph in France owed much to the way its centralized design reinforced the highly centralized administration of the French state. Not only did central government authorities restrict access to this medium to official uses, but its design enabled the central authorities to communicate easily with the provinces at the same time that it made it difficult for the provinces to communicate among themselves. The visual telegraph therefore supported what might now be termed vertical top-down, or centre-periphery, communication—as opposed to horizontal communication among the public.

Source: Attali and Stourdze (1977).

more egalitarian. Kenneth Laudon (1977: 16–17) explained this when calling the telephone an 'interactive' or 'citizen technology' by virtue of it supporting 'horizontal information flows among individuals and organized groups on a regular basis'. He noted:

> **The other technology families are largely vertical communication devices: computers allow experts to collect and centralize information about citizens, and the mass media technologies allow a small group of persons to broadcast to millions. Interactive technologies allow, on the other hand, ordinary citizens to talk with one another . . . to discuss issues and plan actions—in short, to communicate with like-minded people and groups on demand.**

Today, Laudon might add the Internet to his list of citizen technologies. But his point would remain valid. The facilitation of horizontal communication reinforces a more democratic distribution of communicative power.

Creating or Eliminating Gatekeepers

Technological change can also alter the role of gatekeepers in the dissemination of information. In an age of 'spectrum scarcity', with a limited number of television channels, newspapers, and books, gatekeepers play a critical role in deciding who gets on television, what is news, and what is fit to print. A proliferation of channels, news outlets, and desktop publishing operations can alter the role of gatekeepers, and possibly shift greater control to audiences. However, if the proliferation of options becomes overwhelming, audiences might well look again for gatekeepers to filter, prioritize, and select information for the consumer. In either case, the role of gatekeepers is likely to change.

Gatekeepers can play a powerful role in business and personal telecommunications, and not just in the mass media. The secretary was once the prime gatekeeper who screened and prioritized calls for an executive, but this role is becoming less common as electronic mail and voice mail tend to reduce the need for secretaries. Innovations in telecommunications switching systems are therefore increasingly shaping 'who says what and to whom'?[2] These changes can have consequences beyond any gains in the technical efficiency of the service.

For example, the invention by Almond B. Strowger of an automatic step-by-step switch to replace telephone operators was one of the most important advances in telecommunications.[3] This enabled a caller to use a finger-wheel dial (later replaced by the push-button phone) to make a call without the help of a telephone operator. The personal motivations which stimulated Strowger (see Box 3.4) to achieve this are an illuminating example of the diverse factors that contribute to the outcome of an ecology of games shaping tele-access.

When Strowger's switching system was first installed in 1892, the Bell telephone companies did not think their subscribers should dial numbers themselves and chose not to adopt his system. However, the switch was sold to independent telephone companies in the USA and telephone companies in other countries. It was eventually adopted by the Bell companies in 1919 (Pierce and Noll 1990: 170–2).

By 1928, the manufacturer of the Strowger switch, Automatic Electric Inc., had grown to seven factories in the USA and abroad, employing 10,000 people. Strowger's invention became one of the most important technological innovations in the history of the telephone, lasting nearly 100 years—a remarkable period for an electromechanical technology (Noll 1991: 103). Not until the 1980s did most telephone companies begin replacing electromechanical switches with computer software on digital technology.

Box 3.4. Personal Motivatations Behind the Strowger Telephone Switch

Almond Strowger was an undertaker in Kansas in the late nineteenth century. He grew up in New York, completing a university education that was interrupted by the Civil War, in which he served as a bugler. He began a teaching career but moved to Topeka, Kansas, where he started his career as an undertaker. He moved again to Kansas City, where he established his business.

In the course of running his business, he became incensed by the mistakes, delays, curt answers, negligence, and interruptions of telephone operators. He also became convinced that a particular telephone operator, the wife of a competing undertaker, was diverting business to her husband by reporting the Strowger line as being 'busy' whenever prospective customers rang him up. He vowed to eliminate human operators.

His work was supported by his nephew, Walter D. Strowger, whose father (Almond's brother), was a nurseryman and inventor, who had discussed the idea of an automatic telephone switch with Walter as early as 1880—nearly ten years before Almond patented his invention. Walter came to visit his uncle and ended up moving into his uncle's house to work with him on the development of the switching system. In 1889 Almond filed for a patent on his switching system.

On 3 November 1892 a special train took a group sixty miles from Chicago to the City of La Porte, Indiana, to see a demonstration of the Strowger Automatic Telephone Exchange. According to the press release at the time, the purpose was to demonstrate:

> to the electrical experts and to the scientific world, that through this system a degree of perfection has been reached that will be of great commercial value and enhance many fold the uses of the telephone. We will demonstrate that this system entirely obviates the many annoyances to which subscribers are at present subjected, besides being much more economical than the system now in use.

Sources: Pierce and Noll (1990: 169–73); Noll (1991: 102–3).

Redistributing Power between Senders and Receivers

Innovations in electromechanical switching not only allowed individuals to place their own calls, but also made the calls anonymous. Whether by accident or design, this anonymity contributed to the privacy and simplicity of the phone call. Now the new technologies threaten to take away this anonymity.

For example, e-mail was designed from its earliest years to identify the person sending a message. Much discussion surrounded the standards that would be set for e-mail networks within the US Department of Defense's Advanced Research Project Agency in establishing the ARPANET, which evolved into what became known as the Internet (see Box 4.6). One agreed standard was that the header of every message would identify the account name of the person sending a message, the account name of the recipient, the date the message was sent, and the subject of the message. All of the information about messages can be used by recipients as a means to prioritize and screen messages in order to manage better their communications and to become their own gatekeepers. But this system does not promote anonymity.

Call line identification or caller ID (see Box 3.5) is widely marketed as a tool for screening out unwanted calls, but is criticized as an invasion of privacy by disclosing personal information without the caller's consent (Dutton 1992*b*). Whatever your personal opinion about the desirability of this service, caller ID shows how intelligent digital networks and services can substantially and innovatively reshape who gets access to whom and to what information.

Box 3.5. Call Line Identification: Caller ID

Call line identification or caller ID service can remove anonymity in telephony by displaying the telephone number of the caller on a display panel before the telephone is answered. This number can automatically trigger interactions with databases that provide a display of additional information about a caller, such as name, address, and business-account histories.

A telephone call is transmitted as signals sent over a pair of copper wires. These wires and the systems that switched them have always been viewed as scarce resources, not to be wasted. Older telephone systems used the twisted wire pair to ring the telephone as well as to transmit any conversation. In a long-distance call, for example, this could tie up many circuits and switching resources while the phone rang, even when no one was talking.

Newer systems created a separate data signal to ring the telephone. Only when the phone is answered is an end-to-end circuit completed using the twisted wire pair. This was economical—promised to cut seconds off the ring time—provided the basis for third parties, like a long-distance provider, to bill users; and it provided other information which could help to enhance business competitiveness and improve and extend customer services. It also enabled the provision of caller ID services.

This technology is also illustrative of the complex mixture of motivations that underpin the technological choices that shape access. Caller ID was made possible by changes designed to improve the efficiency of the telephone network. Other things being equal, such improved efficiency in telecommunications networks is in the public interest. But one unanticipated outcome of these innovations was the loss of anonymity, as a person's telephone number was carried with his or her call. This was not driven by customer demand, nor even a strategy for offering the new caller ID service, but by efficiency concerns—a technical rather than a social rationality. Nevertheless, this technology has led to changes that have major social implications for tele-access.

Who Controls ICT Content?

An ICT innovation in broadcasting was high on the agenda in US President Bill Clinton's 1996 State of the Union Address: the violence-chip (V-chip). He saw this technical innovation (see Box 3.6) as central to his Administration's approach to protecting children, as it provides parents with a tool to help reduce their children's exposure to sex and violence on television.[4]

Box 3.6. The Violence-Chip (V-Chip)

The V-chip is an electronic device that can be installed in TV sets to identify TV programmes which parents deem objectionable, and to block display of the video portion of those programmes. In order for this system to work, a rating, much like movie ratings, is required for each programme. (Programmes can be rated or categorized in many ways, such as 'designed for adults; unsuitable for under 17'.) These ratings are encoded and transmitted within the vertical blanking interval of an analogue television signal. Parents or other members of the household could set the chip to screen out the programming with ratings they did not wish their children to view. The passage of a US telecommunications reform bill in 1996 mandated that a V-chip be installed in every new TV set sold in the USA, which could occur in 1999.

In fact, the technology used for the V-chip was developed through efforts to support the hearing impaired. For example, a chip could be installed to decode subtitles transmitted within the vertical blanking interval for display on the TV screen. The unanticipated development of a universal censoring device based on ICTs designed to provide greater access for a disadvantaged minority illustrates the unpredictable impacts of technical change on tele-access.

The V-chip is similar in some ways to a cable TV 'lock-box'. The lock-box was introduced in the 1980s to help parents prevent their children from watching adult movie channels on their cable systems. However, a lock-box does not require the transmission of any content that the programmer does not already provide. In contrast, the V-chip depends on content being rated and the ratings being transmitted with every programme, leading some to view this as a violation of the first amendment to the US Constitution, which protects against government control over the press and other political speech.

American broadcasters have long fought against any government-mandated ratings. For example, the film rating system used in the USA and and many other countries has remained voluntary. However, the popularity of President Clinton's proposal amid widespread public concern over violence on TV constrained any concerted action by the broadcasting industry.

The V-chip is a clear example of a technology that does not have an inherent bias with respect to the content it is dealing with. With the appropriate rating schemes, the same technology could be instructed to let through or screen out any type of programme, whether it is about religion, news, or violence. In practice, public policy objectives have now been designed into this device in the USA. This shows how technology is not only analogous to, but can also be an extension of, public policy. At the same time, it illustrates the indeterminant nature of ICTs. The V-chip will take many years to diffuse to most households. Parents might not use it. Children might circumvent the technology, just as many have done by gaining control of the TV remote control and video recorder.

Similar discussions about parental control over what their children do on the Internet have also raised complex issues where the nature of technical capabilities intertwine with public policy objectives and a diverse ecology of games to challenge fundamental social and political notions of freedom and control, personal responsibility, and shared community values. In the Internet arena many also believe that ICTs, like software that can screen out information from particular people or sites, can empower the user to control content. While many parents look towards the content providers to protect their children, many in the Internet community regard the use of screening devices as simply amounting to censorship.

The Future of Tele-Access

The contemporary and historical examples given in this chapter and elsewhere in this book illustrate some of the many ways in which technical choices can alter the relative communicative power of different actors. In fact, the historical relevance of tele-access is a major reason why this concept is of value in thinking about the future role of technological advances.

In addressing the consequences of technological change, this book is concerned with the 'social impact' of ICTs. Unfortunately, social scientists have often dealt with the social impact of technologies as if they could be extrapolated rationally from an understanding simply of the design of systems (see Chapter 2). As a result, many social scientists now avoid even the use of the phrase 'social impact' lest they be branded a technological determinist.

This is an important academic dispute because it highlights the theoretical controversies surrounding the relationships between technology and society (MacKenzie and Wajcman 1985; Mansell and Silverstone 1996). However, avoiding discussions of the impact of ICTs can blind the public to the many ways in which technical choices make a crucial difference to vital aspects of their lives. ICTs shape tele-access, but, as you will see in the following chapters, their impact in real social and organizational settings will depend as much on the strategies and social choices of consumers, users, providers, experts, regulators, and politicians as on the designs of the ICTs.

Notes

1. This statistic was provided by Acting US Trade Representative Charlene Barshefsky (Gerstenzang 1997).
2. I am paraphrasing Harold Lasswell (1971: 84), who argued that you could describe any act of communication by answering the questions: 'Who Says What In Which Channel To Whom With What Effect?'
3. My thanks to Sheldon Hochheiser with AT&T Archives for providing documentation from AT&T's collection on Strowger's invention. This section draws from reporting in *Sound Waves* (Nov. 1907), *Literary Digest* (7 Apr. 1928), *Science and Invention* (Apr. 1928), *Automatic Telephone* (Jan.–Feb. 1929).
4. It also provided a 'technological fix' to the legal and political issues surrounding first-amendment protections of speech (see Weinberg 1966 for a discussion of the technological fix).

Essays

3.1. Barriers to Convergence

Nicholas Garnham

Digital technology allows all forms of information to be recorded, processed, and distributed in a common electronic form. But Nicholas Garnham argues that it is wrong to assume that this technological convergence will lead automatically to a convergence of the industries developing and using digital technology.

3.2. ICT Innovations in Services

Ian Miles

The take-up of ICT capabilities varies considerably between the different areas that make up the services sector of the economy. Ian Miles analyses the likely future role of ICT-based innovation across a broad spectrum of services.

3.3. Information Politics: The Study of Communicative Power

Nicholas Garnham

Nicholas Garnham contends that research and policy-making should not be targeted on information and communication technologies, otherwise it will be difficult to avoid an overwhelmingly technological bias. Much more attention should be given to the ends to which technologies could be employed—the distribution of communicative power.

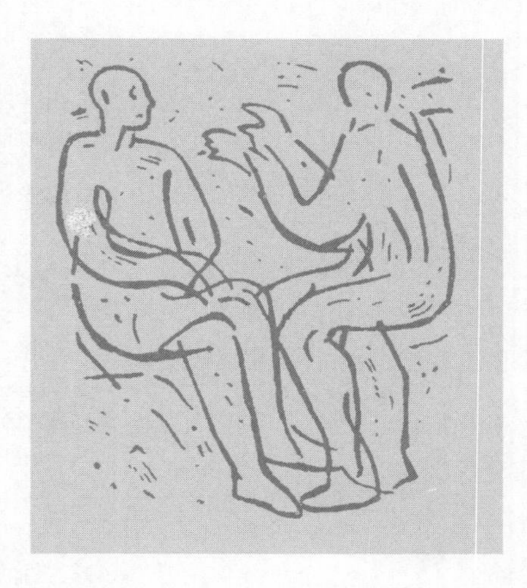

3.1. Barriers to Convergence

Nicholas Garnham

All communication industries have resulted from the need to overcome the problems of communicating across time and space. These include the separate industries of print, film, records, telecommunications, radio, and TV broadcasting. Each has distinct histories and development paths which have exploited the particular advantages of their own special technologies of production and distribution. In doing so, they developed distinct market niches, value chains, industrial structures, and related patterns of regulation.

Talk of convergence among these industries generally assumes a technologically determined scenario. This starts with the belief that a switched, broadband network will be rapidly provided to the majority of homes, either through adding bandwidth to the existing switched telecommunications network or by adding switching to the existing cable TV network. The end result is typically seen as a single communication 'pipe' delivering all media services electronically to homes and business premises. In some versions, all these services will be received by the user on a converged multimedia computer terminal. It is this vision that lies behind the term 'information superhighway'.

The existence of such a physical infrastructure, it is argued, will enable more than only conventional telecommunications and broadcasting services to be distributed over the same network. In addition, a range of new interactive entertainment and information services—such as video games, teleshopping, and other forms as yet undreamt of—will become possible and will eventually supplant more traditional one-way services. All of this will be paid for directly by consumers through some form of usage-based payment system. Many see the Internet as the realization of this model.

The optimistic version of this scenario argues that the result will be an increase in information and entertainment diversity, thereby enhancing individual choice and freedom. The pessimistic version stresses the dangers of monopoly control, of social isolation and fragmentation, and of the further decline of the public sphere. However, the scenario itself fails to address the technological, economic, cultural, and political barriers to a convergence which could create a new and seamless multimedia industry out of historically distinct communication industries (Garnham 1996).

It is clear that, where the regulatory system allows, telecommunications operators will continue to roll out technical capacity in their networks to deliver video to the home and cable TV operators will build in capacity to provide switched telephone and data services to their customers. Since the provision of a fixed-link local loop is likely to retain natural monopoly characteristics, this means that there is likely to be severe competition between telephone and cable companies to control that link. How this struggle works out in practice will depend largely on regulatory intervention based on competition criteria.

Nevertheless, there remain major technological barriers to smooth convergence. It is not clear, for instance, that it makes sense to merge networks designed for largely one-way delivery of broadband services with networks designed to optimize the delivery of switched two-way narrowband services.

Many examples of convergence involve just shared ducting, such as through the underground conduit running from the street to a person's home. This is because civil engineering, not transmission and switching facilities, remains the largest cost element in building a network. There is also uncertainty as to whether the most appropriate form of service delivery is via the provision of more bandwidth or by compression techniques which can exploit narrower bandwidths.

There are likely to remain significant regulatory barriers to convergence. On the one hand, fear of the natural monopoly characteristics of switched networks is likely to lead competition authorities to favour multiple-delivery networks and the separation of content and carriage. At the same time, political fears concerning media influence and threats to privacy are likely to favour the retention of various types of content regulation.

Even if we assume that the technological and regulatory problems are surmountable, the economic and cultural barriers to convergence remain significant. First, as falling transmission costs represent an increasingly small share of the costs of delivering a service, particularly an audio-visual one, any possible cost advantages of a converged network are unlikely to be the main determinant of market structure. Meanwhile, alternative delivery mechanisms will continue to exist and economic power will shift from control over scarce distribution channels to control over scarce intellectual property rights. The broadcasters, film companies, video-game makers, teleshopping retailers, and others who own that property are unlikely to allow any single distribution chain to dominate the market.

With the introduction of digital broadcasting, over-air transmission will become less scarce and therefore relatively cheaper. Conventional over-air broadcasting is therefore likely to remain the most cost-effective way for programme producers to reach economically viable audiences, financed by either advertising or subscription. Indeed, the problem for the telephone companies is to find services which will generate enough revenues to upgrade their networks. To justify broadband residential connections, for example, a significant revenue stream has to be found from a market tightly constrained by the limited availability of discretionary household expenditure and deeply ingrained consumption habits.

Another major barrier to convergence is the huge gulf between the cultures of the telecommunications and media business. Telecommunications operators are accustomed to dealing with the sale, largely to business customers, of a small set of standardized services based on the flows of large quantities of undifferentiated bits. This allows for little elasticity in the way services are priced. Mass-media markets, on the other hand, are quite different. They involve constantly creating new prototypes and selling them into a very uncertain residential market. The problems are ones of coordinating creative labour, controlling rights, and effective marketing. It is difficult, if not impossible, to combine these very different skill sets and associated reward structures within one organization.

As a result, we are likely to see new multimedia products and services developing out of the computer software and specialist publishing industries. These will be based on the technical infrastructure of the installed base of personal computers used primarily by professionals, managers, and teleworkers. They are likely to have minimal impact on either the existing telecommunications or mass-media industries.

These cultural differences are reinforced by the different markets on which the industries depend. The corporate market that telecommunications suppliers mainly serve is very different from the domestic entertainment and information market. These differences include the relation between buyer and seller, such as in the different importance attached to reliability in comparison to pure price, and the patterns of investment and skills required.

The problem of constructing a viable corporate structure to serve both markets with equal efficiency may prove to be the major barrier to the effective exploitation of the converged products and services that technological development potentially makes available.

References

Dutton, W. H. (1996) (ed.), with Malcolm Peltu, *Information and Communication Technologies—Visions and Realities* (Oxford: Oxford University Press).

Garnham, N. (1996), 'Constraints on Multimedia Convergence', in Dutton (1996), 103–19.

3.2. ICT Innovations in Services

Ian Miles

Services comprise the bulk of employment in advanced industrial economies. They span a wide range of activities, including: the wholesale and retail trade, repairs, hotels and restaurants, transport, financial intermediation, public administration, education, health, and social work. Given this diversity, any generalizations are clearly going to be problematic. Nevertheless, services are often contrasted with manufacturing in generic terms.

For instance, services are typically perceived as laggards in the use of new technology. They are also often regarded as passive users of that technology rather than innovators in their own right. The reality is very different, for services are actually the dominant users of new ICTs.

For example, services account for over 75 per cent of the investment in IT hardware in the UK and USA (Miles *et al.* 1990). Some service sectors are highly advanced in ICT use, as in financial businesses. Others lag behind, like the multitude of corner-shop retailers who may not even use electronic scales and cash registers.

In the future, cheaper and more versatile ICTs, such as mobile communications, are likely to be applied more widely in services (Ducatel 1994). Large numbers of ICT professional staff, including systems analysts and electronic engineers, are also being employed in services (see e.g. Miles *et al.* 1990; Kimbel and Miles 1993). At least 25 per cent of corporate R&D in some developed countries is undertaken by services companies (Miles 1995). Service firms are also centrally involved in software production, telematics systems, and other ICT-intensive activities which develop or transfer technology for the whole economy.

Manufacturing and services industries are becoming more similar in many respects. However, many unique features traditionally attributed to services still apply to many of them, particularly in relation to the specific forms that ICT innovation takes in terms of the service product, the production processes used, and the nature of the services market.

The service product is often largely intangible and frequently information intensive. Some services are produced and consumed at the same moment, like nursing; some are hard to store, transport, and trade. If you want your teeth repaired, you cannot order this by mail, or leave your mouth at the dentist, while you get on with your job. Process and product (and thus product and process innovation) are hard to distinguish. The customer is also often involved in some co-production. Even without this, services are often tailored to consumer requirements and there is much interaction between producer and user.

ICT innovations are often employed to enhance the capabilities of services. Telematics can be used for ordering products, reserving hotel rooms, and designing and delivering services that primarily involve information—for example, by using optical disks and other new media. ICTs can make traditional limitations of time and space less relevant—cash machines can operate around the clock, and teleservices can be accessed without going to the firm involved. ICT capabilities like electronic data interchange (EDI) can be used to input client requirements remotely and match these to service products. This offers opportunities for more self-service capabilities using telephones, PCs, and newer technologies and more user-friendly software interfaces.

The production of services is typically described as featuring heavy investment in buildings and less in technology. The labour involved in services includes both highly professional work—especially that requiring interpersonal skills—and relatively unskilled jobs, often using casual or part-time labour. The service labour process frequently consists of craft-like production with limited detailed management control of work. It is frequently non-continuous and has limited potential for economy-of-scale benefits. The organization of services is highly variable: some are large state-run operations, while, at the other extreme, there are many small-scale service companies, family firms, and self-employed individuals.

ICT innovations play important roles in services production. Building costs can be reduced by using teleservices, toll-free phone numbers, and other ICT capabilities. Key operations can be relocated in areas of low labour costs using telecommunications to maintain coordination. Reliance on expensive and scarce skills may also be reduced by using expert systems and related innovations.

More general reorganization of work practices can be facilitated by employing ICTs to monitor work—for example, with tachometers and mobile communications for transport staff. And flatter organizational structures can be facilitated by ICTs which allow data from field and front-office workers to be entered directly into databases and management information systems.

Finally, some unique characteristics of service markets mean it can be difficult to demonstrate products in advance. This can be overcome by increasing the use of techniques such as demonstration software and trial periods of use to enable the customer to get a feel of the performance before placing an order. Some products are also supplied as public services with little attempt to base prices on actual costs. ICTs are enabling new modes of charging to be introduced in what is becoming called the 'pay-per' society, together with more pricing flexibility to be introduced into retail systems. It is also worth noting that public and commercial services which are subject to regulation can use ICT capabilities such as databases and performance monitoring to provide the diagnostic evidence required by regulatory institutions, as well as being useful to service providers themselves.

There are great differences in the ways ICTs are used and developed by different types of service providers (Miles *et al.* 1996). For instance, hotels and tourist industries are relatively plant-intensive and have a low use of ICTs in core activities, except for a few pioneers. On the other hand, R&D, design services, and the providers of customized training packages have the transfer of ICT-related knowledge to clients as their core role, and are often pioneers in the use of ICTs. Generalizations about ICTs in services must, therefore, be made with caution.

References

Ducatel, K. (1994) (ed.), *Employment and Technical Change in Europe* (Aldershot, UK: Edward Elgar).

Kimbel, D., and Miles, I. (1993), *Usage Indicators: A New Foundation for Information Technology Policies* (ICCP, 31; Paris: Organization for Economic Cooperation and Development, Information Computer Communications Policy).

Miles, I. (1995) 'Innovation in Business Services: Knowledge-Intensity and Information Technology', paper presented at PICT International Conference, 'The Social and Economic Implications of Information and Communications Technologies' The Queen Elizabeth Conference Centre, Westminster, London, 10–12 May.

—— Brady, T., Davies, A., Haddon, L., Matthews, M., Rush, H., and Wyatt, S. (1990), *Mapping and Measuring the Information Economy* (Library and Information Research Report, 77; Boston Spa, UK: British Library).

—— Kastrinos, N., Flanagan, K., Bilderbeek, R., den Hertog, P., Huntink, W., and Bouman, M. (1996), *Knowledge-Intensive Business Services: Users, Carriers and Sources of Innovation in the EC* (DG13 SPRINT-EIMS; Luxembourg: Office for Official Publications of the European Communities).

3.3. Information Politics: The Study of Communicative Power

Nicholas Garnham

A technological fetishism has affected most research and policy-making concerning ICTs. However critical the research, however far towards the 'social-shaping' end of the agenda it seeks to go, if it starts with the problem defined in terms of technology, then it cannot easily, if at all, avoid the debilitating effects of that fetishism.

For example, the PICT research, on which this book is based, suffered from this bias. From its inception in 1985, the PICT research agenda was determined by the question it was set up to answer: 'What is the social and economic impact of ICTs?' The approach adopted, the nature of the questions asked, and the potential solutions available would have been very different if the basic question had been about cultural production and consumption, or about democratic participation, or about the changing nature of work. ICTs would then have been seen as part of either the problem or the solution, but the approach, the nature of the questions asked, and the potential solutions available would have been different.

ICTs have raised questions of social power ever since their birth with the invention of forms of writing. Once communication expanded beyond face-to-face interaction and the natural endowments of speech and gesture, the question of who commanded the cultural and material resources for communication—and for what purposes—became central to an understanding of the social order. I refer to this as 'information politics'.

The differences between individuals and groups in their ability to mobilize communication power in pursuit of their goals have always been intertwined with ICTs. Since we also know historically that those patterns of power distribution only change slowly, rarely, and with difficulty, it would be safe to assume that the so-called new ICTs are unlikely to be either as new or as dramatic in their impact, for good or ill, as the technologically focused approach assumes. And we should not let this focus distract us from attending to more fundamental questions concerning the unequal distribution of communicative power.

Thus, we must avoid the trap of thinking that an increase in the number of available TV channels will solve the deep-seated problem of the relationship between cultural producers and cultural consumers, or between different cultures, in societies characterized by division of labour and structured inequalities of wealth and status. It will also not do anything to illuminate the question of the social purpose and value of what has come to be called art. And it is unrealistic to imagine that the Internet will do much to solve the crisis of democracy and the profound philosophical and practical problems involved in attempts to specify and create a viable democratic political order.

We can hope to understand what impact, if any, new ICTs are having—or are likely to have—on the processes and structures of our society only by first focusing on understanding by whom and to what ends communication has been, and is, currently mobilized. For instance, changes in printing technology tell us something, although not much, about the economics of the press. They tell us nothing about its social and political role. Changes in broadcasting technology tell us nothing about the role of public-service broadcasting in fostering political and cultural democracy.

It is these questions of information politics that should be at the heart of the research and policy agenda. Research should focus not on how we can remove barriers to the introduction of multimedia services, but on how we can ensure a wider social spread of the cultural and political competencies and opportunities necessary for full and equal participation in social life. Without such an effort, the notion of the information society will be scarred by the same problems of inequality, injustice, and inefficiency as the society it is being designed to replace.

4 The Social Shaping of Tele-Access: Inventing our Futures

Predicting the Unpredictable: Technologies are Socially Shaped

The discussion in earlier chapters of the social processes shaping tele-access explains why no single set of actors can steer the development of ICTs along a chosen path. The unpredictable interactions of social choices shaping technological innovation and its outcomes place major constraints on efforts to guide and anticipate the course of ICT development.

As a result, forecasters have had a generally poor track record in anticipating public responses to innovative products and services.

Exciting forecasts of major markets for technologies such as interactive cable television and the video phone were not fulfilled. On the other hand, pessimistic forecasts of a limited future for advances such as the facsimile machine and wireless telephony proved also to be wide of the mark. As a result, policy-makers have often failed to accomplish their intended objectives, and many industrialists have lost money on ICT investments.

Alan Kay—one of a handful of key players in the development of ICTs—argues that the 'best way to predict the future is to invent it' (Kay 1994). Indeed, the personal preferences of pioneering entrepreneurs and inventors have influenced the design of ICTs, such as the personal computer (Cringely 1992). But the role of creative inventors is circumscribed by the impacts of many other actors. For instance, Alan Kay and his colleagues at Xerox PARC in the 1970s had a considerable influence on the computer industry (see Box 4.1), but they did not persuade their own corporate

Box 4.1. Xerox's Palo Alto Research Center (PARC)

The Computer Science Laboratory at Xerox PARC near Stanford University, California, brought together some of the most creative individuals in computing. It was founded in 1970 by Bob Taylor, a psychologist who helped fund the development of the ARPANET in the mid-1960s while working at the Department of Defense (see Box 4.6). Taylor and one of the first people he hired, Alan C. Kay, assembled a group of about fifty computer scientists who advanced key technologies for supporting human interaction with computers (Kay 1994). They were inspired by the work of Douglas Engelbart at the Stanford Research Institute (SRI), who strove to make computers a more useful tool for people to think and work with. The PARC group introduced many influential innovations, including graphical user interfaces (GUIs) employing overlapping windows, object-oriented computer languages such as Smalltalk, high-speed computer networks such as Ethernet, and the laser printer. Kay's 1972 proposal for a 'hand-held' 'stand alone' 'interactive-graphic computer' called Dynabook incorporated many features of the future PC (Brand 1987: 96). Some Dynabook features were incorporated into the first PCs, such as the Alto, which was completed at PARC in 1973. The Alto used GUIs and a mouse to make it easier for the user to operate. Xerox funded PARC largely out of a concern over the impact of computers on its copier business, rather than a strategy for entering new product areas. PARC scientists were unable to persuade Xerox to commercialize many of their innovations, like the Alto, but they later inspired the products of other companies such as Apple's Macintosh and Microsoft's Windows (Cringley 1992: 80–92).

sponsor to commercialize many of their inventions, although other companies did.

Inventors must convince their peers of the wisdom of their designs, since engineers and scientists have differences of opinion on the best way to do things. Technical experts must compete with rivals to persuade industry to refine, manufacture, and market their inventions. Producers must enrol users and the media, among a number of other constituencies (Molina 1989, 1990). And some large and powerful users, such as governments or the military, must often enlist the support of producers to create a particular technical standard or infrastructure.

In this process, technologies are invented, reinvented, designed, produced, diffused, and consumed over time by groups of people who fundamentally change the nature and paths of a technology in unpredictable ways. Nevertheless, some useful insights can be gained from an understanding of the rich history of responses to innovations in ICTs (Greenberger 1985; Noll 1985, 1992; Silverstone and Hirsch 1992; Elton 1991, 1992; Dholakia *et al.* 1996).[1]

Field Trials and Experiments

In the 1990s, new visions, like that of an information superhighway, combined with real technical advances and huge multimedia mergers to rekindle interest in all sorts of innovations affecting tele-access (Burstein and Klein 1995; Emmott 1995; Negroponte 1995). Although services such as digital video, multimedia, virtual reality, and the Internet may incorporate many technological advances, they are based on many of the same assumptions about user behaviour that guided the application of earlier technologies.

Many so-called emerging ICTs are, therefore, familiar to those who have tracked field trials and experiments over many years. Some key early initiatives include wired city experiments, with Hi-OVIS and QUBE being particularly important landmarks (see Box 4.2).

For example, the 'request video' which was a central feature of some Hi-OVIS channels for about five years provided something near to what is now called 'video-on-demand' (VOD). Hi-OVIS utilized a robot in the cable studio to retrieve a requested video cassette and plug it into a video tape monitor for play back to the home (Kawahata 1987). Yet in the 1990s, initiatives in the USA and UK were still vying for recognition as the 'first' VOD service in the world, including a fibre-to-the-home (FTTH) trail in a suburb of Los Angeles which depended on a human being fetching video cassettes for play back to just the four homes included in the trial, which later shifted to schools.

Box 4.2. Wired City Trials and Experiments in the USA and Japan

Tama New Town, a Tokyo suburb. Sponsored by the Japanese Ministry of Posts and Telecommunications (MPT), the Coaxial Cable Information System (CCIS) project began in 1973 and focused on local community programming (Murata 1987).

US National Science Foundation (NSF) Experiments. A series of interactive cable trials took place in Reading (Pennsylvania), Spartanburg (South Carolina), and Rockford (Illinois) from 1974 on, with university–industry collaboration (Brownstein 1978; L. B. Becker 1987).

QUBE. In 1977 Warner-Amex introduced a thirty-channel interactive cable television system in Columbus, Ohio—'average city' USA—to test the commercial viability of new programming formats. Viewers could signal the cable station, for example, to answer a poll or order a pay TV programme. QUBE could monitor how TV sets were being used to test-market new programming. It closed in 1984 (Davidge 1987).

Hi–OVIS. The Highly Interactive–Optical Visual Information System was supported by Japan's Ministry of International Trade and Industry from 1978 to 1986. It experimented with interactive TV, video conferencing, request video, electronic shopping, character and still picture information services, and news (Kawahata 1987).

Full Service Network. Time Warner's interactive TV trial outside Orlando, Florida, ran from 1994 to 1997 (Burstein and Kline 1995: 131–46; Snoddy 1997).

Interactive TV has also become a new media buzzword of the 1990s. However, there were live and interactive programmes on QUBE over a decade before, including sports and game shows. But these innovations were not profitable because of the higher cost of two-way networks and the lower profitability of producing for only a local audience. QUBE's closure in 1984 became a symbol of the demise of interactive cable TV, despite the fact that the experiment gave birth to successful national channels such as Nickelodeon for children and MTV for popular music fans (Davidge 1987).

Innovation Amnesia

The history of the electronic media is often forgotten, even though much of it happened within the last several decades. One reason is that technological change often brings entirely new groups of technical experts and business executives into the development of ICTs.

In the 1970s, cable enthusiasts included many investors and journalists who were unfamiliar with broadcasting and had a somewhat naïve view about the prospects for promoting local and interactive television (Smith with Cole 1987). In the 1980s, many telephone company executives had very limited knowledge of the history of broadcasting and cable TV services, and so became overly optimistic about the prospects for VOD. Many computer 'techies' of the 1990s were not even alive when their predecessors advocated the construction of computer-based public information utilities, and exhibit some of the same attitudes about television—that it is too passive—as the early pioneers of interactive cable communications (Gilder 1994: 66). Bill Gates, for example, championed visions of interactive television in the 1990s that echoed calls for interactive cable communications in the 1970s (Burstein and Kline 1995: 209).

Two key lessons from this history of trials and experiments are explored in the remainder of this chapter:

1. The responses of various groups to visions of the future can be as important as the responses to the actual products and the services that they promote. They can, for instance, create a coalition of groups in support of innovation, while also raising unrealistically high expectations.
2. Those who envision and produce ICTs sometimes base their designs on misconceptions of the user. When faced with this reality, many developers ignore or seek to change the way people do things, rather than rethink their design of ICTs.

Visions of the Future of ICTs: Enrolling Multiple Players

The public at large does not only respond to concrete technical products, gadgets, and services. Often the proponents of ICTs attempt to enrol the public and other critical players long before technical innovations reach businesses or households. They do so by articulating a vision of the future that will be enabled by ICTs. While gaining the support of a variety of constituencies, the success of such a vision can undermine those who favour alternative technologies. Visions, and public responses to them, are important factors shaping the paths of technological change.

This kind of 'politics of ideas' strongly influences the ecology of games affecting tele-access. Debates over new technology are like a game in which there are no limits on the number of players on each team and spectators can choose to join in the game at any time. The political scientist E. E. Schattschneider (1960: 2) once compared such conflicts to a fight: 'Every fight consists of two parts: (1) the few individuals who are actively engaged at the center and (2) the audience that is irresistibly attracted to the scene. . . . The outcome of every conflict is determined by the extent to which the audience becomes involved in it.' Contestants with a clear lead in such a conflict might well try to defend the status quo by keeping their side intact and the spectators in the stands. In contrast, losing contestants can change the odds by dividing their opponents or drawing spectators into the game. By determining what the conflict is about, key players can shape the ways in which individual players choose sides and the number and type of spectators drawn into the conflict. There can also be a struggle among competing visions of the future of ICTs, especially in the commercial market place. If the conflict is understood to be about access, for example, rather than merely technology, then a wider diversity of players might become involved.

Before the Internet became a household word, a tremendous amount of mass-media coverage and public debate was given to the idea of an 'information superhighway' (see Box 4.3). Much of this was stimulated by the way strong leadership in the USA from the White House associated

Box 4.3. The Information Superhighway

The 'information superhighway' is an ICT network which delivers all kinds of electronic services—audio, video, text, and data—to households and business (Gore 1991). It is usually assumed that the network will allow two-way communication which can deliver narrowband services like telephone calls as well as broadband capabilities such as video-on-demand, teleshopping, and other multimedia applications. Services on the superhighway can be one-to-one (telephones, e-mail, fax, etc); one-to-many (such as broadcasting and video conferencing); or many-to-many (typified by bulletin boards and forums on the Internet).

There are a number of competing views on the most appropriate architecture or network model on which to build the information superhighway. It could be based on models derived from the traditional practices and experience of different industries, including the telephone, cable, or computer data communications. The Internet computer communications model of an integrated 'network of networks' has proven the value of enabling the smooth and efficient movement of information across many types of public and private networks, despite their use of different equipment and standards.

the superhighway with a vision of the social and economic importance of new telecommunication infrastructures (Gore 1991). In addition, the information superhighway was a powerful image in its own right, crystallizing issues surrounding telecommunications infrastructures into a concept which could be readily understood by non-specialists. The popularizing effect was reinforced by the subsequent growth in worldwide popularity of the Internet, which gave the mass media a tangible peg on which to hang stories about what the superhighway could mean in practice.

In the context of organizations, visions can be especially powerful as they need only enrol relatively easily identifiable key constituencies in order to shape management responses to ICTs. For example, the idea of a 'virtual organization' has had an impact on how many firms have thought about the role of ICTs, even when the vision has not been realized in practice (see Essay 4.1). Of course, visions are likely to fade if they are not continuously reinforced by actual developments which embody the vision.

In addition to organizational visions, which are examined in depth in Chapter 5, democratic and technocratic visions have been important to tele-access.

Democratic Visions

The concept of an information superhighway is a democratic one of enabling all citizens to get access to any information and to any person, at any time, from any place. It is one in a history of visions that promote more democratic tele-access. As indicated in Box 4.4, the ideas of 'public information utilities', 'wired nations', and 'network nations' share many of the same ideals as the 'information superhighway'. The underlying conception of supporting more democratic access to huge electronic stores of information has remained remarkably similar over the decades, although the supporting technologies have changed dramatically. For example, the domination of large, centralized mainframe computers has given way to distributed networks interconnecting a myriad of different types of computing systems.

The politics of ideas surrounding tele-access is critical to enrolling a variety of different constituencies in the promotion of a technological innovation.[2] US Vice-President Al Gore and President Bill Clinton captured the attention of politicians and journalists around the world in the 1990s with their evolving concept of the information superhighway. The idea gained legitimacy in debate within industry and policy circles (Dutton 1996*a*; Sawhney 1996), and played an important role in shaping

Box 4.4. Democratic Visions of the Future of Communications

- *Public information utility.* On-line computer systems accessible over telecommunications networks triggered initiatives from the early 1960s to enable the mass public to access electronic information and communication services (Sackman and Nie 1970; Sackman and Boehm 1972).
- *Wired nation.* In the 1960s, the idea of marrying the new 'information highway' of coaxial cable and computer technology spawned discussion of two-way cable television providing all kinds of information and communication services (R. L. Smith 1970, 1972; Goldmark 1972).
- *Network nation.* Computer mail and conferencing stimulated conceptions of a networked society in the 1970s which envisioned computers as facilitating the emergence of 'vast networks of geographically dispersed persons'—a new form of human community (Hiltz and Turoff 1978: p. xxv; Valle 1982).
- *Information superhighway.* The Internet embodied the 1990s vision of a 'nationwide, invisible, seamless, dynamic web of transmission mechanisms, information, appliances, content, and people' to 'improve the lives of individuals, reinvigorate education, expand buinsesses, and strengthen communities' (US Advisory Council 1996*a*, *b*; Gore 1991).

public policy in many other nations (Kubicek *et al.* 1997). Public officials have subsequently knocked down all sorts of barriers between cable, telephone, and broadcasting that might constrain the construction of a modern information infrastructure.

Many of the earlier democratic visions of the future of ICTs also captured a great deal of attention but were short lived and failed to achieve their initial high hopes. For example, a belief in the 'wired city' was propelled by links with President Lyndon Johnson's 'Great Society', an agenda for social action outlined in his 1964 campaign against Republican Senator Barry Goldwater (Johnson 1971: 104). A journalist, Ralph Lee Smith (1970, 1972), popularized the wired-nation idea in ways that helped convince key people at the US Federal Communications Commission (FCC) that the long-term social and economic benefits made it worth taking the known risks to the broadcast industry involved in removing regulatory barriers to the development of cable TV (Dutton *et al.* 1987*a*). However, the market failure of local and interactive cable TV, such as QUBE, undermined the credibility of this vision. Changes in technology and policy then led to the growth of satellite-linked cable systems that delivered one-way entertainment programming to ever wider audiences.

Technocratic Visions

Democratic visions often originate outside the technical community, as with journalist Ralph Lee Smith on the wired nation and the role of Vice-President Gore—who has a background in journalism—in popularizing the superhighway. Technical experts have more often promoted visions that are not tied to any social or political ideal, but to a technical logic of efficiency. These technocratic visions can be just as influential and enduring, even if they remain far ahead of the realities of technological change (see Box 4.5).

Box 4.5. Technocratic Visions of the Future of Communications

- *Interactive cable TV.* Coaxial cable could be an electronic highway for two-way communications between cable service providers and businesses and households (Baer *et al.* 1974).
- *Videotex.* TV sets could be used for display of text and graphics delivered over telephone lines that link a remote computer to a terminal attached to user's TV (Martin 1977).
- *The chip.* The microprocessor chip became a powerful symbol in Britain during the early 1980s, which stimulated the development of government policies supporting the production and use of information technologies (Evans 1979; ITAP 1982, 1983).
- *Multimedia digital convergence.* Digital technologies, such as multimedia PCs and digital TV, would transform all analogue media, such as telephone and broadcasting (Martin 1977; Gilder 1994; Negroponte 1995).
- *Integrated Services Digital Network (ISDN) and fibre to the home (FTTH).* ISDN is a wired multimedia network. The optoelectronic-based FTTH is sometimes called 'Broadband-ISDN' (Elton 1992).
- *The Internet.* This computer network of networks, using a standard Transmission Control Protocol/Internet Protocol (TCP/IP) to interconnect many different ICT systems and services, has become a model for future networks (NRENAISSANCE Committee 1994; see also Box 4.6).

One common element of these technocratic visions has been the pursuit of 'convergence'. Cable systems in the 1970s, videotext in the early 1980s, and digital technologies such as the Internet and the multimedia PC in the 1990s embodied the concept of convergence—using a single infrastructure to integrate the once separate technologies and industries of print, broadcasting, and telecommunications. 'Digital convergence' promises not only gains in efficiency but also the ability to provide many old services in new ways that fundamentally change the way we use the media, such as providing more personalized information services as more intelligence can be embedded in telecommunication networks (Brand 1987; Negroponte 1995). Nevertheless, convergence is a technocratic vision that might have positive or negative social implications. It also entails industrial and regulatory change that has been far more difficult to achieve than technical enthusiasts generally admit (see Essay 3.1).

Technical visions can help to support or undermine the influence of particular players within the technical community. For instance, the lack of initial interest in the much hyped ISDN vision of digital convergence became a focus of jokes, with commentators saying ISDN stood for 'Incredible Services we Don't Need'. Yet many services highlighted by the promoters of ISDN were strikingly similar to those later promoted in the multimedia technical vision, which has been received more positively.

Promoters of FTTH (see Box 4.5), mainly in the telecommunications industry, emphasized the technological superiority of fibre optics (Elton 1992). However, they have promised few new services beyond the claim that VOD would be a new source of revenue which would enable phone companies to construct a broadband infrastructure to replace the narrowband wire pairs strung to households and businesses. In the USA, the lack of a compelling social and economic vision for FTTH left the telephone companies with little support within the FCC and Congress, who saw no reason to jeopardize the cable television industry to create what seemed to be nothing more than a 'video juke box'.

Most of these FTTH trials were limited to technical trials with fewer than 200 subscribers (McGilly *et al.* 1990). Driven by a few telephone companies and manufacturers, over two-thirds of American trials focused on voice telephony services, comparable to the services provided over twisted wire pairs. Less than one in five experimented with cable TV as well as telephone services. Only about one in ten experimented with broadband telecommunication services (McGilly *et al.* 1990).

Even when technical visions do not capture the public's imagination, they can direct a great deal of investment and development in telecommunications. In an ecology of games, as discussed above, it is often useful to keep spectators out of the contest. If these visions leave the general public mystified or uninterested, they can buttress the influence of the technical community. For example, despite the failure of ISDN to capture the public's imagination, ISDN has developed further and become an important option to support the growing market for better Internet connections to households and businesses.

One of the most successful and influential technical ICT visions has been the Internet (see Box 4.6). Its main technical innovation was to divide

Box 4.6. The Evolution of the Internet, 1958–1998

1958	The Advanced Research Projects Agency (ARPA) of the US Department of Defense is created in response to the Soviet Union's launch of Sputnik.
1968	ARPA designs the first packet-switched computer network (ARPANET), with the first node installed at the University of California at Los Angeles in 1969.
1970s	Data communication protocols are developed for packet-switched networks.
1983	ARPANET moves from its original Network Communication Protocol to Transmission Control Protocol (TCP), developed in the USA and UK.
Later 1980s	Public and private university, regional, national, and international networks are developed.
1989	ARPANET is decommissioned, with the National Science Foundation Network (NSFNET) taking on its role as a worldwide backbone 'network of networks'.
1990s	The focus is on consolidating and interconnecting networks worldwide around the Internet, exploring its commercial potential, and building Internet2.

Source: Denning and Lin (1994: 133–6); NRENAISSANCE (1994: 237–43).

data into small 'packets' that could be individually addressed and routed to the right destination via computer 'switches' located anywhere in the world. This idea of a 'packet-switched' network based on the TCP/IP protocol allowed network links to be made between 'heterogeneous' computers and services, which may have been based on different and previously incompatible standards.

The Internet seemed to burst upon the public consciousness suddenly, although packet-switching technology had been developed since the 1970s. This is typical of the way apparent 'overnight' ICT technical breakthroughs are actually the result of processes that unfold over years or decades (Hafner and Lyon 1996).

This illustrates how exciting visions and real advances created by technologists have triggered an exploding supply of ICTs that have driven many new tele-access products and services. However, user attitudes in business, industry, and households have often had a major braking effect on technical change. For example, the growing capacity of telecommunications networks in the 1960s motivated many telephone companies to develop the video telephone to make use of increasing bandwidth capacities (Dickson 1974). But subsequent failure of demand from potential users has had a major impact on the course of technological change (Noll 1992).[3]

Technical Choices and Conceptions of the User

The history of ICT trials and experiments provides evidence of some recurrent patterns in the ways that producers tend to misjudge responses to ICTs (Neuman 1985; Elton 1991; Dutton 1995*c*). It is useful to look at these misconceptions at this point because they cut across all tele-access arenas discussed in subsequent chapters.

Energizing the Couch Potato

One area of application that has become the focus of a growing supply of technologies and services is the home, with what Ian Miles has called 'Home Informatics' (see Essay 4.2). While the early market for many innovative ICTs is in business and industry, the largest markets are often in the household, making the private consumer one of the primary long-term targets of ICT producers.

Many producers of home informatics assume that consumers should (and want to) play a more active role. Among them are critics of traditional broadcast TV, who believe the public should interact with television, not be passive couch potatoes, and that ICTs will effect this change (Gilder 1994; Gates 1995).

Interactivity has regained a great deal of currency. Interactive television, interactive cable—even interactive dashboards and 'Teddy Bears'—have been among the multitude of innovations flowing from ICT advances. Nevertheless, as with other supply-driven developments, interactive services have a chequered history which raises serious doubts about the public's interest in participating more actively in televised entertainment (Dutton *et al.* 1987*a*). QUBE was typical of 'wired-city' experiments (see Box 4.2) which demonstrated that a variety of public and commercial services could be provided over two-way cable systems, with interactive offerings showing no more than marginal advantages over conventional broadcasts—but at a significantly higher cost (Elton 1980; L. B. Becker 1987; Davidge 1987: 84–5).

Despite these disappointing results, the interactive vision continues to motivate new service offerings. For example, a new generation of interactive cable TV trials were launched in the 1990s, such as a trial by British Telecommunications (BT) in Colchester, north of London in Essex, and Time Warner's 'Full Service Network' in Maitland, Florida, outside Orlando. Time Warner spent as much as $100 million on the Full Service Network from 1994 to 1997, when it was closed, but reached only about 4,000 subscribers (Snoddy 1997).

In 1997 Vice-President Gore urged the cable industry to move faster in modernizing its networks for interactive communications. Nevertheless, the new trials have bumped up against many of the same difficulties in attracting a mass audience for interactive services as in the earlier decades. They have also found a limited market for VOD, expected to be the 'killer app'—the crucial application that would be the turning point in opening up a new mass interactive TV market (Burstein and Kline 1995: 131–45). Time Warner and other cable system operators have refocused on less expensive 'near VOD' that permits viewers to the same movie on any one of several channels depending on the time at which they wish to view it—one-way broadcasting of entertainment programming.

Other 'interactive' television projects have used hybrid technologies to piggy-back on existing broadcasts—for instance, to allow viewers, say, to play along with game shows from their homes using a hand-held terminal that can receive a radio signal simulcasted with the broadcast. On-line chats and information have been organized over the Internet in conjunction with live broadcasts of celebrities, such as the chess match between IBM's Big Blue and Gary Kasparov in 1997. These hybrid systems lower interactive costs by adding value to existing broadcasts, which could possibly bring financial viability even if the interactive service engages a small proportion of the total audience.

Revitalizing Community in a Global Market

Many different policy approaches have been adopted around the world towards balancing local, regional, and national needs in communication. However, policy innovations aimed at promoting localism using new electronic media have generally failed in the face of audience responses to local programming.

A link between technical advances and local communities was nurtured in part by the early history of cable systems, which began as local community antenna television (CATV) systems. Tele-access was biased towards the local because the attenuation of electronic signals limited the geographical coverage of early cable systems. Moreover, the provision of more channels opened up the potential for offering local programming as well as many other educational, cultural, and governmental services. However, experiments with community cable TV in many countries failed to establish a proven market for local services.

In some respects, there has been a shift in the geographical orientation of new media from local to more national and global networks. This shift was furthered by the arrival of satellite-linked cable systems in the late-1970s and later by direct broadcast satellite systems (DBS). However, this shift is best exemplified by the Internet and Web. Yet the advocates of electronic networking for the twenty-first century continue to emphasize the central role that ICTs can play in local communities. The US Advisory Council (1996*b*: 9) on the National Information Infrastructure (NII)—the original information superhighway policy statement—identified the creation of 'stronger communities, and a stronger sense of national community' by 'getting America online' as one of its main goals.

This American attention to community in building a national information superhighway appeals to a strong tradition in the USA of promoting localism in broadcasting.[4] The Clinton Administration also had clear ambitions to pursue the NII concept at a more global scale through what has been called 'Global Information Infrastructure' (GII) initiatives (Gore and Brown 1995). This global approach to information superhighways was driven by a variety of factors, including the greater value attached to any network that could provide similar levels of service around the world, such as the Internet in conjunction with ordinary telephone networks.

There are some exceptions to this global orientation, such as the emergence of some electronic bulletin board systems (BBS), where a local orientation is a central feature of the services. Another exception is the emergence of successful initiatives to launch 'neighbourhood television', which build on the growing ranks of local volunteers that have experience and interest in video production (Dutton *et al.* 1991). Localism also remains one of the few strategic advantages of wired over wireless systems, such as in the competition between cable communications and DBS. But the economies of scale behind the provision of the same programming to millions rather than hundreds of households creates major disincentives for devoting resources to productions aimed only at a local audience, whether on TV or on the Internet.

Privileging Face-to-Face Communication

The developers of ICTs often take face-to-face communications as the ideal which mediated forms of communication should seek to replicate as closely as possible. For example, the idea of developing a 'telephone' that transmits and receives images on a screen is as old as television. A number of major telephone equipment manufacturers designed video phones in the 1960s, including Nippon Electric Company, Stromberg-Carlson, the British Post Office (now BT), and AT&T, which developed the Picturephone (see Box 4.7).

Box 4.7. AT&T's Picturephone

Bell Labs began experiments with video telephony in the 1950s, leading to the display of a prototype at the 1964 World's Fair in New York City. AT&T commercialized the Picturephone in the late 1960s, forecasting a market of up to 1 per cent of all domestic and 3 per cent of all business phones by 1980. It was an analogue system that used a phone, linked with a video screen, camera, speaker, and control unit to receive and transmit sound and a 5½ inch by 5 inch black-and-white image. Three wire pairs were required, but the existing telephone transmission facilities could be used. A light signalled if the person called had a video phone. Developers hoped the Picturephone would become a business status symbol, with users putting pressure on their friends and colleagues to purchase and use one. But the very limited response from business led AT&T to withdraw the product in 1973 after investing between $130 million and $500 million in its development.

Sources: Dickson (1974), Martin (1977: 113–31).

More recently, France Telecom found that only a minority of households in an experiment in Biarritz (see Box 4.8) used the terminals they were provided as a video phone—and then only after a fairly long period of adaptation. Subsequent video phone initiatives have also failed to generate customer demand to match the developers' expectations, at least in the short-term. This was the case, for example, in the early 1990s with AT&T's new VideoPhone 2500 and similar products from Mitsubishi, Sony, BT, and other companies.

Critics have taken the market failures of video phones as evidence that the public simply does not need or want video telecommunications (Noll 1992). Proponents argue that video telephony is destined to arrive, once the cost and design are right, perhaps by providing video communications within a window on a personal computer, which can be expanded or closed by the user.

Box 4.8. France Telecom's Biarritz Experiment

Biarritz is a resort and tourist centre in the south-west of France. In 1979 France Telecom—the nation's public telecommunications monopoly—chose Biarritz as the site for an experimental fibre optic network, providing voice, video, and data services via a multimedia terminal. France Telecom's objectives were to develop the technical know-how for building and installing fibre optic systems, to showcase French technology, and to explore the market for broadband services. The first fifty households subscribed in 1984, and all 1,500 residences covered by the project were connected by the summer of 1986.

The Biarritz system provided broadcast services, involving fifteen cable TV channels and twelve stereo sound channels, as well as switched point-to-point services, including video telephony. Subscribers were provided with a multi-service, integrated multimedia terminal with a video monitor, camera, keyboard, and telephone handset. The terminal could be used as a video telephone to transmit still or full-motion images. Subscribers could consult on-line image and databanks and access thousands of videotext services provided over the *Télétel* system, better known as Minitel (see Box 4.10). Subscribers were charged a monthly fee for cable service and both a monthly and usage-sensitive rate for switched services. Usage by subscribers, however, proved disappointing and the experiment was discontinued.

Source: Gérin and Tavernost (1987).

The Technical Mirage

When I first demonstrated digital video to my students in the 1980s, I was astonished by their response. I had just shown them over 30 seconds of motion video with text and sound—all stored on a 3½ inch diskette. All they had to say was: 'Television is better.' Years later, when I showed my students a new interactive CD-ROM designed by Bill Gates and his Microsoft colleagues (1995), most of my undergraduate students were interested only in seeing the tour of his home! Many trials of new services motivated primarily by a technological breakthrough experience the same kinds of unenthusiastic responses from potential users and customers.

The success of the personal computer beyond its initial hobbyist 'nerd' market was based on 'killer apps', such as word processing and spreadsheets, which inspired a wide range of people to buy their own PC. Many of the new media have yet to identify equivalent services to trigger this vital acceleration in demand.

The hype that surrounds most technological breakthroughs often creates expectations of improved services that do not square with the reality, as with interactive cable initiatives. Trials of VOD in the 1990s found that households demanded fewer than three movies per month (Burstein and Kline 1995: 144), which was fewer than the number rented from video stores and not enough to support a profitable VOD service. Reasons for this included an inability to compete with video shops in terms of picture quality and price, despite the convenience of being able to download films directly to a TV set.

In the early 1990s, the multimedia personal computer was also more of a technological than market breakthrough. A multitude of applications were promoted for this technology, such as business training, educational CD-ROM software, games, and surfing the Internet. But it was the advent of the Web on the Internet later in the decade that provided one of the main triggers for market expansion (see Box 4.9).

Box 4.9. Hypertext and the World Wide Web (WWW, the Web, W3)

The Web is a body of multimedia information accessible over networks such as the Internet. A key, central feature of its organization is that it permits 'hypertext' searches, which allow a user to follow chains of electronic pages of information without regard to where they are physically stored. Users can access information housed on any networked Web server as easily as that stored on their own desktop computer. The Web achieves this through a set of software and standards for representing and cataloguing electronic information. As a user browses an electronic document on the Web, highlighted keywords, symbols, or images indicate that there exists a link to related text, sounds, or graphics located elsewhere in the document or in any other document on the Web. By directing the mouse pointer to the highlighted area and clicking the mouse button, the user can go to where the link is pointing. This provides a simple way of moving through information stored on computers. Information included in the Web must be specially coded using the hypertext markup language (HTML) to identify the keywords and the electronic addresses used in hypertext links. The Web emerged from a set of networking projects at the European Laboratory for Particle Physics (CERN) in Switzerland, with Tim Berners-Lee as one of its driving forces (Segal 1995). Berners-Lee's 1990–1 prototype was influenced by Ted Nelson's visionary work on the Xanadu project from 1966, which aimed at creating a commercially viable network that would 'give you a screen in your home from which you can see the world's hypertext libraries' (Nelson 1974: 56–7). Xanadu was inspired by Douglas Engelbart's 'oN-Line System' (NLS) at SRI (see Box 4.1).

The desperation surrounding the search for killer applications is evident in the claim that prurient interest will propel the multimedia market. Sex and violence are often suggested as the trigger for multimedia, with claims that an enormous proportion of CD-ROMs and Internet accesses are devoted to pornographic material—called 'high tech peep shows' in the *New York Times* (9 Jan. 1993). One estimate is that an adult entertainment Web site, called *ClubLove.com*, received 1.4 billion hits in 1997 (*Time*, 29 Dec. 1997, 43).

Others argue that the impact of this market is far less significant (Greenberger 1992: 142). Experience with other media underscores this point. Adult films were among the most popular pay-per-view movies on QUBE (Davidge 1987), but they were not popular enough to generate a profit for QUBE or a market for interactive cable TV. Adult sites were among the most popular on the Web in its early years, but this usage is way out of proportion with the scale of the mass market sought by the industry.

Media Habits

Certain ICT innovations essentially make it easier to do what users are already in the habit of doing, such as have a new form of screen for a PC. They are compatible with existing technology (Rogers 1986: 116–49). However, many key developments demand that we change how we do things. This usually requires social and institutional changes which lag behind the technological advances.

Many trials of new media have been constrained by media habits. To summarize the views of Hazel Kahan (1984), who worked with the QUBE system: You should never underestimate the difficulty of getting people to do something they have never done before, such as 'talk to their TV', particularly if it is something they don't like to do. In US and Japanese interactive cable trials in which providers could constantly monitor channel selections, it was found that viewers took years to change from a habit of watching one channel all evening to 'hopping' more frequently between channels (Gonzalez 1986). Now, channel-hopping is taken for granted in multichannel environments, with many people grazing through programmes—'watching TV' more often than concentrating on individual programmes. But the TV grazing habit took over a decade to develop and is not the engagement required to participate in an interactive programme.

So, while interactive TV pioneers believed they were building on the TV-watching activity people already liked to do, they found they were helping to create fundamentally new 'viewing' cultures. For instance, a fundamentally different kind of behaviour is involved in interacting online with a game show compared to the traditional way viewers shout out answers intermittently when they can outwit TV contestants.

Another example of how developers of new services challenge existing communication habits even when they believe they are building on them is provided by the video phone. In seeking to emulate face-to-face communications, developers of video telephony actually created a quite different medium that called for new rules, a new etiquette, and a change in communication habits. Even early studies of the AT&T's Picturephone found that it altered normal face-to-face communications as well as normal telephone use (Dickson 1974; Noll 1978). The video phone moved the image of the other person closer than the conventional social distance that Americans like to maintain and created an abnormal level of direct eye contact between the participants in a conversation.

Tele-Access: More than Information

Providing access to vast stores of information is given great emphasis in most discussions of the information superhighway and other multimedia developments. The assumption that there is a widespread interest in getting access to information *per se* lay behind early visions of the public-information utility and remains strong in the ICT industry and elsewhere. However, this assumption might well be a misleading guide to the development of ICTs, as indicated by many videotex trials (see Box 4.10). These found that the public might be more interested in interpersonal and group communication and specialized services than in just having lots of information at their fingertips (Hooper 1985: 190).

For example, Times Mirror's Gateway found a growing interest in e-mail among its subscribers. But Times Mirror was a publisher interested in the future of the newspaper business. Realizing that videotex presented no immediate threat to the newspaper, the company saw no reason to continue with Gateway, as it did not see itself as being in the electronic communications market. A few years later, the emergence of on-line newspapers and then the popularity of the Internet led the company back into this arena, but once again because it wished to be prepared for changes in the newspaper business should more people seek information on-line.

The history of the Internet (see Box 4.6) strikingly highlights the importance of the 'C' in ICTs. ARPANET was originally designed to permit students and researchers at one university to use computing facilities at others (Hafner and Lyon 1996). In practice, however, ARPANET evolved instead to become primarily a medium for interpersonal communications —'e-mail'. Only in the 1990s has growing interest in the World Wide Web led to more remote access to computing facilities, although e-mail and

Box 4.10. Videotex Trials and Services since 1979, including Minitel

Videotex originally referred to systems modelled on 'viewdata', invented by the British Post Office, later BT, in 1971. Viewdata employed a TV set, telephone, and modem to access a computer database of thousands of 'pages' or 'frames' of news and other electronic information for a wide variety of services. Text and colour graphics were transmitted to and from households or businesses via telephone lines and displayed on the TV screen. Users selected pages using a remote control. This hybrid of TV, phone, and computer influenced the design of videotex systems round the world, including:

- *Télétel* from France Télécom. First tested in 1978 using a compact 'Minitel' terminal that avoided any link with the TV set. Over five million free terminals were eventually provided to French households for an electronic telephone directory. Thousands of other service providers emerged by 1986.
- *Prestel*, introduced in 1979 by BT as a commercial viewdata service.
- *CAPTAIN* (Character and Pattern Telephone Access Information Network), launched in 1979 in Japan by NTT.
- *Regional US* systems, such as *Viewtron*, commercially launched in south Florida in 1983, and Times Mirror's *Gateway*, started in southern California in 1984 and stopped after several years and multi-million dollar losses.
- *Bildschirmtext*, launched by Deutsche Bundespost in 1984.

Sources: Greenberger (1985); Aumente (1987).

the exchange of documents and opinions with others at local and distant locations remain the most popular uses of this capability of the Internet.[5] The success of the Internet has been driven primarily by avid communicators, not information-seekers.

E-mail and teleconferencing between individuals were also central aspects of successful computer-based services that pre-dated the spread of the Internet outside academia. These included public-interest applications, such as SeniorNet in the USA designed to support communication among the elderly (Arlen 1991), and global commercial business facilities, like CompuServe. The significance of such communication capabilities underlines the importance of the concept of tele-access, which captures the centrality of people, services, and technologies as well as information.

The value of any telecommunications network increases in a non-linear fashion as more people are connected to the network. Given the importance of communication in tele-access, it is, therefore, clear that a central problem facing new media is to reach a point at which there is a large enough number—the 'critical mass'—of users to provide an effective and profitable service (Valente 1995). Many new media trials, like France Telecom's Biarritz experiment (see Box 4.8), found that the value of innovative communication technologies is severely diminished if a critical mass of users is not accessible. This problem cannot be solved just by new design or technological solutions, but demands innovative thinking about how to subsidize users until a critical mass is established.

A critical mass—not necessarily a mass market—has been achieved with the Internet. It was nurtured in part through public research support such as the ARPANET and university funding that provided a broad base of users by the early 1990s, when graphical user interface designs such as Mosaic greatly facilitated use of the Internet by simply pointing and clicking a mouse. It has since diffused at a dramatic rate to encompass private and public networks round the world.

E-mail is widely used for social and family communications, as well as in business, government, and educational contexts. These major strides have been made despite the fact that the user interface to most e-mail networks remains non-intuitive for a large proportion of potential users, who can still be deterred by the costs, complexity, and need for special equipment and advice (Mossberg 1996). There are also regulatory difficulties with respect to issues such as privacy, freedom of speech, and security when moving out of the academic environment which nurtured the Internet (Dutton 1996*b*). These problems raise cautions in predicting the Internet's future, and suggest the need for more systematic research on actual usage patterns of ICTs over time.

Social Concerns at the Bottom Line

The kinds of ICT trials and experiments discussed in the previous section highlight the significance of enduring social concerns. Video telephones, for example, are often perceived to be an intrusive technology that invades the privacy of the user (Dickson 1974: 102). Many people express concerns about being seen on a video phone at inconvenient moments, such as when they step out of the shower. Although design changes—like locating video phones mainly in offices using small windows in the screens of workstations—might allay these concerns, they could also make telecommunications less versatile, portable, and convenient (Dutton 1992*b*).

In the Hi-OVIS project (see Box 4.2), a video camera installed on top of the household's TV set permitted video from the home to be transmitted to the cable head-end and then broadcast live to the other Hi-OVIS households (Kawahata 1987). Families cleaned their homes and dressed up to appear live on this local cable TV service. Critics argued that this kind of participation clashed with cultural traditions in Japan that discouraged individuals from attracting such attention to themselves (Ito and Oishi 1987: 212–13). Similar fears about the intrusive potential of video communications in the household are applicable to most other cultures and newer media. For example, live video has emerged as an Internet security service, permitting individuals remote access to live video cameras at a day-care facility—to see that their children are safe.

Cost can also be viewed as a social concern, not just an economic issue. In the main offices of a major US aerospace firm involved with the commercial development of computer simulations for training, a sign stating that 'Reality Costs' reminded staff that simulations become more expensive as they move closer to reality. This message has been relevant in the broader context of many tele-access trials. For instance, AT&T's VideoPhone 2500 was introduced in 1992 at about a hundred times the cost of an ordinary telephone. Yet real costs are only part of the picture. The perceived costs of ICTs are just as important, if not more so. This is illustrated by the way in which the perception in the UK that residential telephone calls are expensive has contributed to the relatively low popularity of bulletin board systems (BBS) compared to the USA, where local calls are perceived to be free.[6] In this case, perceptions exaggerate real differences in cost. In contrast, the popularity of the Internet has been based in part on the perception that use is free. Yet this 'culture of free access' has made it difficult for many commercial services to price services at a level necessary to profit (Shaw 1997).

A variety of rationales beyond the success or failure of a new service can also influence business choices. Take Warner's decision to drop QUBE. This was based in part on losses in the video-game business by another Warner subsidiary (Davidge 1987). The complex interactions between business rationales and the responses of users creates intrinsic uncertainties in predicting the future of ICTs and therefore related issues of tele-access. If the uncertainty of this ecology of games is acknowledged, industry and government can develop realistic and flexible long-term plans and policies. These should include varied and extensive field trials which encourage more experimentation, accept more failures, and support continued R&D—rather than pursuing the vain hope of targeting sure winners.

The Social Shaping of Tele-Access

This chapter has highlighted studies of how the public's response to ICT visions, products, and services can shape, stimulate, and constrain the actual development of ICTs. It has illustrated how many everyday consumer choices about tele-access—like whether individuals want to see the person they are speaking with—affect the course of technological change and, thereby, the degree to which ICTs provide tele-access.

Public responses to new media offerings truly affect the 'bottom line' (Silverstone 1991), so they must not be ignored by the ICT industry. Public responses to the visions and realities of ICTs may not be predictable, but the history of experiments, trials, and commercial offerings suggests that the public—as consumer, audience, user:

- is uninterested in technological breakthroughs unless they offer clear and substantial advantages over existing services;
- uses media out of the force of habit as much as by rational weighing of its practical costs and benefits;
- is more interested in communication than information *per se*;
- cares about the cost of services; and
- thinks about privacy and other social issues in ways that can be critical to the take-up and long-term viability of information and communication services.

Social science has a major role to play not only in understanding the impact of technology, but also in illuminating the ways social factors shape technological change and at every point in its development cycle (see Essay 2.2). The important general lessons identified in this chapter from the history of efforts to introduce ICTs, including the roles that producers and users can play in the development of ICTs, are built on later in this book by focusing on more specific social contexts, such as the household.

Notes

1. This chapter draws on two earlier efforts to draw lessons from the history of new media experiments (Dutton 1995*c*, 1997*a*).
2. The need to create a coalition of diverse 'socio-technical' constituencies in support of technological innovations, such as in parallel computing, was one general finding of PICT research on ICTs in the European context (Molina 1990).

3. Nicholas Garnham (1994*b*) has characterized this general issue as 'demand failure' that extends to many ICTs beyond the video telephone.
4. Throughout the 1960s, the support of localism in broadcasting was behind the restriction of cable system development in the major US television markets. Later, in the 1970s, this same goal became a major justification for the promotion of cable communications (Sparkes 1985).
5. A US survey asked Americans who go on-line what activities they performed 'at least one day per week' and found that 53% mentioned e-mail, which was the most frequently mentioned activity (Times Mirror 1995: 39). Other communication activities were 'conduct research or communicate with colleagues', mentioned by 41%, and participate in discussions, forums, chat groups, which was mentioned by 30%. Information was less often, but still frequently mentioned. These information activities included getting news (30%), getting entertainment information (19%), and getting financial information (14%).
6. A BBS is a means for an individual to use his or her PC for providing information and conferencing services to other PC users who call up the system operator's modem and connect with the individual's PC, which houses BBS software.

Essays

4.1. Interpreting Conceptions of the 'Networked Organization'

David Knights and Hugh Willmott

The ambiguity surrounding many aspects of 'information systems' leaves them open to different interpretations—and the interpretation which gets embedded in actual information systems strongly shapes perceptions of organizational norms and behaviour. Here, David Knights and Hugh Willmott use the widely discussed notion of the networked organization to underscore the power of such ideas.

4.2. Home Informatics: New Consumer Technologies

Ian Miles

Consumer markets for ICTs are potentially huge. Ian Miles focuses on the household to explain how major technological trends are tied to the emergence of consumer product applications that could have dramatic implications for the way we live our lives. His essay highlights the role of consumer values and choices in shaping the social implications of ICTs.

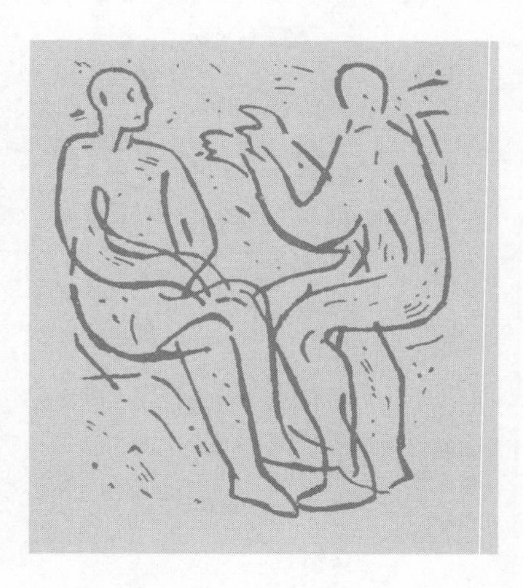

4.1. Interpreting Conceptions of the 'Networked Organization'

David Knights and Hugh Willmott

The term 'networked organization' has become increasingly popular as ICTs have created new means of interconnecting individuals, groups, workplaces, and organizations. It has been used to imply the emergence of a new and more flexible mode of structuring organizations. In this mode, mutually supportive relations of negotiation and trust within and among organizations are meant to supplant those of domination and dependence, which have typified traditional structures based on contract or hierarchical authority.

Networking is supported by the growing availability of ICTs that enable horizontal coordination and mutual monitoring of activity to replace forms of vertical control and surveillance (Hammer and Champy 1993). This is a major reason why ICTs have been favoured as a means of meeting commercial and competitive pressures which are driving managers to find new ways of reducing costs, increasing efficiency, and improving service to customers. However, it is questionable whether there has been any real shift away from practices supported by relations of domination, dependence, and control—or even whether the networked organization is a radically new approach. An alternative scenario is to see networking as a means of tightening horizontal forms of control in the service of established, hierarchically defined objectives (Knights *et al.* 1997).

There are two basic perspectives which could be adopted when studying the networked organization. One approach is to assume that the concept of networking reflects distinctive elements of a changing reality that can be captured by the use of instruments such as questionnaires or structured interviews to measure features like new communication flows. Although this approach produces plausible accounts, it disregards—or is blind to—the rhetorical sense in which the term 'networked organization' is used to suggest the possibility or desirability of change, rather than to reflect change itself.

A second approach explicitly recognizes the rhetorical dimension and seeks to encourage a greater sensitivity to the use of concepts as politically charged means of defining and constructing a particular reality. From this perspective, terms such as 'networking' and 'the networked organization' are understood as aspirational concepts that are developed and widely adopted because they resonate with a mood which their use helps to amplify (Senge 1991). That raises questions about the value of measuring features like communication flows, such as who e-mails whom, as these may have little meaning in terms of how far organizational reality has actually been transformed.

Attention is instead focused on the question of how the concept of networking is deployed rhetorically to create the sense of a different, changing reality. This approach uses the term 'rhetoric' to challenge whether the supposed reality depicted can be mapped or measured directly, but not to imply that it conceals or dissembles the reality. It is the process of organizational change, rather than the presumed outcome, which is regarded as being most important.

A framework for interpreting processes of organizational change from this perspective is needed, such as that developed by Michel Callon (1986). He identified what he called four 'moments' in the process of introducing, promoting, and institutionalizing a distinctive way of seeing the world: problematization, interessement, enrolment, and mobilization.

During the moment of problematization, one or more agencies are involved in exploring what is perceived to be a difficulty or opportunity. With networked organizations, the problems are typically interpreted as being caused by the failure of an established model of organization to be able to respond quickly enough to changes and emergent opportunities. In a study we did of the UK insurance industry, networking ICT was seen as a solution waiting for a problem that was eventrally found when independent financial advisers became disadvantaged relative to direct sales because of the greater sophistication of ICT available to the latter (Knights *et al.* 1997).

The moment of interessement occurs when those who believe themselves to possess relevant solutions to acknowledged problems feel obliged to persuade others who do not share their way of seeing the world. When this moment bears fruit, enrolment occurs. Then, key actors become tied into the proposed arrangements by symbolic mechanisms, such as a public commitment to a new organizational structure, and/or material means, like an investment in new ICT-based networks. Finally, the moment of mobilization involves deploying methods to sustain commitment—for example, holding regular meetings, updating relevant software, or increasing the costs of withdrawal from an agreement.

Callon's moments are recurrent rather than sequential. In the UK insurance industry case study, for example, moments of problematization recurred in diverse guises as novel features of the problem were identified, while interessement and enrolment required continuous renewal in the face of periodic crises of confidence. There was also comparatively little integration or commonality among the companies who wanted to do something about what was widely perceived to be an uneven playing field, which left open the possibility of alternative approaches being developed. However, as the network succeeded in enrolling more players, it gradually became the principal means of developing an electronic data interchange (EDI) capability within and between the major insurance companies and their independent distributors.

The second approach to interpreting 'the networked organization' contests and subverts the assumption that social science can, and should, provide more accurate and reliable accounts of the reality of organizations in a way that directly parallels the contribution of the natural sciences. It reminds us that apparently impartial descriptions of the worlds being studied are actually definitions and prescriptions which could be divisive as well as unifying.

A focus on the rhetorical aspects of networking expresses the tendency to use apparently scientific descriptions to conceal the political manœuvring involved in enrolling participants in activities and relations that become solidified and taken for granted to the point of irreversibility. It allows us to see the concept of the networked organization as a device for securing the position of some dominant actors and threatening other subject positions, rather than treating it simply as a structure which can be unproblematically celebrated.

References

Bloomfield, B. P., Coombs, R., Knights, D., and Littler, D. (1997) (eds.), *Information Technology and Organizations: Strategies, Networks, and Integration* (Oxford: Oxford University Press).

Callon, M. (1986), 'Some Elements of a Sociology of Translation: Domestication of the Fishermen in St. Brieuc's Bay', in Law (1986), 196–233.

Hammer, M., and Champy, J. (1993), *Reengineering the Corporation: A Manifesto for Business Revolution* (London: Nicholas Brealey Publishing; repr. 1994).

Knights, D., Murray, F., and Willmott, H. C. (1997), 'Networking as Knowledge Work', in Bloomfield *et al.* (1997), 137–59.

Law, J. (1986) (ed.), *Power, Action and Belief: A New Sociology of Knowledge?* (Sociological Review Monograph, No. 32, London: Routledge and Kegan Paul).

Senge, P. (1991), *The Fifth Discipline: The Art and Practice of the Learning Organization* (New York: Doubleday).

4.2. Home Informatics: New Consumer Technologies

Ian Miles

Intense efforts are being made to capitalize on ICT applications in consumer markets (Miles 1988; Cawson *et al.* 1995). New products are multiplying—from the utilitarian (pocket calculators and programmable sewing machines) and socially responsible (appliances designed to assist the physically disadvantaged or protecting the environment), to the playful (video games and synthesizers) and frivolous (singing birthday cards and computerized jogging shoes).

Widespread rejection of the technology itself is unlikely (Miles and Thomas 1995). Earlier technological revolutions were similarly built into consumer goods and services. For instance, cheap steel from the early Industrial Revolution eventually underpinned numerous household goods, from cutlery to sewing machines. Gas and electricity powered a wide range of new appliances. Electronics capabilities led to the emergence of a distinction between 'white goods' (the typically enamelled kitchen appliances such as fridges and cookers), and 'brown goods' (such as wood- or bakelite-cased record players, radios, and TVs).

By the 1990s ICTs were being applied both to familiar goods, like video cassette recorders, and to new types of products, such as video games and multimedia CD-ROMs. Although the applications of such new ICT capabilities are part of a technological revolution, they are not themselves the drivers of a revolution in domestic life. However, they may complement such transformations—just as the previous revolutions in domestic technology were used to speed and shape earlier reorganizations in family life and the nature of the home.

Current ICT applications look much less likely to be labour-saving to housework than were, say, electrical goods. Serious domestic robots remain on the far horizon; in the immediate future we are likely to see little more than the application of motors and control to some unfamiliar areas, such as helping to pull curtains, and to the further enhancement of some established brown goods.

More interesting is the possibility of shifts in the household division of labour. For instance, there is controversy as to whether the microwave cooker, which is controlled by microprocessors, has facilitated these shifts.

Three main trends have developed in the application of new ICTs to consumer technology. These cover changes to existing products and systems, the creation of new ones, and the development of more integrated systems.

There are numerous examples of how ICTs have brought about changes in the quality or functionality of products. For example, cheaper, more powerful, and smaller microelectronic components have been combined with other components to open many varied opportunities through the exploitation of capabilities such as: improved miniaturization and portability; better interfaces, controls, and displays; more programmability; greater information storage potential; higher quality and more reliability; and enhanced multimedia telecommunications and broadcasting technologies.

Totally new products and services are also proliferating. These include: home security systems that coordinate and control various sensors and alarms; multimedia home computers; virtual-reality facilities; portable electronic books, communicating notepads, and videophones; and new information and communications services, such as e-mail, the Internet and World Wide Web, and the use of cable and satellite channels for 'interactive TV' services such as teleshopping and telebanking.

ICT-based products can also be linked into integrated systems by a variety of communications connections, including infra-red, wireless, cable—and even signals sent along the electricity mains. This makes it possible to connect together household devices into integrated home systems or 'smart' houses which could, for example, have computerized controls that automatically conserve power, refresh the air, and provide appropriate levels of light and heat or safety systems which prevent power points delivering shocks, children operating cookers, and burglars gaining access to the home. What is more, it will be possible to communicate with the smart home from a distance—for example, to find out who is tampering with your front door or to turn on the hot water so that you can have a bath when you get home.

The above are just a few of numerous possibilities created by the way new ICTs offer a vast and growing toolbox which can be used in the continuing transformation of the social and material world. Even if some early forms of particular innovations are disappointing in terms of price and performance and some applications, like home automation, take longer to get accepted and established than their proponents hope, novel ICT-based products and services will continue to emerge onto the market. How these are further developed and used is the big question.

There is justified foreboding about a world saturated with interactive video games, multichannel soap operas, and pornography on the Internet—a world in full retreat into the tackiest of virtual realities. But media can bring people together as well as coming between them. And new ICTs offer considerably more opportunity for many-to-many communications and participatory activity. ICTs are also almost certainly part of the solution to environmental problems. Everything depends upon our broader social goals and the generation of awareness about the ways in which individual and collective action, using new technologies where appropriate, can begin to achieve them.

References

Bauer, M. (1995) (ed.), *Public Resistance to New Technologies* (Cambridge: Cambridge University Press).

Cawson, A., Haddon, L., and Miles, I. (1995), *The Shape of Things to Consume* (Aldershot, UK: Avebury).

Miles, I. (1988), *Home Informatics: Information Technology and the Transformation of Everyday Life* (London: Pinter Publishers).

—— and Thomas, G. (1995), 'User Resistance to New Interactive Media', in Bauer (1995), 255–75.

Part III

Tele-Access in Business, Management, and Work

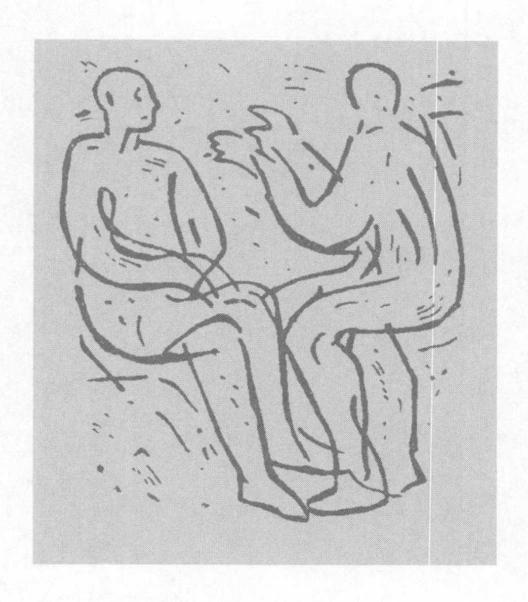

5 The Reach and Boundaries of Management and Business: Virtual Organizations

Access Management

The access revolution described in this book has created new electronically mediated opportunities for people to exchange information, meet, collaborate, work, and make decisions.[1] This is of profound significance to all organizations because the new forms of access enabled by ICTs challenge many conventional management, organizational, and business structures and practices. For instance, ICTs question the need for 'line-of-sight' supervision in which managers monitor and supervise activities by being located close to subordinates, if not actually within view.

ICTs can overcome many constraints of time and distance to place organizations in the presence of critical information, people, and services. This can liberate creative, administrative, management, production, and distribution resources and processes from the constraints imposed by traditional organizational structures. It also reduces the role played by the need to share information in locating work at particular physical sites. That freedom should lead managers to rethink how they organize and run their businesses in order to be competitive in a networked economy.

The scope given by ICTs for re-engineering business processes and organizational responsibilities has opened up opportunities to create what management expert Charles Handy (1996: 212) describes as 'virtual organizations' that 'do not need to have all the people, or sometimes any of the people, in one place in order to deliver their service. The organization exists, but you can't see it. It is a network, not an office.' Many highly successful networked organizations—some with no physical centre—demonstrate the potential for organizations to use ICTs to enable major change in the firm (Wells 1996). They illustrate how essential ICTs have become to creating new products and services, enabling productivity improvements, and managing tele-access not only within and among firms, but also to consumers.

However, all organizations cannot expect to succeed by emulating models of the virtual organization, which, by virtue of their networking, appear to outsiders to be like most other traditional organizations that operate from a major headquarters and satellite branches—but in reality have little centralized physical presence (see Box 5.1). There are other strategies for using ICTs to manage tele-access that can also competitively advantage certain organizations, such as supporting the vitality of large integrated firms. This chapter identifies the key opportunities and limitations of the virtual organization in the context of a broad analysis of the most effective ways in which tele-access can either be a barrier for a business or enable it to revitalize everything it does.

Box 5.1. The Virtual Organization: A Technical, Structural, and Cultural Vision

Networking through ICTs

- substitutes electronic media for physical files and face-to-face meetings to encourage more efficient sharing and gathering of information;
- focuses on face-to-face communication for building cohesion and trust;
- changes business processes to take advantage of the Internet and other ICT networks.

Restructuring to a decentralized network of companies

- uses networks to re-engineer and evolve structures of communication within and outside the organization;
- emphasizes horizontal coordination across functional, divisional, and company boundaries;
- creates jobs that span formerly separate functions and locations.

Building a team culture of trust and responsibility

- devolves authority and initiative;
- encourages workers to operate as trusting and cooperative team players;
- places high value on learning, change, and reorganizing to be responsive, rather than on rules, procedures, and hierarchy.

Sources: Hammer and Champy (1993); Nohria and Berkley (1994: 115); Fulk and DeSanctis (1995: 340); Murray and Willmott (1997).

Enabling New Organizational Forms: The Virtual Organization

Electronic communications and the use of fixed and portable computers provide highly adaptable work spaces (see Chapter 6). These are the foundations on which virtual organizations can be built and rebuilt in a continuously shifting pattern of alliances over time and across space. Virtual enterprises could integrate business processes which encompass people working for many different groups within and outside the company, including direct links to suppliers and customers. Boundaries of the organization and the units that compose them are becoming so permeable that a unit of a firm today could be replaced by an outside contractor from another part of the world tomorrow.

The enormous potential competitive power that can be unleashed by this enhanced flexibility and responsiveness depends largely on the strategic use of ICTs to support new patterns of tele-access. However, the excitement generated by tele-access innovations like the Internet has led to an overly deterministic portrayal of the virtual organization as a 'silver bullet' to solve all business problems. This is based primarily on a technical logic, derived from the escalating use of ICT networks, which ignores social and organizational constraints on fundamental change in the design and use of ICTs in organizations. These other factors mean that tele-access is also often used successfully to facilitate the growth of large integrated firms. Nevertheless, for many enterprises the virtual organization can be a valuable approach to realizing the benefits of tele-access in order to reach new markets, open new channels to existing customers, transform existing products and services, and innovate in entirely new areas.

Imagine, for instance, entering a bookshop with over 2½ million volumes you can browse through before deciding what you want delivered to your door. This unlikely physical prospect is feasible on the Internet, as shown in 1995 when 'Amazon.com' opened a virtual bookshop on the Web—an innovation that other firms have emulated. An earlier example, based on telephone rather than computer networking, is provided by First Direct, a sort of a virtual bank established in 1989 by Midland, one of the UK's major banks. First Direct captured hundreds of thousands of customers, primarily among wealthier households, by delivering most of its services over the telephone, twenty-four hours a day, seven days a week (Goddard and Richardson 1996: 209–10).

In addition to these technical changes, there are three related social and organization innovations that are essential to discussions of the virtual organization (see Box 5.1):

1. **Networking.** Telecommunication and computer-based communication systems are such a crucial element of virtual organizations that the overall approach is often referred to as 'the networked organization' (see Essay 4.1). Networks can, for example, help managers develop new connections as individuals, generate more communication across, up, and down the organizational hierarchy, and create opportunities for new gatekeepers between the 'on-line' and 'off-line' worlds within the enterprise (Hiltz and Turoff 1978; Rice and Associates 1984; Culnan and Markus 1987; Charan 1991; Sproull and Kiesler 1991). ICTs can be used to create new cross-functional 'virtual teams' spanning traditional departmental boundaries (Hammer and Champy 1993; Mankin *et al.* 1996) and improve tele-access connections across firms (Johansson *et al.* 1993: 9). Both internal 'intranets' (see Box 5.2) and inter-organizational networks, like the Internet, are important tele-access media. Less than

10 per cent of companies in the UK, for example, were using an intranet by 1997, but this proportion is growing rapidly.[2] These open up new 'electronic corridors' as channels of access, while they also create some electronic barriers to close off others, such as 'firewalls' (explained in Box 5.2).

Box 5.2. Intranets: Private Computer Communication Networks

A popular corporate innovation in the 1990s has been the private 'intranet', based on Internet and Web technology, such as the TCP/IP protocol and Web-like hypertext links and browsers. Intranets provide the benefits of the Internet in interconnecting diverse hardware and software. Its private ownership also enables much higher levels of security by creating electronic barriers such as passwords and security 'firewalls'—links to outside networks that are programmed to limit what kinds of access will be permitted by whom.

Booz Allen and Hamilton Inc., a major management consulting firm, has used a corporate intranet since 1995 to interconnect over 6,000 employees in 80 offices round the world. Whatever equipment is used locally, employees can get access to the same repository of information, containing thousands of documents cross-referenced by geography, industry, and topic. Its intranet also contains a directory of employees and their areas of specialization and skill, which allows quick identification and location of experts needed to form a consulting team for a particular client's requirements. E-mail and teleconferencing allow employees to engage in private and public discussions, for brainstorming, to ask questions, and to promote new ideas.

Source: http://home.netscape.com/ provides information on corporate intranets.

2. **Restructuring the organization.** The virtual organization provides a strategy for downsizing, outsourcing, and transforming the business processes of the firm. Contracting out functions that are not among the core competencies of the firm can create more flexibility and lower costs. Likewise, the use of direct electronic data interchange (EDI) links between a manufacturer and a supplier to carry orders, invoices, and other key business transactions can facilitate the outsourcing of supply or manufacturing (see Box 5.3). Outsourcing also presents a structural approach to managing the changing geography of the firm, creating more autonomous units that do not need to be managed directly (Heckscher and Donnelon 1994).
3. **Building an appropriate 'learning-organization' culture.** Many advocates of virtual organizations stress that their approach is most likely to succeed within a culture which promotes cooperative teamwork at all organizational

Box 5.3. Electronic Data Interchange (EDI)

Inter-organizational linkages, such as that between a manufacturer and a supplier, can be faciliated by EDI systems that offer electronic interchange of data between computers, within or across organizations. EDI message standards permit companies with different types of hardware and software to exchange data. By creating a system in which networked machines communicate directly without requiring information to be re-entered by hand, the speed and accuracy of routine processes can be enhanced.

Colgate-Palmolive implemented an EDI software system in 1995 to link and organize data from all aspects of the company, including product shipments, schedules, balance sheets, and orders. For example, the software tracks when a shipment has left the factory, updates any data related to it, and alerts managers and technicians located hundreds of miles away. The goal is to track detailed demand in order to fine-tune production and delivery schedules to keep its inventory to a minimum. Tracking is done by monitoring the customers of Colgate-Palmolive's products, like Wal-Mart. As noted in the *Wall Street Journal* (18 Nov. 1996): 'If it works, no corner of its business will go untouched.'

levels among individuals and groups who each have a great degree of local autonomy. Incentive systems, such as competitive contracting and pay rises based on team performance, have been developed to encourage this culture (Mankin *et al.* 1996). Given a dependence on creating, disbanding, and recombining teams with new projects and strategies, proponents also emphasize the need for continuous learning and change (Drucker 1993: 177–90; Handy 1993: 243–5).

Technology, the Structure of Organizations—and Business Pay-Offs

Innovations like the virtual organization have reopened discussions about relationships between technology and structure. These were early themes of the kind of research on ICTs that reflected broader debates over technology and society (see Chapter 2). Three general perspectives have emerged as an aid to understanding these relationships: technological, organizational, and emergent (Sampler 1996: 6–12).

The technological perspective emphasizes the influence of ICTs on the structure of organizations. At its technologically deterministic extreme, proponents of this view claim that new organizational forms, like virtual organizations, arise primarily from technological change (Tapscott 1995). The organizational viewpoint focuses on how the strategies and characteristics of organizations influence the design and use of ICTs (Danziger *et al.* 1982; Child 1987). The emergent perspective incorporates the idea of an ecology of games in arguing that both organizational and technological change unfold in unpredictable ways as a consequence of ongoing debates among multiple actors and institutions (Fulk *et al.* 1990; Mansell and Silverstone 1996; Bloomfield *et al.* 1997). Actors in this process are regarded as being constrained by existing technologies and structures, while also actively interpreting and using them to influence others in pursuing their own interests. While this emergent perspective is in line with the notion of an 'ecology of games', it does not focus on issues of tele-access.

In fact, by focusing attention on the interplay between technological and organizational change, the debates among adherents of these alternative approaches have, unfortunately, deflected attention from the need to identify more precisely how organizational and technological changes combine to make a difference to 'bottom-line' outcomes, such as the productivity of the firm, or who is in and who is out of the firm. They have also failed to highlight change in the geography of the firm, another issue of tele-access, that is enabled by technological and organizational change (see Chapter 6).

The tele-access concept shifts the focus away from the relations between technology and organization towards a concern with how both technical and organizational designs interact to shape access. In addition, tele-access exposes a flaw in the argument that the revolution in ICTs inevitably enables 'communication of richer, more complex information' (Fulk and DeSanctis 1995). While it is true that ICTs can enable access to people, information, services, and technology that would not be possible in others ways, they can also reduce, screen, and change the content and flow of information.

Economic pay-offs and competitive advantages from ICTs are not inevitable (Kraemer *et al.* 1981; Scott Morton 1991). Firms need to reorganize business processes in order to take full advantage of ICTs. Technical innovations also require social and organizational innovations which take longer and require more political will than that involved in acquiring new equipment. Moreover, the complex technical and organizational interdependencies of ICTs can create a type of 'electronic concrete' (Quintas 1996: 85–9) that can constrain organizational flexibility even more and leave organizations isolated from information or customers (see Box 5.4).

Box 5.4. Electronic Concrete

PICT researcher Paul Quintas (1996: 85–9) coined the term 'electronic concrete' to counter the conventional belief that computer-based software and systems are always far more malleable that physical structures and systems.

In early stages of development, software and its requirement specifications are indeed very flexible. However, most large-scale software development projects evolve over many years, even decades, of design and programming change, involving teams of developers working on different parts of the system.

As Joseph Weizenbaum (1976) pointed out, such systems can become so complex as bugs are repaired and new capabilities added that they evolve into something that is literally 'incomprehensible'—the system may do what it was programmed to do, but no individual fully understands exactly how the overall system works. As Quintas (1996: 89) put it: 'The sunk investment in such legacy systems, together with their complexity, means that organizations find it difficult to replace them or to make significant changes in their information and communication technology strategies.' They become set in the electronic equivalent of concrete.

Technological Underpinnings of Tele-Access in Business and Industry

All ICTs play a role in the politics of information in an organization. They shape tele-access by creating, rather than only reporting, 'organizational reality' (see Essay 5.1). Even the most routine budgeting and accounting system will shape knowledge about the organization by defining the ways in which a firm categorizes and views its costs and revenues. Computer models, such as a financial projection calculated on an executive's spreadsheet program, can also define how managers see their reality.[3]

Managers seldom think of an accounting system or forecasting model as an ICT that shapes their view of reality. The indirect and often unintended role of ICTs in shaping an organization's knowledge of the world might be the most profound implication of tele-access for management control. Yet increasingly managers are taught to use models of a business's operations to 'manage-by-wire', as it has been argued by Haeckel and Nolan (1993: 122–3): 'expert systems, databases, software objects, and other technical components are integrated to do the equivalent of flying by wire. The executive crew then pilots the organization, using controls in the information cockpit of the business.' Inadvertently, or purposively, ICTs are being applied more to manage tele-access in four interrelated areas: electronic meetings, corporate communications, tele-mediated collaborative working, and the creation and tele-marketing of new products and services.

Electronic Meetings

Until the 1970s, most computer systems focused on administrative support and clerical applications. This meant the technology was rarely treated as a matter of priority by top management. Likewise, the telephone was viewed as a cost of doing business rather than an element of corporate strategy. Telecommunications managers were hired to keep the phones working and the costs down, not to help the firm rethink its business processes, products, or services.

One of the first applications that made top managers aware of the broader uses of ICTs was the introduction of electronic-meeting services, such as video teleconferencing. These ICTs—sometimes called 'groupware'—are providing an ever-expanding range of capabilities to support face-to-face meetings and direct teleconferencing systems for electronic group meetings involving participants in different locations (see Box 5.5).

Box 5.5. ICTs Supporting Executive Meetings and Decision-Making

For face-to-face meetings

- presentation software, multimedia PCs, and projectors to aid speakers
- group decision-support systems for brainstorming, ranking alternatives, and polling or voting on options
- computer-based models, spreadsheets, and gaming simulations to provide analyses prior to, but also during, a meeting or strategy session

For electronic group meetings of people that span time or distance

- audio conferencing systems (speakerphones, high-fidelity audio systems, and conferencing bridges for linking multiple voice lines) for bringing together distributed groups
- text (facsimile, e-mail, computer conferencing systems) that permit individuals to participate asynchronously when they are available
- audio-graphic conferencing systems (electronic black boards, combinations of facsimile and audio conferencing) that permit visual aids to be used
- video conferencing (desktop video and workstations, or permanently installed video conferencing rooms) to emulate face-to-face communication, and to permit the introduction of motion video, or other visual presentation aids

In face-to-face meetings, a computer running a program like Microsoft's PowerPoint is now commonly used to enable the graphical layout, production, and presentation of visual material to be handled relatively easily by one person on a PC. Animation, video, and other multimedia information can be incorporated into such presentations. Many major firms, like the Bank of America, have set up 'electronic boardrooms' to hold electronic meetings on a routine basis. These can include advanced ICT capabilities such as audience-response and electronic brainstorming systems, high-quality audio, group conference calls, computer-based models, simulations, and access to the Web. Sometimes these ICTs are used simply to show the trustees and directors of a company that management is technologically up to date, but their use can reshape tele-access, purposefully or unintentionally, as discussed below.

The most notable growth in person-to-person and group electronic communications in the 1990s has come through the boost given by the Internet to e-mail and computer conferencing systems, which were previously confined mainly to private internal networks in a few organizations. The Internet combined with the growing availability of PCs to help create a critical mass of people who can be contacted electronically within and outside organizations. An important benefit of these systems is that they are asynchronous, so the people do not have to be in the same place at the same time to meet, and can respond at a convenient time from distant locations to messages stored in an 'electronic mailbox'.

Growing use of the Internet in corporations has generated new interest in video communications, such as multi-point video conferencing. The first video-conferencing systems made available to the public by the major public telecommunications operators (PTOs) in the 1970s and 1980s (see Box 5.6) generated great expectations, but insufficient user demand (Fulk and Dutton 1984; Noll 1986: 69). Nevertheless, falling prices and technical advances in video-compression technologies led many major companies to install private video-conferencing facilities in the 1990s to support business meetings and routine collaboration among multiple, far-flung activities that could benefit from visual presentations.

Despite improved technical telecommunications quality and declining costs of equipment, video conferencing has remained more of a corporate status symbol than an everyday feature of organizational life (Johansen 1984, 1988; Noll 1986). Technical advances that allow a video window to be opened (or closed) on the screen of an executive's multimedia PC could help to make that breakthrough.

Box 5.6. Early Public Teleconferencing Systems and Trials

The high cost of TV-quality video conferencing meant that early efforts focused on the design of publicly accessible rooms that could be leased for executive meetings. Early systems included:

1. *Confravision.* Set up by the British Post Office in the 1970s, five Confravision studios could be linked for two-way conferences over video monitors that were similar to normal TV screens. Commercial trials of Confravision were disappointing.
2. *Picturephone Meeting Service (PMS).* In the late-1970s, AT&T shifted the marketing of its Picturephone technology to its use for executive business meetings by establishing the PMS trial. This initially employed several Picturephones around a conference table. PMS soon moved to full motion, TV-quality pictures between any two of its multiple sites that could also show slides, overheads, and videotapes. Each video-conferencing site was located in an AT&T room that could be used by the company, but also leased to the public for a price well below the cost of a fully commercial service. PMS failed to generate great demand and was phased out following the divestiture of AT&T in 1984.

Similar systems were trialled by NTT and Bell Canada with similar results.

Sources: Short *et al.* (1976: 1–10); Martin (1977: 128–31); Noll (1997: 22–3).

Corporate Communications

Companies have increasingly augmented their internal corporate communications channels with ICT-mediated systems covering four main types of activities:

1. **distribution of corporate memoranda and notices,** for example through automated facsimile distribution, computer bulletin boards, and e-mail distribution lists;
2. **corporate newsletters** produced by desktop publishing systems, multimedia CD-ROMs, and other ICT capabilities;
3. **interaction through intranets** that provide electronic mail, Web sites and other means of broadcasting corporate information and getting feedback on it;
4. **business television** via satellite-distributed closed circuit TV and video cassettes.

The escalation in Internet use fostered corporate moves towards establishing private intranets and acquiring lower-cost, but more specialized PCs, such as 'the network computer' (see Box 5.7). Other forms of corporate communications, such as business TV, have also been greatly expanded. For example, some of the largest companies are using satellite communications to provide dozens of closed-circuit TV channels for corporate communications and training.

Box 5.7. The Network Computer and Tele-Access

The idea of a network computer (NC) was promoted by Larry Ellison of Oracle in the mid-1990s to replace the PC as the standard for corporate computing. The NC accesses software over a network, rather than from programs stored on the computer. This means that the NC does not require as much expensive storage capability, like a hard drive, as a PC. All NC software can be centrally provided and updated for all users linked to the corporate network, which avoids the risks associated with user-installed PC programs. The NC is in some ways a reinvention of 'dumb terminals' linked to time-shared computer systems, which were used extensively in the pre-PC era.

The NC could reshape tele-access in two very different ways. For users, it could advantage corporate management by recentralizing control over software and data within the organization. Rather than creating unique configurations of software on their own PCs, individuals would use standardized NC software and databases. They would be housed on the network and maintained and used by the organization as a whole, perhaps creating a standard company 'desktop'. Among ICT producers, the NC could influence the market access of different hardware and software vendors. For example, the NC has been one strategy for by-passing Microsoft, which dominates the market for PC operating systems—and which initially resisted new programming languages like Java that permit the development of software which, unlike Microsoft Windows, can run on any type of computer. In fact, Java's developer, Sun Microsystems, collaborated with Microsoft rivals IBM and Oracle.

All these capabilities permit organizations to link top managers with their employees in every corner of the globe. This provides the dual capability to centralize some components of organizations, such as corporate strategy, while decentralizing others, like manufacturing.

Day-to-Day Collaboration and Networking

The rapid growth in the use of mobile phones, voice-mail messaging systems, electronic mail, and other ICT-mediated person-to-person communications has been critical to rethinking the role of telecommunications in the day-to-day management and operation of organizations. These facilities have been supplemented by 'groupware' software (Johansen 1988) and organizational information management capabilities which provide networked computer-supported collaborative working (CSCW) and group decision-making (see Box 5.8). The marriage of these technologies and the Internet has transformed conventional assumptions about the nature and extent of collaboration and management control that can take place at a distance. This is discussed further in Chapter 6.

Box 5.8. The Uses of ICTs for Collaborative Work in Organizations

Person-to-person communication and networking

- e-mail, videotext, facsimile, voice mail
- PCs, modems, and desktop video
- mobile phones, pagers, and hand-held messaging devices

Collaborative work and groupware

- exchanging files, electronic file transfer, transferring discs
- telephone, facsimile, e-mail, and Web access
- collaborative systems for screen sharing, group authoring
- computer models, simulations, and forecasting tools, such as spreadsheets
- electronic data interchange

Organizational databases and archives

- hand-held and laptop computers, personal digital assistants
- hypermedia, such as the Web and randomly accessible CD-ROMs
- data and information warehouses, accessible over corporate intranets

The flexibility and fluidity of networking create new opportunities for collaborative developments of products and services, including multinational 'relationship enterprises', a type of virtual organization consisting of 'temporary networks of strategic alliances that shift among firms in different industries, locations, and even countries' (Whitman 1994: 35–6). ICT-mediated links and common goals enable and encourage allied firms to behave more like a single entity. Tele-access can also bring even very small local companies into profitable contact with customers at a great distance, as discussed below.

Electronic Products, Services, and Markets

Advances in ICTs have become bound up with the way an organization defines its products and services—not only in the way an organization does business and markets itself (Miles 1993; Penzias 1995; Bloomfield *et al.* 1997). A company's ICT infrastructures can be an intrinsic part of the products and services offered to customers, such as airline reservations, twenty-four-hour banking from automatic teller machines (ATMs), home-based shopping, and computer consulting over the Internet (see Essay 3.2).

The ATM cash machine illustrates how direct information access can simplify and reduce the labour involved in the provision of a service. It also shows how electronic networks can incorporate customers within the boundaries of the organization by giving them direct access to carry out some tasks, like data entry, that were previously done by the firm's employees. Airline reservation systems, like American Airlines' SABRE System, are another classic example of providing services more directly to consumers over the telephone or via PCs in the home or office. A United Airlines advertisement put it this way: 'With United Connection and your mouse, it's easy to make your own travel arrangements.' This is being extended all the way to the runway as many airlines automate passenger boarding.

A critical issue is whether electronic access to services over such facilities as the Internet will wire customers into businesses, or permit them greater flexibility in picking and choosing the best airline, bank, or insurance firm. Proprietary systems, such as used in the initial years of electronic banking, could make it more difficult for a customer to switch providers, such as by obliging a customer to obtain different software, or learn new procedures. More standardized systems for the preparation of finances on a home PC, and banking over the Internet, could undermine the force of proprietary systems, as well as location, in locking in access to customers.

Freed from the constraints of geography, or unstandardized systems, ICTs also enable competitors to find customers. For example, ICTs allow customers to be targeted more precisely through tele-marketing systems, such as by using computer-based mail-merge programs, detailed consumer data, profiling, and direct marketing tools available on the Web.

Key Factors Shaping Management Strategies

Three of the factors shaping choices about tele-access which were outlined in Chapter 1 are of particular relevance to developments like the virtual organization: economic resources and constraints, ICT paradigms and practices, and organizational structures and business processes.

Economic Resources and Constraints: Virtuous or Vicious Productivity Spiral

The most productive nations and firms are investing larger proportions of their resources into ICTs, which is advantaging these players by delivering greater levels of productivity (Dertouzos *et al.* 1989; Kraemer and Dedrick 1996; Melody 1996). This indicates why ICTs are central to defining the 'state of the practice'. A significant level of access to ICTs is, therefore, essential to any enterprise or country wishing to be viewed as a serious player in the world of business and industry. This can advantage the larger and wealthier nations and firms also, since they can better afford access to these ICTs, creating a virtuous spiral (Castells 1996; Melody 1996). It can also create a vicious spiral for many organizations than cannot afford access to technologies (Scott Morton 1991; Freeman 1996*b*).

But access to ICTs does not translate automatically into productivity gains (Freeman 1996*b*). The inability to achieve significant productivity gains from a tool that offers tangible improvements at a personal level has become known as the ICT 'productivity paradox'. One reason for the apparent paradox is anchored in the poor design and implementation of ICTs (Kraemer *et al.* 1981; Fincham *et al.* 1994). As one survey of management specialists noted:

> **In the majority of cases new systems do not meet their objectives and are counted at best as partial successes, at worst as failures. Most systems are delivered late and over budget. Failures in this area are rarely purely technical in origin. Most organizations are poor at evaluating the effectiveness and impact of their investments in this area.** (Clegg *et al.* 1996: 13)

Another reason why ICTs have often failed to achieve gains in productivity is that organizations have not had sufficiently strong incentives to cut their costs. They have often been able to support increased costs by raising revenues, either through price increases or through growing market share in an expanding economy. It was only when faced with increasing global competition and lower rates of growth that many organizations were forced to cut their costs. The ICT infrastructures in place have permitted them to do so (Penzias 1995: 2; Kraemer and Dedrick 1996).

A third explanation for the paradox is the long period that is needed to make the work-design, cultural, and structural changes within organizations which are necessary to allow the full benefits of technological change to be realized (Kraemer *et al.* 1981; Moss 1981; Freeman 1996*b*).

Field research highlights the degree to which organizations resist change (Bloomfield *et al.* 1997). In the early decades of computing, the development of management information systems (MIS) was viewed as a means for technically reforming structural deficiencies of organizations (Kraemer *et al.* 1981: 4–7). Since the 1980s, however, change in organizational structures and business processes have been seen as important to realizing the productivity pay-offs of ICTs (Scott Morton 1991; Coombs and Hull 1996). For instance, the increasing importance of ICTs to corporate strategies has helped to promote innovations in organizational design and business process re-engineering (BPR) (Hammer and Champy 1993).

ICT Paradigms and Practices

Radical technological innovations, such as in ICTs, require major social and organizational change—in fact, they are inseparable. One major socio-technical change lies in a shift in the paradigm for how we do things. For instance, much enthusiasm for the virtual organization of the future is based on a vision steeped in rhetoric about the information society and the Internet. Although this vision can be misleading, such new paradigms are often necessary to convince an organization that it makes sense to do things in a new way, which is essential before innovations can move effectively from the research lab into real organizational life.

Two interrelated paradigms play a critical role in the visions and realities of new organizational forms like the virtual organization:

1. **The new ICT paradigm.** As discussed in Chapter 2, ICTs have fostered a new way of thinking about the management of business and industry that makes a break from the 'Fordist' mass-production orientation of industrial organizations (Perez 1983; Miles and Robins 1992; Freeman 1994). The virtual organization fits many characteristics of the new ICT paradigm as it is information intensive, relies heavily on ICTs, and promises to reduce hierarchical patterns of communication (see Table 5.1). Practical manifestations of the strength of the ICT paradigm include the way in which many companies that are more committed to ICTs have also become: more specialized—for example, by mechanisms like the spin-off of non-core businesses; smaller in terms of the number of full-time employees (Whitman 1994: 33); and more geographically distributed (Goddard and Richardson 1996).
2. **The telecommunications-substitution paradigm.** ICTs can be an adequate substitute for face-to-face communication for some tasks, such as

Table 5.1. Changing Paradigms and the Virtual Organization

Features	**Fordist**	**Post-Fordist**	**Virtual organization**
Technology	Low technology, energy intensive	High technology, information intensive	Organization is a network
Focus of ICTs	Automation, mechanization	Systematization	Electronic data interchange (EDI)
Design and production	Sequential	Concurrent	Collaborative
Mass production	Standardized production of fixed product lines	Customized production of changing products	Organized for innovation, mass customization
Production resources	Dedicated plant and equipment	Flexible production systems	Outsourced to maintain flexibility
Ownership	Single large integrated firm	Networks, profit centres, internal markets	Outsourcing, spin-offs to maintain lean organization
Management control structures	Hierarchical, vertical chain of command	Flat horizontal structures, lateral communication	Coordination through the market place, competition
Work	Departmentalized	Integrated in teams across departments	Project and *ad hoc* teams
Knowledge, learning, memory	Centralized	Distributed	Distributed strategically to protect core tacit knowledge
Job skills	Specialized, bureaucratic	Multi-skilling, professional, entrepreneurial	Employees involved in more aspects of the business

Sources: Womack *et al.* (1990); Miles and Robins (1992); Freeman (1994).

in giving and receiving information (Short *et al.* 1976; McGrath and Hollingshead 1994). However, despite the compelling economic logic of using tele-access to reduce costs, telecommunications have been applied mainly to enhance, rather than to substitute, person-to-person communication (Elton 1985; Dutton 1995*c*). Substitution is efficient, but carries several risks. First, it might not happen. Substitution has been forecast for decades, but executive travel has not diminished, as most executives like to travel (Johansen 1984: 16–34). Secondly, many executives and others seem to be governed by an alternative 'tele-enhancement' paradigm in which telecommunications become a means to complement, rather than replace, face-to-face communications. For example, people most often communicate electronically with those they also travel to see person to person.

These emerging paradigms can be criticized as being overly deterministic by failing to point out that, for example, ICTs can help to make both mass or customized production more efficient. Tele-access suggests a non-deterministic way of thinking about ICTs because it highlights the value of using ICTs to shape a variety of strategic choices about access. The tele-access perspective explains why it would be wrong, for example, to view the 'virtual organization' as a prescription for always substituting ICTs for person-to-person communication. Face-to-face meetings may actually be superior in some situations, such as in getting to know someone, or exercising the authority of a higher status position (Short *et al.* 1976; Rice and Associates 1984; Daft and Lengel 1986). An effective tele-access strategy would seek to find the appropriate balance between using ICTs to enhance or substitute for other forms of access to information, people, services, and technologies. Individuals would view such choices as critical to what they will learn, who they will meet, and how well they are served.

Organizational Structures and Business Processes

Organizational structures and business processes can be viewed in the same way as ICTs. For example, management structures, like the design of reporting relationships, are used to shape access to information. Executives limit the number of department heads reporting to them as a means of screening and filtering information. Changes in ICTs are therefore difficult to separate from change in organizational structures and processes.

Awareness of the interplay between ICTs and business and organizational issues started in the 1950s, when computers first became a serious business resource. Computers were regarded primarily as a tool for enhancing centralized, hierarchical control until the early 1980s. Since then, computer-based ICTs have been increasingly seen as leading to the flattening of organizational hierarchies by permitting drastic reductions at middle-management levels (Leavitt and Whisler 1958; Dertouzos *et al.* 1989: 122–4; Bloomfield *et al.* 1997). This is illustrated by the case study of TRW (see Box 5.9).

Forecasts were made as early as the 1950s that ICT capabilities could be used to flatten organizational structures by reducing the need for middle managers whose jobs entail the management of information (Leavitt and Whisler 1958). Yet computerization actually accompanied growth in the

Box 5.9. Video Conferencing at TRW: Restructuring Access

TRW is a diversified automotive and electronics equipment firm based in Cleveland. In 1981 its space and defence operations, based in Redondo Beach, California, experimented with video teleconferencing a quarterly forecast report for its East Coast operations in New Jersey. This report is a formal mechanism for reviewing units by focusing on year-to-date revenues and expenditures, and forecasts for the remainder of the fiscal year and beyond. In this case, the Vice-President (VP) of one operating group within TRW's electronics sector in Redondo Beach was responsible for reviewing the activities of a unit in New Jersey. The Redondo Beach VP and his colleagues travelled to a video-conferencing facility in Los Angeles, while the New Jersey group travelled to a facility in New York City to hold the meeting. Normally, this report involved the New Jersey VP, General Manager, and Director of Financial Control flying to Los Angeles to meet with the group VP and five or six of his senior staff for meetings lasting about three hours. Participants got over 'being on the screen' within a few minutes and felt the content and tenor of the video conference was much like their face-to-face meetings. However, the New Jersey VP brought six (rather than two) staff to the table and a dozen more managers sat behind them. Los Angeles had direct, less filtered access to more specialists without going through the hierarchy and vice versa. One LA participant noted: 'If you had a question for [a specialist] you could get him to come forward. You didn't have to ask his boss what he meant; you could ask him.' TRW subsequently installed private video and uses conferencing facilities extensively to link its 300 locations across fifteen countries.

ranks of middle managers through the 1970s, in line with the tendency for organizations to apply ICTs in ways that reinforce existing power and control structures (Danziger *et al.* 1982). It was not until the growth of organizations could no longer be sustained by new revenues that traditional management paradigms and business and organizational structures and processes came under great challenge by new ICT-paradigms (Penzias 1995; Kraemer and Dedrick 1996).

The relatively high cost and the lack of on-line links of early large mainframe computers were particularly suited to their use in the prevailing centrally administered organizational structures (Winner 1986: 19–39). Subsequent innovations, however, have made ICTs flexible enough to support any structure which is deemed to be appropriate for an organization's internal and external interactions. Research has found that this flexibility has been applied largely to encourage greater decentralization (see Box 5.10). As telecommunications become less and less a constraint on organizational design, other factors such as wage differentials and the

Box 5.10. Dimensions of Change in Organizational Forms Tied to ICTs

Intra-organizational: vertical control

- decentralized, less hierarchical control over decisions
- delayering or flattening hierarchies: eliminating middle management layers
- reducing proportion of administrative support staff and costs
- decreasing formalization of behaviours and job requirements

Intra-organizational: horizontal coordination

- organizing work groups by electronic workflow, not physical location
- concurrent engineering: simultaneous design by distributed teams
- stockless production, such as just-in-time (JIT) systems

Organization and unit size

- downsizing and delayering to reduce permanent staff numbers
- spinning off or outsourcing activities
- creating federated organizations: decentralizing within a centrally coordinated framework

Inter-organizational coordination

- Inter-organizational coupling, such as in electronic data interchange (EDI)
- strategic alliances across industries to share information or networks
- other linkages, such as through coordinating associations or directorates

Sources: Jablin (1987); Fulk and DeSanctis (1995); Rice and Gattiker (forthcoming).

need for face-to-face interaction are playing a greater role in locational decisions (Downs 1985; see also Chapter 6).

One way in which ICTs have promoted decentralization has been through their use in supporting new forms of strategic alliances in many companies, including collaborations between organizations in formerly distinct industries. In 1996, for example, the Bank of America and the Lucky chain of supermarkets in California agreed to a partnership which involved placing the bank's ATMs in Lucky's stores. Tele-access can also support improved coordination between trade associations and other informal links among companies, such as a group of companies which depend on each other for the development of certain products and services.

Despite the growing popularity of alliances, such organizational forms provide no easy answer to vital problems of communication, coordination, and trust that are central to their success (Weber 1993). Many factors beyond ICTs are critical to the outcome of strategic alliances (see Essay 5.2).

The same is true of outsourcing, where a critical issue is the risk a company runs of losing some core competencies by outsourcing activities vital to its business.[4]

For instance, IBM's decision to outsource the provision of its IBM PC microprocessor to Intel and operating system to Microsoft was judged critical to the speed with which it was able to bring the IBM PC to market successfully. But that decision also contributed to the subsequent ability of Microsoft to become a powerful competitor to IBM.[5] When Virgin Atlantic followed the path of many companies in outsourcing its information-processing through a facilities-management (FM) contract[6] with British Airways, it found that BA employees had used Virgin Atlantic computer data to 'poach customers and analyze route profitability' (Dempsey 1993).

Some firms have also looked to ICTs to extend centralized control over far-flung activities, such as by utilizing advances in mobile communications that allow contact to be maintained among individuals distributed across many locations. This kind of tele-access need not be used just as a means of enforcing centralized monitoring and control. For example, a San Francisco company has contracted clerical work to a small firm, Kinawley Integrated Teleworking Enterprise Ltd. (KITE), that operates out of a 'tele-cottage' in a rural community of Northern Ireland (Bryden *et al.* 1995). ICTs facilitate access from San Francisco to a skilled workforce at a lower cost and a lesser security risk than it could otherwise access, while allowing the tele-cottage firm to maintain great autonomy in deciding how to meet its contractual goals. In 1997 over 20 per cent of the 'top 500 US businesses' were reported to be using 'software services provided in India' (*Financial Times,* 30 June 1997).

Where Tele-Access Could Take Some Organizations

Organizations must think strategically about how they can align tele-access, structural change, and the business processes of the firm. This must be done without becoming fixated on the virtual organization or any other single approach to organizational design. The strategy should aim to use ICTs to enhance innovation, business processes, and interpersonal communication within the organizational structure most appropriate to the enterprise.

Account should be taken of the great benefits that can be derived from exploiting tele-access using relatively basic technologies, such as e-mail, fax, and telephony capabilities such as group conference calls and the Internet. It is not necessary to wait for something more technologically exotic, such as video phones or Webcasting, before starting to investigate what can be achieved by reshaping access. In fact, such basic technologies will remain critical to most tele-access efforts until more advanced ICTs are accessible to a large proportion of companies and households.

This chapter has underscored the degree to which ICTs have become a key aspect of modern business practices. It has illustrated a few of the ways in which managers can use ICTs imaginatively and effectively to manage access in ways that improve business practices and processes to deliver better overall organizational performance and customer satisfaction. This places great incentives and responsibilities on government and industry to provide access to ICTs, including essential education and training know-how. Success will depend not only on the paradigms and strategies of management, but also on effectively resolving the variety of political, social, and cultural issues related to tele-access—such as those resulting from widespread organizational 'delayering' and staff 'downsizing' cuts—that I turn to in the next chapter.

Notes

1. This chapter is based on a study supported by the Fujitsu Research Institute (Dutton 1997*b*, 1998).
2. This was based on an independent survey of 1,000 British companies conducted by the Future Foundation for BT (P. Taylor 1997*a*).
3. Computer models are used by organizations to facilitate negotiation and bargaining among individuals in ways that often help resolve conflicts and lead to a social definition of the real world (Dutton and Kraemer 1985).
4. One study of UK businesses found that outsourcing was not used frequently, and, when used, it had a 'low success' rate (Waterson *et al.* 1997: 7). As Bruce and Littler (see Essay 5.2) found, inter-organizational alliances and collaborations are problematic.
5. Overviews of the decisions and strategies behind the IBM PC are provided by Cringely (1992) and Chesbrough and Teece (1996), who use this case to discuss the strengths and weaknesses of the virtual organization.
6. An FM arrangement allows an organization to contract with an outside company to handle certain functions, such as all billing, payroll, or computer facilities.

Essays

5.1. Information Systems and Management Control

Rod Coombs

The relationship between information systems, management control, and power is a critical aspect of the role of ICTs in organizations. Rod Coombs draws on detailed case studies of organizations to illuminate the factors shaping this relationship.

5.2. Collaborative Product Development

Margaret Bruce and Dale Littler

ICT-based product development is becoming increasingly complex as the pace of technological change creates pressures to reduce product development periods, while increasing the demand for new skills as markets and technologies converge. This essay examines the formation of collaborative arrangements designed to meet these challenges.

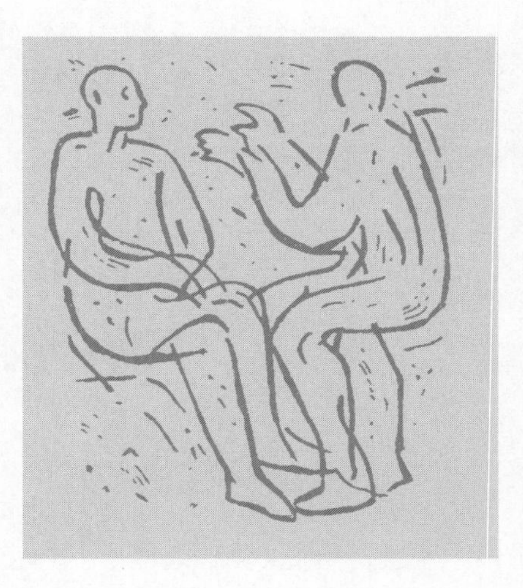

5.1. Information Systems and Management Control

Rod Coombs

Much specialist literature on management information systems treats power as a sort of commodity which exists in a zero-sum framework (Bloomfield and Coombs 1992). Within that framework, information systems are viewed simply as a new resource in the struggle for increments of power.

The interpretations of ICTs as a means of extending and enforcing control depend partly on emphasizing technological capabilities for gathering information on organizational phenomena in one location and reporting on them in another. Such reporting can be done in a more intimate, detailed, and timely way than with non-ICT systems. This concept of control therefore focuses on the surveillance potential of ICTs.

However, another distinctive property of information systems is their capacity to create, rather than merely report, organizational reality (Bloomfield *et al.* 1997). Placing this capacity of ICTs in the foreground helps to explain why the nature of management control may change as a result of attempts to implement it though the medium of ICTs.

This alternative approach is based on a disciplinary concept of power inspired by the work of Foucault (1979). His treatment of power focuses on the role that self-discipline can play in organizations and other social settings. As individuals internalize the norms of an organization's culture, they are likely to discipline themselves. For instance, a wealth of technical and everyday terms, such as 'the networked organization', is carefully defined and argued over in the design of information systems (see Essay 4.1). When they are used in live systems to explain what is said to be happening in the organization, they can decisively influence the understanding of members of the organization and define the terms of debate within it.

Every management report deploys terms such as costs, overheads, customers, quality levels, performance, assets, and cross-subsidy. All of these have to be defined within the information system and organizational phenomena reconciled with them. Yet their apparently benign and neutral appearance hides their carefully constructed nature and their power to include and exclude.

The internalization of such concepts and the willingness of an organization's members to construct their world and work in terms of these concepts is an instance of the exercise of disciplinary power. This style of management control can also be present without ICT systems. However, it is frequently achieved in greater measure when the concepts are also embedded in information systems.

The disciplinary concept of power is, therefore, the most effective way of accounting for some of the special properties of information systems, because it emphasizes how the categories are defined and embedded in the information system and their ability to constitute reality for users (see Bloomfield and Coombs 1992; Coombs *et al.* 1992).

This concept of power focuses on the compliance of individuals with procedures of self-discipline as well as more traditional forms of discipline. However, it is important to emphasize that such power is seen neither as an omnipotent form of control nor as being purely negative. Power is always subject to resistance and can open up a discursive space that brings opportunities for new arrangements, although these are constrained by the organizational contexts in which negotiation and action can take place.

For instance, when a new resource management information system was introduced into the UK's National Health Service (NHS), initial resistance by doctors to the system gradually changed as many of them began to see how the system could improve efficiency, patient treatment, and medical research without impairing their notions of clinical freedom (Coombs and Hull 1996). This occurred within a context where some managers regarded the new ICT system as being a success if it integrated doctors more tightly into new management processes, irrespective of how well it worked in purely technical terms. As in this case, addressing the political dimensions of change involved in ICT innovations is essential in general to understanding the vital relationship between information systems and management control.

References

Bloomfield, B., and Coombs, R. (1992), 'Information Technology, Control and Power: The Centralisation and Decentralisation Debate Revisited', *Journal of Management Studies*, 29: 459–84.

—— Coombs, R., Knights, D., and Littler, D. (1997) (eds.), *Information Technology and Organizations: Strategies, Networks, and Integration* (Oxford: Oxford University Press).

Coombs, R., and Hull, R. (1996), 'The Politics of IT Strategy and Development in Organizations', in Dutton (1996), 159–76.

—— Knights, D., and Willmott, H. (1992), 'Culture Control and Competition: Towards a Conceptual Framework for the Study of Information Technology in Organizations', *Organization Studies*, 13: 51–72.

Dutton, W. H. (1996) (ed.), with Malcolm Peltu, *Information and Communication Technologies—Visions and Realities* (Oxford: Oxford University Press).

Foucault, M. (1979), *Discipline and Punish* (Harmondsworth: Penguin).

5.2. Collaborative Product Development

Margaret Bruce and Dale Littler

As industries become more international and global in their operations, many products are being marketed in a number of regions simultaneously. Rapid technological innovation and intense competitive pressures are also combining to create shorter windows of opportunity. Gaining an advantage in this environment demands stringent cost and risk controls and the ability to remain responsive to quickly shifting market needs.

The formation of collaborations, especially in the area of ICTs, has been heralded as the most effective route to spreading the costs and risks of product development to meet these tough demands. The main forces driving this interest in collaborative ventures are:

1. access to new technology and expertise to complement existing competencies and to share or acquire know-how (Dodgson 1993);
2. technological convergence that makes it harder for individual companies to exploit promising opportunities on their own, particularly in an area like ICTs, which encompasses so many different specializations;
3. marketing advantages that provide access to overseas markets and knowledge of local customs, which can extend the scope of potential products beyond what was possible via internal development alone;
4. lowering the risks and costs of product development; and
5. shortening the time to market.

Nevertheless, significant disadvantages may arise with collaborative product development because of factors such as:

1. the high costs associated with the time taken to set up the collaboration and monitor its progress;
2. unsatisfied expectations of the collaborators;
3. fear of losing proprietary knowledge; and
4. failure to ensure that an equitable contribution is made by all parties.

Although this can lead to a cautious approach to collaborations, organizations can learn to manage collaborative product development effectively (Bruce *et al.* 1995; Littler *et al.* 1995). Companies which have had experience of collaborations in ICTs identify a number of influences critical to determining the outcome of ventures (Leverick and Littler 1993). The most significant of these include:

1. choice of partner (compatibility of culture, operational and management styles, creation of a high degree of trust);
2. establishment of ground rules (agreeing specific objectives, responsibilities, and project milestones which state clearly what is expected from each partner);
3. nature of procedures used (frequent communication and progress reviews);
4. ensuring equity of contribution (mutual benefit, equality in power sharing);
5. people (presence of a project or collaboration champion, top-level management commitment, effective personal relationships); and
6. environmental factors (such as monitoring markets to establish the need for a product).

All forms of ICT-related product development are taking place in difficult, varied, and quickly changing environments. Collaboration adds an extra degree of complexity that demands special management consideration. It is particularly important to build flexibility into collaborative agreements in ICT markets, which are extremely dynamic in terms of their changing competitive arena and the uncertainties surrounding appropriate technologies, potential customer values, and profitable market segments. Inadequate contingency planning for managing such inherently risky and unpredictable markets is one of the main reasons why collaborations fail.

The key elements outlined above as critical influences on collaborative environments highlight the importance of giving priority to internal issues, such as accountability and control, as well as to market developments and other external issues. They also emphasize that this is another vital context where an understanding of the social and cultural factors shaping technological development—such as the need to balance trust and fear when disclosing proprietary information between partners—is essential to the effective management of organizations.

References

Bruce, M., Littler, D. A., Leverick, F., and Wilson, D. (1995), 'Success Factors for Collaborative Product Development: A Study of Suppliers of Information and Communication Technology', *R&D Management*, 25/1: 33–44.

Dodgson, M. (1993), *Technological Collaboration in Industry* (London: Routledge).

Leverick, F., and Littler, D. A. (1993), *Risks and Rewards of Collaboration: A Survey of Product Development Collaboration in UK Companies* (Manchester: University of Manchester Institute of Science and Technology).

Littler, D. A., Leverick, F., and Bruce, M. (1995), 'Factors Affecting the Process of Collaborative Product Development', *Journal of Product Innovation Management*, 11/1: 1–18.

6 Redesigning the Workplace: Challenging Geographical and Cultural Constraints on Access

Revolutionary visions of the paperless office and the automated factory, along with real advances, have combined to make the workplace one of the major focal points of efforts to apply ICTs. Such grand-scale visions, however, can obscure more modest incremental ICT innovations which are creating fundamental changes in tele-access. Large or small, these changes are reshaping workplaces in terms of their communicative power, and, thereby, their physical design and location, and their culture. This chapter discusses the opportunities and risks that come with a greater reliance on ICTs for tele-access.

The Design of the Workplace

The nature of the workplace is intrinsically connected to the technologies used by people to do things, as illustrated by the many jobs and offices that were once designed around typewriters. The new forms of tele-access afforded by advanced ICTs are enabling organizations to redesign workplaces along three interrelated dimensions:

1. increasing the centrality of ICTs in the workplace;
2. altering the composition of the workforce; and
3. managing the location of work across time and space.

The new ICT paradigms and management structures discussed in Chapter 5 are behind many of these changes, whose feasibility is greatly enhanced by the potential of tele-access to allow organizations more flexibility in distributing work. Yet workplace flexibility depends on more than technical matters. Social and cultural change is also required to take full advantage of ICTs and to avoid the risks inherent in a more open and accessible workplace. Business and industry can waste money, fail to recruit talented personnel, and lose customers if they fail to understand the ways in which social and cultural factors, such as the values and attitudes of managers and other personnel, facilitate and constrain the use of ICTs in the workplace.

Tele-Access in the Workplace

The application of new ICT paradigms by business and industry has been driven by what Chris Freeman (1987) has called a 'technological-cum-economic imperative'. Organizations that have designed business processes to take advantage of ICT capabilities have gained substantial improvements in productivity and in the range of products and services they offer (see Chapter 5).

This pursuit of competitive advantages through ICT innovations has raised major social concerns about the indirect effects of technological change on the quality of the work environment, ranging from fears over rampant unemployment to workplace surveillance (see Box 6.1). Tele-access provides a more consistent way of thinking about these concerns, while encompassing new and more fundamental changes in the workplace, such as those entailed in relocating jobs and functions.

Box 6.1. Workplace Concerns Raised by ICTs

- *Unemployment*: replacing people by machines in particular types of jobs, or across the economy as a whole (Bell 1974; Freeman 1996*a*).
- *Deskilling*: computerizing functions to enable the same jobs to be performed by people with lower skill levels (Braverman 1974; Attewell 1987*b*).
- *Stress*: speeding up work and heightening time pressures, creating more psychological stress (Attewell 1987*b*).
- *Surveillance*: increasing the monitoring and electronic surveillance of the workplace (Attewell 1987*a*; Zuboff 1988; Bellotti; 1997).

Senior executives in leading organizations are deploying ICTs to restructure management processes and relocate work in ways that improve access to information, skilled labour, and customers (Hepworth 1989; Goddard 1994; Goddard and Richardson 1996). The goals and strategies of other actors, particularly users of ICTs within the organization and among the public at large, also contribute strongly to shaping the direction of change in the workplace, sometimes in ways that counter the strategies of top managers.

The Networked Office

A sense of how the diffusion of new ICT paradigms and practices has shaped every corner of business and industry can be gained from a look at a central locus of technological change—the office (OTA 1985).

Early electronic data processing (EDP) systems did not fundamentally affect the core activities of the office since most technical and clerical functions, like keypunching, tended to be centralized in the EDP department. When advances in word and text processing in the 1970s made the office a prime target of the ICT industry, the high cost of equipment led to the establishment of centralized units, such as word-processing departments in which a 'pool' would type letters and documents dictated by phone or sent via internal mail.

The increased use of on-line computers with text-processing capabilities helped to spawn many visions of a 'paperless office of the future' utilizing digital media to integrate and distribute information and activities freed from geographical constraints (Uhlig *et al.* 1979). As early as the 1970s, the Information Sciences Institute (ISI) at the University of Southern California—which established one of the initial nodes of the ARPANET —sought to develop a paperless office through facilities like e-mail for the distribution of internal memos. It took about twenty years for such practices to become widespread.

Most early ICT-based office applications were focused on the efficiency gains that could be achieved by substituting electronic systems for paper, 'snail-mail', filing cabinets, and manual typewriters. Debate focused on whether ICTs would deskill and 'Taylorize' secretarial and clerical work —making office staff mere cogs within large information assembly lines —or create a new highly skilled cadre of information workers who would play more pivotal gatekeeper roles in the office (see Box 6.2). In practice,

Box 6.2. Frederick Taylor's Contribution to Scientific Management

Frederick Winslow Taylor was an American industrial engineer whose *Principles of Scientific Management* provided a foundation for managers to apply advanced technologies and systematic analyses, such as time and motion studies, to improve the efficiency of production processes. The scientific-management school placed responsibility for inefficiency on managers, who needed to break a production process into its component parts and segment these tasks across more specialized workers, such as in an assembly line. Fordist paradigms (see Table 5.1) were anchored in Taylor's work, but principles of scientific management, such as redesigning work to take advantage of advanced technologies, are also seen in contemporary moves towards business process re-engineering. One difference is that scientific management stressed efficiency, while later approaches place greater emphasis on the quality of the work environment and the responsiveness of an organization to its customers (Handy 1997). The popular children's novel *Cheaper by the Dozen* (Gilbreth and Carey 1948) poked fun at Taylorization.

Sources: F. W. Taylor (1911, 1916); Knights and Willmott (1988); Handy (1993: 21); Hayward (1996).

the automation of offices has reduced the number of repetitive, assembly-line jobs (Kraemer *et al.* 1981: 73–106; Danziger and Kraemer 1986), but reinforced the existing structure of secretarial and other office roles (J. Webster 1996*a*).

Rethinking Office Work

Since the PC became a tool for word processing in the 1980s, many managers and employees have begun to take advantage of ICT capabilities to rethink the way they do their work. Some executives began 'typing' their own letters, creating their own spreadsheets, and sending their own mail. This technological shift, supplemented later by ICT-based networks, enabled extensive redesigning of the way work is done in the office, although we are still a long way from the paperless office.

In the late-1980s I worked with a university department in the USA going through an upgrade of its word-processing equipment at a time when dedicated word processors had replaced the electric typewriter and were still a preferred option to the PC. Indeed our secretarial staff decided to stick with dedicated word processors when they were offered the choice of moving to PCs. They reasoned that the word processor technically reinforced the secretary's job, which they saw as being concerned mainly with typing letters and documents. If they had PCs on their desks, they feared that academic staff would ask them to do work that went beyond their job descriptions, such as maintaining databases or checking e-mail. In fact, when this department eventually moved to PCs, many secretarial jobs were cut to reduce personnel costs, and academic staff started to do many tasks previously done by secretaries.

In such ways, ICTs have challenged traditional roles of both the executive and the secretary. At the same time, ICTs have created tele-access infrastructures for a more networked organization in which more and more information is in electronic forms, and opportunities are opening for growing ranks of new information teleworkers.

One manifestation of this development has been the emergence of new forms of 'alternative officing' that break away from traditional conceptions of the office (see Box 6.3). This trend suggests that tele-access will radically alter the design and status of offices and office work. In 1997 a London billboard for Nokia captured this idea in the snappy catchline: 'Remember when more power meant a bigger office?' in advertising a handheld device incorporating phone, fax, and Web access capabilities.

Box 6.3. Alternative Officing: Re-Engineering Office Work

- *Economizing and standardizing offices*: reducing costs per employee through smaller standardized offices, such as modular open-office designs or cubicles.
- *Sharing facilities*: two or more individuals, or a team, sharing an office space. This could be done on a first-come-first-served basis or by hotelling —creating a scheduling, reservation, and check-in system that may employ a 'concierge' to prepare the space and equipment for each individual or team.
- *Using space outside the firm*: working off the company's premises, such as from home, telecentres, satellite offices, customer premises, mobile offices, or out of an 'electronic briefcase' at a temporary location like an airport.

Sources: HOK Facilities Consulting (1994); Johansen and Swigart (1994); Nilles (1994).

Intelligent Buildings and Regions

Another way in which ICTs have influenced the workplace is in the location of office buildings and industrial developments. In the 1980s, for example, the growth of competition in telephone equipment and services focused attention on the use of ICTs to enhance the value of real estate, marketed through such terms as the 'intelligent building' and 'landtronics' (Downs 1985). Designs for office buildings began to incorporate more shared telecommunications facilities, such as receptionists and private automated branch exchanges to be shared by tenants. Land developers viewed ICTs as an additional competitive edge.

Many factors influence the locational decisions of firms, so telecommunications can have only a marginal—but by no means negligible —impact. In the 1990s, intelligent buildings and regions, which incorporate state-of-the-art ICT systems in their design and construction, have continued to be one device for attracting firms to particular developments (see Box 6.4). The founders of DreamWorks SKG[1] originally chose a site miles from the core of Hollywood in part because they wanted to build a film studio that had state-of-the-art ICT infrastructures and facilities. Their plans to build a new facility were held up by a variety of planning, zoning, and environmental issues,

Box 6.4. Intelligent Buildings, Smart Places: Asia, North America, and West Europe

- *Multifunction Polis (MFP), Australia.* A joint plan of the late 1980s by the Australian and Japanese governments has envisioned advanced ICTs as the key to the infrastructure of a new global city that integrated high-tech industry with attractive urban and environmental amenities (Droege 1997*a*: 6–8).
- *Multimedia Super Corridor, Malaysia.* A government-funded development corporation has been used to attract global multimedia firms, by liberalizing some policies, such as government censorship, and installing advanced telecommunications infrastructures, intelligent buildings, and environmental amenities in a 9 × 30 mile corridor south of Kuala Lumpur.
- *Smart Valley, California.* Government and industry in Silicon Valley have teamed up to encourage the use (not just the production) of ICTs in ways that will support and retain business, industry, and communities in the area.
- *DreamWorks, Los Angeles.* Dreamworks SKG, with incentives from the City of Los Angeles, have planned a 100-acre state-of-the-art digital studio in Playa Vista to become a 'center of excellence' and 'incubator' for multimedia firms in LA.
- *Smart Isles, UK.* A consortium of educational, industrial, and government agencies have collaborated to utilize ICTs more effectively to support education, training, and community services in the British Isles.
- *Science Parks, EU.* Properties located near research and higher education institutions have provided attractive settings and offices for high-tech businesses to encourage academic–industry collaboration (Monck *et al.* 1988).

but also by pressures to remain in closer geographical proximity to the pool of talent, such as animators, that contract with many other Hollywood studios.

ICTs have also blurred the boundaries between the household and workplace. From the first wave of PCs for the household that went beyond the hobbyist market, PCs were used primarily for work-related activities (Dutton *et al.* 1987*b*). Since then, homes have been designed to incorporate 'offices' based on the PC, and other ICTs—for example by installing additional phone lines to accomodate faxes and Internet access. In such ways, tele-access to ICTs is shaping the design of physical spaces to accomodate equipment and new ways of living and working.

The Geography of the Firm

The widespread use of ICTs has enabled other changes in the geography of the workplace. The location of the firm and the people within it has always been tied to access, as illustrated by the concentration of manufacturing plants near major transportation arteries and the co-location of individuals who collaborate on projects or use the same relatively expensive technology.

Earlier innovations, such as the telephone, mail, and air travel, enabled companies to relocate manufacturing and other labour intensive activities, including information work, long before computer-based networking became central to their businesses. But, as white-collar information work becomes the target of digital technologies, all organizations will need to undertake a fundamental review of the assumptions underlying their locational choices.

The way advances in tele-access have loosened the need to tie functions and tasks to specific geographical locations has challenged traditional assumptions behind locational decisions. Increasingly, tele-access enables companies to accomplish more across time and space, with individuals being able to communicate electronically across the globe in the same way as they communicate across offices (see Chapter 5). Even small firms are able to improve their competitive edge by distributing collaborative virtual work groups across time zones to permit work to continue round the clock.

Since the 1960s some have suggested that, once work is done on a computer or over the phone, there are fewer reasons for employees to travel for miles to central city locations (Memmott 1963; Healy 1968). Instead, they could 'telecommute' from locations distributed around a metropolitan area (Nilles *et al.* 1976; Short *et al.* 1976). This capability for remote work has led some to argue that geography will no longer be a factor in business and industry. However, although tele-access dissolves some time and space constraints, other factors—such as prevailing wage rates and the availability of skilled personnel—will matter even more in the locational decisions of business and industry (Downs 1985; Goddard 1994; Goddard and Richardson 1996). For example, firms might choose to locate individuals where they gain access to 'tacit' knowledge that cannot be easily codified and distributed electronically (Faulkner and Senker 1995: 200–12; Lamberton 1997). This is one reason why many industries remain highly centralized and regional.

The creation of animated cartoons and feature films, to cite one example, remains very labour intensive.[2] Despite advances in computer-based animation techniques, the animation process at the major studios requires thousands of drawings and much colouring by hand. In order to reduce costs, major Hollywood animation studios began subcontracting the most labour-intensive aspects of the animation process in the 1970s, increasingly to firms in East Asia. Advances in ICTs have allowed more flexibility in pre- and post-production processes, such as in editing the final film, thereby providing the potential to re-centralize more aspects of the production process. This began to bring more parts of the animation process back to some of the pioneering Hollywood studios who had contracted much of this work out. This trend is reinforced by the importance of personal networking in the establishment of virtual organizations for each new production, as well as by the tacit information gained from having a workplace in the midst of a major centre for this industry.

In the following sections I will explore how tele-access is shaping the geography of the workplace through its inter-related use for meetings, telework, and distributed work.

Electronic Meetings

As discussed in Chapter 5, electronic meetings are one of the oldest and simplest means for supporting the decentralization of a firm or linkages among firms. Most levels within an organization, from top executives to clerical staff, can use various types of audio, video, or computer conferencing systems to facilitate group decision-making and communication at a distance (Johansen 1984, 1988).

In practice, teleconferencing has had a much more limited impact on the geography of the firm than expected. For example, video conferencing has been marketed to firms since the late 1970s as a substitute for travel to executive meetings. Yet, as explained in Chapter 4, publicly accessible video-conferencing facilities, such as AT&T's Picturephone Meeting Service (PMS), failed, initially at least, to find a viable base of repeat users, even when full motion broadcast-quality systems were provided well below cost.

One reason for the relatively slow and limited take-up of teleconferencing is that only a small proportion of executives feel they travel too much and few wish to substitute a teleconference for travel unless it is to an undesirable location (Johansen 1984). Also, business executives and managers do not perceive that visual information is a critical component of most meetings (Johansen 1984; Noll 1986). Despite improvements in video-conferencing capabilities, most non-supplier evidence suggests that teleconferencing continues to be used to enhance communications, such as in advance or as a follow-up to face-to-face meetings, rather than as a substitute for executive travel. This benefit has led many firms, such as TRW (see Box 5.9), to build permanently installed, private video-conferencing facilities within the key offices of the corporations.

Telework

ICTs enable people to work at a distance from their employers in a wide diversity of jobs (see Essay 6.1). For example, teleworking from home could be an option for managers, professionals, or secretarial and clerical staff using home-based systems. Some might require high-performance multimedia computers, while others might work effectively with simpler PCs and fax machines.

Provided telework is deployed appropriately, it can bring advantages in terms of increased productivity, lower office and travel costs, and a higher morale among employees (Nilles 1992, 1994). However, there are also barriers to home-based teleworking. Some executives fear they will lose management control over employees not physically present at the office. Employees' fears of being isolated or being passed over for assignments or promotions because they are less visible can also lead to resistance. Nevertheless, telework is becoming more prominent, as virtual business processes increase and ICTs become more ubiquitous and richer in information and communication capabilities. One 'middle-of-the-road' US estimate suggests that home-based tele-work almost doubled between 1990 and 1997, from 5 million to 9.2 million employees 'who work from home at least a couple times a month'.[3]

A clear example of the potential of telework in the household is provided by iLAN (see Box 6.5). This company is atypical because the personnel are more technically proficient with ICTs than most users. However, the company's experience provides a plausible scenario for many other types of businesses.

The enormous potential of telework has been responsible for the development of publicly funded or private local 'telecentres' that offer individuals and organizations access to PCs, the Internet, technical support, and other ICT resources. These centres provide an alternative to teleworking from home, or additional resources for home-based teleworkers, such as those who lack particular capabilities, like a colour copier, in their home office.

Since the 1980s many telecentres have been established to help reduce travel and traffic congestion in major cities. In Los Angeles, for example, over twenty telecentres were established along these lines. Only one of these, however, was clearly successful in attracting 'telecommuters' (Wilson and Dutton 1997*b*). A variety of factors contributed to the failure of the others (see Box 6.6). Those who promote telecentres

Box 6.5. iLAN Systems Incorporated: A Network Producer and User

iLAN is a Southern California computer-services company founded in 1992 by Tom Reynolds to design, install, and maintain ICTs, including local-area (LANs) and wide-area networks (WANs), desktop computers, and network file servers. By 1997 the owners and all seventy employees worked from home and their customers' facilities using iLAN-supplied computers, Internet access, cell phones, pagers, and whatever other ICTs were needed to support their work. Employees are expected to work independently, be flexible, and manage their own time. Four employees and the co-owners oversee operations, coordinate installations, and supervise the teams—with no support staff. iLAN grew from earnings of $180,000 in the first year to over $6 million four years later. In this highly specialized company, employees are organized into teams within particular areas of expertise and iLAN provides training opportunities and new equipment to enable its employees to keep pace with advances in technology. Most communication among iLAN employees is by means of ICTs. Employees are encouraged to use ICTs to keep in touch with one another and are expected to spend a large proportion of their expenses on communications.

Source: Wilson and Dutton (1997*a*).

Box 6.6. Factors Accounting for the Failure of Telecommuting Centres

- *Location*. Many sites were chosen first and marketed later, failing to capture a major employer with many potential 'telecommuters'. Such employers should be enrolled early so that telecentres can be sited near employees' homes.
- *Management*. Most telecentres were funded, and often run by, public-sector managers who were not entrepreneurial and market oriented.
- *Paradigms*. Supervisors and managers of most firms continued to believe in line-of-sight management of employees. They were therefore reluctant to permit middle- and lower-level staff to work remotely.
- *Conceptions of the user*. Most telecentres designed standard, functional office spaces for lower-level clerical personnel, but those with the autonomy to telecommute in fact tend to be managers and professionals who want 'executive suites', or can afford the equipment and space to work at home.
- *Policy*. The easing of environmental regulations and the growth of transportation facilities diminished incentives for employers to support telecommuting options.

Sources: Nilles *et al.* (1976); Wilson and Dutton (1997*b*).

as a means for extending tele-access to distressed rural and urban areas must also consider the lessons learned from efforts to support telecommuting.

Mobile work is a growing form of telework. Mobile workers can carry with them what is effectively a virtual office in less space than the traditional travelling salesman's suitcase of goods. The mobile office can include pagers, cellular phones, faxes, portable multimedia computers, printers, and many other capabilities, enabling employees to work anytime, anywhere. Networking innovations are also making it possible to 'teleport' all the software that a person would use at an office computer to the home or a portable computer 'on the road'.

Combined with the use of sophisticated project management software, a mobile office can make each business trip more productive. Hand-held computers can facilitate routine tasks, such as sales and delivery, by offering links to databases and intranets containing up-to-the-minute information on inventory, and offering tele-access to other staff. Top executives and professionals also work productively on the move. For example, the headquarters of a major US law firm uses case-management IT systems to inform an attorney arriving at a particular court what action can be taken on any cases at that specific time and location.

Distributed Work

The most significant impact of ICTs on remote work might be in facilitating the 'spatial reorganization' of the functions and personnel of the firm (Moss 1987; Goddard 1994; Goddard and Richardson 1996). This started in the 1970s with banks, insurance companies, and many other firms relocating clerical functions from 'back offices' at their headquarters in major cities to areas outside the central business district where human and building resources are available at lower costs (Moss 1987; Richardson 1994).

A similar logic is being played out in larger geographical arenas using a growing number of management options. For example, companies in industrialized nations often employ programmers in less developed nations, like India. Many enterprises now use 'call centres' to provide telephone-based services and support to customers around the world (see Box 6.7). In 1997, for instance, one of the largest tele-service providers, Matrixx Marketing, employed 15,000 customer-service representatives in twenty-three facilities in North America and Europe. They handled over half a

Box 6.7. Call Centres and the Relocation of the Workplace

Financial institutions, insurance firms, the travel industry, and a growing number of other businesses are using the telephone and ICT networks like the Internet for an increasing range of services and business processes, such as opening a new account, or making credit-card verifications. This escalation in tele-services has been accompanied by the development of 'call centres' designed to handle large volumes of interactions in an efficient manner. The geographical location of call centres can permit them to provide twenty-four-hour service, respond flexibly to the language of the caller, and realize the benefits of locations with lower wage rates and better availability of staff with appropriate skills. The centres use a variety of intelligent networks and systems to route and answer calls and automatically handle requests. Virtual call centres can be created by networking distributed offices so that they function as if they were located in one location. The Internet permits similar geographical flexibility in handling electronic commerce. The Home Shopping Network, for example, processes hundreds of thousands of telephone calls a day in a central location, round the clock, from viewers of TV shopping channels in the USA. In the UK, the use of call centres by British Airways (BA) and First Direct bank has led both to relocate teleservices to areas that promised lower-cost labour, with less labour turnover, than the companies experienced in London and the south-east of England.

Source: Richardson (1994).

million calls for service every day for a collection of firms, including thirty-three of the top 100 of the Fortune 500 companies (Fishman 1997: 14–15). Some geographically distributed firms have decided to centralize administrative functions, such as accounting and billing, sometimes by outsourcing the work (see Essay 6.2).

Leading-edge companies are seeking to use the flexibility afforded by these kinds of tele-access approaches to choose the most cost-effective places for carrying out necessary tasks, then networking them together in changing virtual configurations (see Chapter 5). This could make local and regional economies more significant as localities compete in the global economy to provide the human and physical resources which are in most demand. Tele-access also allows new players from anywhere in the world to enter a market place, creating increased competition for local customers. An example of a shift in tele-access involving all these dimensions is a move towards mobile work in Fujitsu Business Communication Systems (see Box 6.8).

Box 6.8. Fijitsu Business Communication Systems (FBCS)

FBCS develops, manufactures, installs, and maintains intelligent networks and equipment, such as private branch exchanges (PBXs), to support functions such as voice processing, automatic call distribution, and video conferencing. FBCS's headquarters are in Anaheim, California. It had thirty service facilities throughout the USA in 1997, when it decided to make 'mobile workers' of its sales and account representatives. Its aim was to decrease the time staff spend in, and commuting to, company offices—while increasing the time they spend in the field with customers. Tele-access is being used in this case to relocate sales staff to where the social rapport and the tacit knowledge gained from personal, face-to-face contact is more critical to the success of the job. These mobile workers use laptop computers, intelligent exchanges, the company's intranet, and other ICTs to support work from their homes, cars, and customers' premises. The gain in mobility achieved by this was one reason for a larger reorganization from geographical to 'vertical' markets, such as health care, education, or entertainment, which cover much larger geographical areas of southern California. This shift permitted a consolidation of offices and regions, allowing FBCS to cut the numbers of middle management and support staff in sales offices. It was complemented by other changes of strategy. Instead of getting to know all telecommunication managers in a limited territory, such as in all major companies in the central business district, sales representatives needed to gain expertise in their 'vertical', including the key products and decision-makers, by searching data- bases, attending conferences, and networking.

Source: Ellison and Dutton (1997).

High-Tech Cultural Barriers in the Workplace

Discussions of the impact of ICTs on work often overestimate the role of rational management strategies—such as moves towards a virtual organization—and underestimate the role of users, whose habits, skills, and attitudes shape the production and use of ICTs in ways that support or constrain tele-access strategies (see Chapter 4). It is, therefore, critical in the workplace to develop accurate conceptions of users and how they will respond to various ICT innovations, and recognize that access to a trained workforce could depend on the openness of the culture of the organization as well as its technological infrastructure.

Assuming a High-Tech, ICT-Centric User

The enthusiasm of many people for new technologies like the Internet has contributed to the way the ICT industry and computer-system developers frequently overestimate the technical agility and enthusiasm of personnel, while underestimating the effects of users' valid concerns over the impact of innovations on productivity, privacy, and the quality of the work environment. The failure of many ICTs in the workplace has been tied to a lack of acceptance by users, whether they be executives who are reluctant to type or use e-mail, or consumers expected to shop or bank over interactive television. This can create problems for the whole organization, not just for individuals who underuse or overuse ICT capabilities, as the productivity gains from capabilities like networking depend on everyone making effective use of ICT systems.

Structural change, as well as ICTs, meets with much resistance. Managers and supervisors are often concerned about losing control over remote staff, which can have a braking effect on the establishment of more distributed organizations. According to Handy (1996: 212), this means that managers will be more dependent on working primarily with individuals they can trust to manage themselves, such as subcontractors who could lose their contract for poor performance. People are also likely to be judged mainly on the outcomes of their work, rather than by their work experience or practices. Many tele-work schemes are focused on sales, in part because the outcomes of their work are easily measured and sales staff often have personal incentives to perform well, such as receiving commissions on sales.

User resistance is sometimes a symptom of flaws in management strategy. For example, the advocates of virtual organizations often fail to recognize that employees may have conflicting expectations imposed on them (Murray and Willmott 1997). On the one hand, they are expected to assume more autonomous and self-directed roles. The media often report on how, for instance, a team of young programmers—all with a stake in the software company—work and play through the night in a converted warehouse to get their product out before the competition. On the other hand, competitive pressures are causing many private and public enterprises to offer employees less secure forms of employment, lower raises, fewer hours, while demanding more from them in terms of travel, personal commitment, time, and managerial skills (B. Crichton 1993: 9). One tele-worker explained that she had lost some of her support staff in the head office through downsizing and quipped, ironically: 'I have been empowered.' And some of the famous exemplars of the new management style, such as the software firm Netscape, are moving towards more traditional management practices as they have become larger and more established companies (Denton 1998).

Gendered Organizational Cultures and Stereotypes

Stereotypes of those who produce and use ICTs can be particularly misleading. The ICT-centric user is only one of many misconceptions. Another common one is that ICTs are produced by men for men (see Essay 9.2). This is anchored in the early history of the PC, but also in the male gender imbalance of computer science departments, computer clubs, electronic discussion groups, video arcades, and many firms. This stereotype ignores the extent to which women work with ICTs in organizations and the decisive influence that socialization in the family and workplace can play in shaping gender affinities towards technical artefacts (Webster 1996*b*). For example, in the public sector, women in white-collar employment are more positive than men about the impact of ICTs on their jobs (Dunkle *et al.* 1994). But the male-dominated culture of some firms, engineering departments, clubs, groups, and arcades can negatively influence the educational and career choices of women (see Essay 6.3).

Producer and user organizations that fail to create a supportive culture for women and other groups, such as younger as well as elder employees, will risk the loss of vital talent and creativity. This is particularly important, as the more pervasive use of ICTs means that the most successful companies are those which will gain access to segments of society that have not been the target of earlier products and services, such as women, the handicapped, and minority-language communities.

The concept of a virtual organization might need to be modified in different cultural settings. National and organizational cultures which place great value on face-to-face interactions can find it difficult to envision how e-mail and computer conferencing can enhance communication, particularly given the domination of the English language on the Internet (Asper *et al.* 1996). This might help to explain why virtual organizations are less prevalent in Japan than the USA, for example. In a global market place that is increasingly dependent on tele-access, it will be important for firms to recognize these cultural differences and recruit and train skilled gatekeepers—individuals who can create an effective interface between the on-line virtual organization and the rest of the firm.

Failures to Manage Tele-Access: Risks in the Workplace

These opportunities to exploit geographical opportunities are reinforcing the vital role of ICT use in all sectors of the economy. But, as tele-access becomes more important to an organization, so do the risks entailed with failures to manage the ICT infrastructure to support that access. The design of ICTs therefore needs to respect the human capabilities and perceptions of users, as well as addressing business and technical issues.

Information Overload

Much of the rhetoric associated with concepts like business process re-engineering (BPR) and virtual organizations emphasizes the need for highly skilled, adaptable personnel who will 'multi-task' across functional divides and respond effectively to rapidly changing business needs. Although there is empirical evidence to show that ICTs can increase demand for skilled staff, rather than deskilling (Kraemer *et al.* 1981; Attewell and Rule 1984; J. Webster 1996*a*), even highly skilled and trained personnel can be overloaded with information by poorly designed systems.

Information overload and difficulties in comprehending complex systems, particularly in stressful environments, have been at the root of a number of major ICT failures and disasters (Peltu *et al.* 1996). This indicates that ICT-based systems should be designed and managed to respect human limits of understanding. As Russell Ackoff (1969) argued in the case of management information systems of the 1960s, users—managers, professionals, and now consumers—seldom need all the information they say they want, or which the system delivers to them, resulting in so-called information systems ending up as 'mis'information systems (see Box 1.2). Systems need to be designed and users trained to screen, prioritize, and regulate tele-access.

Inappropriate Substitutions

Computer-mediated communication, such as e-mail, can eliminate cues that establish a social presence, or define a social context, such as facial expressions and physical appearance (Short *et al.* 1976; Rice 1984; Sproull and Kiesler 1991). This can facilitate communication, such as by reducing the status differentials of individuals, in ways that promote more effective, spontaneous, and open communication. This same effect can impair interpersonal interactions, such as in attempting to reach consensus on a matter in which individuals disagree. That is why great care must be taken to design and use ICTs to meet social as well as technical criteria (Daft and Lengel 1986).

A variety of studies suggest that mediated communications reduce cues about the relative status of members in a group (McGrath and Hollingshead 1994). This becomes less relevant as members of an organization are well known to each other, which is one reason why the findings of studies are mixed—it depends on the real world setting. Nevertheless, this loss of status cues could have positive and negative effects simultaneously by redistributing the communicative power of individuals within the group. On the one hand, it could lessen inequality among participants in shaping the final decision (Short *et al.* 1976; Hiltz and Turoff 1978). On the other, it could result in members with critical information and high status having less influence in a mediated group (Hollingshead 1996). The lack of social context can, therefore, diminish the influence of everyone on the group—including those with expertise.

ICTs can distance the user from colleagues and customers if they are applied more to replace rather than to augment other forms of communications. The most pervasive example of this is the degree to which many organizations inappropriately implement voice systems, such as the answering machine, in ways that prevent customers from speaking with a real person. In order to avoid this danger, activities as diverse as drive-in McDonald's Restaurants, banks' automated teller machines, and loan managers are experimenting with introducing video communications to 're-humanize' their mediated and automated user interfaces.

Human contact is often critical, particularly in work that entails making judgements about individuals. Mediated systems of communication risk creating what Kenneth Laudon has called a 'dossier society' (Laudon 1986), in which we increasingly know individuals by their 'data image' or electronic dossier, such as a credit report, rather than by how they look and what they say. This could be good in some circumstances, such as in reducing discrimination based on a person's physical appearance. However, it can have negative effects by reducing compassion or personal discretion, replacing human judgement with automated calculation.

Privacy and Security

The widespread use of ICTs has generated concerns about security and electronic surveillance (see Chapter 3). In the workplace, fears of electronic surveillance often underestimate the ability of employees to bypass and sabotage systems designed to monitor and control their work (Attewell 1987*a*). However, tele-access developments, such as personal locator systems and mobile phones, can extend management oversight outside the walls of an office or organization; a device like the 'Active Badge' (see Box 6.9) can continuously track the movements of personnel.

Box 6.9. The Active Badge: A Personal Location Device

Researchers at establishments such as the Olivetti & Oracle Research Laboratory (known as ORL) in Cambridge, UK, have developed prototypes of a microprocessor-based communication device that is small enough to be clipped on a person like a badge. The ORL Active Badge, for example, contains an infra-red transmitter that can transmit the indentity of the person wearing it. The transmission is picked up continuously by receivers at infra-red cell stations located throughout any area encompassed by the system. Placing this locational data on a company intranet, for example, would permit anyone in the company to know the whereabouts of every colleague. This would allow personnel to make more informed decisions about when to contact each other. The badge could also enable a variety of services to be automatically provided. For example, security-locked doors could be unlocked, printers enabled, and personal files and computer workspaces downloaded to the nearest PC as a person moved through a building. Badges could be equipped to permit the individuals wearing the badge to signal whether they were working or did not wish to be disturbed. Similar systems can be used to identify the location of a cellular phone user. Other ICTs, such as pagers, can serve some functions of an active badge, by maintaining contact when individuals are away from their office, for instance; but they are less capable of activating other devices or providing information on the location of individuals without disturbing them.

Source: Weiser (1991) and ORL <www.cam-orl.co.uk>, Mar. 1998.

Effective and well-publicized safeguards which minimize threats of unwarranted surveillance and intrusion into personal privacy can help an organization gain the trust and positive backing of its employees and customers in its overall exploitation of ICT-based systems. This is a key issue in tele-access.

Planning an Effective Tele-Access Strategy for the Workplace

Tele-access needs to be a focal point of discussion in organizations. ICTs can be used in ways that enhance personal access to critical information and people, such as customers. But, if appropriate policies and practices are not developed, tele-access can misinform users, isolate the firm, or undermine personal privacy and the security of information in the workplace.

Enthusiasm for new ways of working will not spring up automatically. One *New York Times* bestseller, based on the popular *Dilbert* comic strip, highlights new ICT management practices to poke fun at the social ineptness of engineers, and the weaknesses of their managers (Adams 1996). It captures the degree to which practices, such as downsizing, or alternative officing, are poorly applied. Tele-access strategies, such as relocating work, must be focused on the right jobs, and be well implemented in order to succeed. Dilbert is funny, because so many managers and executives pick up the latest jargon, but fail to understand the difficulty of genuine social and technical innovation in the workplace.[4]

Effective training, career development, and management policies will be ever more important in establishing and maintaining a commitment to meeting common goals in organizations that are likely to become increasingly dispersed and electronically networked. Changes in the design and manufacture of ICT-based applications and products are also needed to create a closer link between producers and users, in order to establish a clear, non-stereotypical, conception of the user's actual requirements.

Notes

1. The SKG part of the name represents the initial letters of the surnames of three Hollywood entertainment giants who founded DreamWorks—Steven Spielberg, Jeffrey Katzenberg, and David Geffen.
2. This depiction is based on a series of interviews with key individuals in Hollywood animation studios conducted by Vito Curcuru (1997).
3. This is based on research by the firm Find/SVP reported on by Stuart Silverstein (1997).
4. Scott Adams (1996) roasts such popular management concepts as re-engineering, team-building, and management consultants.

Essays

6.1. Home-Based Telework

Leslie Haddon and Roger Silverstone

Networked information systems have extended the range of paid work that can be carried out from home. However, as Leslie Haddon and Roger Silverstone explain, socio-economic factors are more significant than technology in deciding the extent to which home-based telework is actually taken up.

6.2. The Impact of Remote Work on Employment Location and Work Processes

Ranald Richardson and Andrew Gillespie

Telework has implications beyond the household environment discussed by Leslie Haddon and Roger Silverstone in Essay 6.1. Here, Ranald Richardson and Andrew Gillespie explore the wider aspects of remote working on the location of work.

6.3. Women's Access to ICT-Related Work

Juliet Webster

Access to labour markets involving ICT-related skills is a crucial issue because of the growing importance of ICTs to all aspects of economic, business, and industrial life. Juliet Webster reports on evidence relating to one important aspect of this market—the gendered nature of the ICT industry.

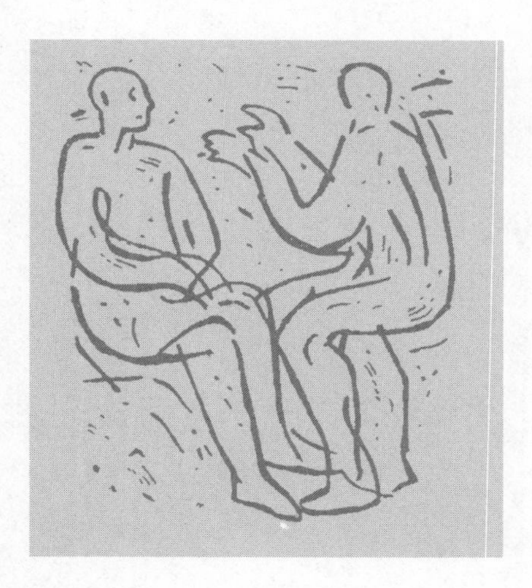

6.1. Home-Based Telework

Leslie Haddon and Roger Silverstone

Individuals work from home using a variety of high-tech and low-tech information and communication technologies and services. Their work can be professional or clerical and involve varying degrees of responsibility, including management. It can include numerous different activities, such as word processing, data collection, and report writing.

Teleworkers may be employees of large organizations or self-employed, men or women. Their teleworking can be a chosen option or result from being made redundant. It can be a permanent life choice or a temporary stage in a career or the life cycle. The commitment to work at home using new technologies can be qualified, and sustained, by regular attendance at a workplace.

The amount of teleworking that takes place at any one time and in any given society is difficult to measure because of differing definitions of telework. An additional problem is that teleworkers are hidden away in the home, giving them low visibility. Although there are teleworker organizations, not all teleworkers join them. Some teleworkers advertise their services, but others do not. Many personnel departments in companies do not keep track of all their teleworking staff, especially when some decisions to work at home are informal.

However, there is evidence to suggest that the phenomenon is increasing, particularly among the self-employed (POST 1995). Teleworking is also getting more attention among policy-makers, employers, trade unions, and the media. This can increase familiarity with the concept and, possibly, its acceptability.

The bulk of research on teleworking has focused on management issues, from the points of view of both the employing organizations and the teleworker. Some research also addresses the social implications from the perspective of the family and household of the teleworker (Haddon and Silverstone 1993, 1994). This indicates that the decision to begin teleworking is often informed by knowledge of the experience of others and frequently supported by self-help organizations and consultants who advise on the implementation of teleworking schemes.

Technology has not been the prime factor in the development of teleworking. Only some forms of telework have been made viable by the availability and affordability of new ICTs. Many types of both clerical work (typing, for example) and professional work (such as preparing reports) were carried out at home for many years before word processing, faxes, and e-mail facilitated this kind of undertaking.

ICTs have, therefore, often been the enabler, rather than the instigator, of home-based work. It is usually socio-economic factors which shape employer, employee, and self-employed decisions to do all or some work at home.

The variation in teleworking is substantial. People work at home for different reasons, with different degrees of choice. Some choose it to be around for the children, while others have few alternatives. The experiences of individual teleworkers are influenced by their personal trajectory into the work—from full-time work, part-time work, unemployment, or looking after a home. That experience and its evaluation range from the life-enhancing, by virtue of the freedoms offered, to the exploitative, when there is no other viable option.

Within this diversity, teleworkers face a common range of issues concerning the handling of the relationship between home and work, which they manage with varying degrees of success. For example, the fact that work is so near causes some to become workaholic, while others dislike the sense that work is always hanging over them.

Without the imposed time regimes of the office, teleworkers have to develop strategies for managing the pacing and timing of work within their overall domestic temporal rhythms. For example, work may intrude on family life through incoming work phone calls, or else the demands of family members may threaten to be disruptive to work. Without the trappings of the office, there can be problems convincing others inside and outside the home that teleworkers are really working.

Negotiation with others and the development of strategies to cope with these factors become crucial. Such strategies may include the use of ICTs beyond just carrying out the job to manage the teleworker's public image and availability—for example in the use of the answerphone or in a conspicuous display of computer equipment as an indicator of an efficient working environment.

The entry of paid employment into the home can also have mixed consequences for the experience of ICTs. Most teleworkers do not require the latest and most powerful equipment and sophisticated multimedia services. For many, therefore, telework is not an unambiguous route for a whole range of new ICTs and services to enter the home.

The greater impact is sometimes felt when telework competes for access to the household's existing technological resources. A family computer then has to be used at times for work, so is not available for game-playing. The domestic phone line also becomes a work link, with the need to make new rules about its use—such as when not to block the phone or guidance on an appropriate manner in which children should answer the phone.

In many cases, telework introduces some ICTs which become part of the household infrastructure, used by others than the teleworker and for purposes other than work. For instance, many households would not have acquired a personal computer or fax had it not been for telework. A new computer or other ICT may then have a displacement effect, making the older systems available for others to use.

In view of the complex and disparate nature of teleworking, both economically and sociologically, it is important to be cautious in making both judgements and predictions about its significance. The implications of teleworking on family life, and on the relationship between home, community, and work, are neither simple nor unambiguous. Nevertheless, there is no doubt that working from home using a range of ICTs and services will become increasingly common and will be seen increasingly to be worthwhile for different categories of workers.

References

Haddon, L., and Silverstone, R. (1993), *Teleworking in the 1990s: A View from the Home* (*SPRU/CICT Report No. 10*; Brighton: University of Sussex).

—— —— (1994), 'Telework and the Changing Relationship of Home and Work', in Mansell (1994), 234–47.

Mansell, R. (1994) (ed.), *Management of Information and Communication Technologies: Emerging Patterns of Control* (London: Aslib).
POST (1995): Parliamentary Office of Science and Technology, *Working at a Distance: UK Teleworking and its Implications* (London: POST, House of Commons).

6.2. The Impact of Remote Work on Employment Location and Work Processes

Ranald Richardson and Andrew Gillespie

Teleworking has generally come to be understood to mean working from home or a small neighbourhood centre using ICTs. Interest in these forms of work has been so great that a telework mythology has developed. We read almost daily of how information professionals are trading their skills over the information superhighway without leaving their home office, generally from some ideal rural setting. We see images on our television screens of tele-villages—'wired' communities rekindling the frontier spirit using new technologies.

The subtext of the telework mythology is that a radical decentralization of work, almost on a pre-industrial pattern, is on the horizon. However, home-based telework has failed to take off on the scale predicted, and remote neighbourhood offices, known in Europe as tele-cottages, have made little contribution to the growth of remote work (Gillespie *et al.* 1995). On the other hand, other forms of remote work have grown rapidly as firms seek to lower costs or improve responsiveness by reorganizing their operations around new technologies.

We should not, however, expect this process to result necessarily in the radically decentralized patterns of work which some techno-futurologists predict. New spatial configurations of work result from the individual decisions of large numbers of firms, each of which seeks competitive advantage from reorganization across space.

Each firm is also still governed to some extent by existing constraints, including labour-process control, labour availability, real-estate costs, transport, and the nature of markets. This means that new patterns of work location will be complex, involving both centralization and decentralization of work. The key point about the new technologies is that they allow firms to exercise a greater number of locational options.

The best-documented example of how ICTs have facilitated the centralization of functions is in global financial services, where the command and control of massive flows of capital have become concentrated in the few 'global cities' which have suitable labour and support infrastructures. Using the new technologies, firms based in these cities—notably New York, London, and Tokyo—can achieve a global reach, remotely servicing markets across the world (Sassen 1994; Leyshon 1995).

In contrast, remote working based on groups of workers located at a distance from each other permits firms to extend their reach in terms of labour markets.

Typically such groups form 'virtual teams', performing interconnected work tasks, either in real time or asynchronously, using ICT applications such as voice and electronic messaging, computer conferencing, shared electronic calendars, shared databases, and technologies that automatically manage the flow of work.

Some of these teams will be permanent, others will come together temporarily for one or more projects. Teams may consist of workers from a single firm or from a number of firms engaged in joint ventures. In all cases, the essence of team telework is the geographical movement of work between team members via telecommunications networks. This substitutes for the movement of workers.

There remain organizational and technical barriers to such teamwork (Feng and Gillespie 1994). Nevertheless, the concept is being embraced by an increasing number of large firms, including Coca-Cola, Digital Equipment Corp., Eastman Kodak, and Exxon (Opper and Fersko-Weiss 1992). Advantages gained from this approach include the ability to overcome, at least partly, the worldwide shortage of scientific and engineering staff with key skills in areas such as R&D.

Firms which have previously been confined to recruiting in a relatively small number of geographical areas, with consequently inflated wage costs, can now range more widely within advanced and less advanced countries. This distributed pattern of team telework can also help with labour retention, as there is less requirement for staff to move to different worksites.

Both global financial services and team telework predominantly involve skills which are at a premium. The locational options which firms can adopt are, therefore, relatively limited. In other types of work which are being organized remotely, such as traditional back-office activities, firms have more varied options because less scarce skills are required. Increasingly, remote working also involves customer-service tasks which are undertaken by telephone rather than face to face.

The reorganization of these work tasks around new technologies may result in the concentration of work to a single site or a few locations, in order to achieve scale economies. For example, in retail banking there has been a trend towards moving many tasks out of the branch network into large offices. However, this process is also likely to lead to a decentralization of work away from high-cost locations to lower-cost areas (Marshall and Richardson 1996). The degree of locational flexibility will vary, depending on a number of factors, such as the type of skills required, the nature of the market, the quality of telecommunications, and occasionally—as in banking—regulations covering cross-border trading.

This means that data-processing functions, for instance, can be fairly easily transferred offshore. It is more difficult to relocate remotely those tasks requiring real-time interaction with customers, not so much as a result of technical constraints, but because of language and cultural differences. Even in this field, there is considerable variation depending on the product sold and the sophistication of the customer.

Nevertheless, by the late 1990s, a significant number of firms, particularly in travel-related services and in computer services, had established single-site customer-service centres in order to service customers in a number of countries (Richardson 1997). Western Europe, Ireland, the Netherlands, and the United Kingdom appeared to be particularly attractive as locations for such firms. For example, British Airways' UK call centres dealt with customers from the USA; the Holiday Inn's central reservation site in Amsterdam handled calls from across

Europe and North Africa; and the US computer software company Quarterdeck International serviced its European customers, by telephone, from Dublin.

ICTs, therefore, clearly help to extend market reach, thus creating the opportunity for firms to reconfigure operations across space. But, despite the proliferation of new ways in which work is undertaken and networked, we see no marked tendency towards radically dispersed work practices, such as the widespread restructuring of work towards home-based teleworkers.

References

Allen, J., and Hamnet, C. (1995) (eds.), *A Shrinking World?: Global Uneveness and Inequality* (Milton Keynes: Open University).

Dumort, A., Fenoulhet, T., and Onishi, A. (1997) (eds.), *The Economics of the Information Society* (Luxembourg: Office of the Official Publications of the European Communities).

Feng, L., and Gillespie, A. (1994), 'Teleworking, Work Organisation and the Workplace' in Mansell (1994), 261–72.

Gillespie, A., Richardson, R., and Cornford, J. (1995), *Review of Telework in Britain: Implications for Public Policy* (Report prepared for the UK Parliamentary Office of Science and Technology; Newcastle: Centre for Urban and Regional Development Studies, University of Newcastle upon Tyne).

Leyshon, A. (1995), 'Annihilating Space?: The Speed-Up of Communications', in Allen and Hammet (1995), 12–54.

Marshall, J. N., and Richardson, R. (1996), 'The Impact of "Telemediated" Service on Corporate Structures', *Environment and Planning A*, 28: 1843–58.

Opper, S., and Fersko-Weiss, H. (1992), *Technology for Teams: Enhancing Productivity in Networked Organizations* (New York: Van Norstrand Rheinhold).

Richardson, R. (1997), 'Network Technologies, Organisational Change, and the Location of Employment: The Case of Teleservices', in Dumort *et al.* (1997), 194–200.

Sassen S. (1994), *Cities in a World Economy* (Thousand Oaks, Calif.: Pine Forge Press).

6.3. Women's Access to ICT-Related Work

Juliet Webster

Women have relatively low rates of participation in the labour market for designing and developing computer-based systems. The precise rates, however, vary between countries.

For example, women have traditionally comprised only about 20 per cent of the IT professions in the UK (Committee on Women in Science, Engineering and Technology 1994) and just 15 per cent in the Netherlands (Tijdens 1991). On the other hand, women in the Nordic countries have enjoyed higher than average representations in IT jobs—24 per cent in Sweden and Denmark and nearly 30 per cent in Finland (Blomqvist *et al.* 1994). Women constitute 25 per cent of computer professionals in Australia (Stobart 1992) and around one-third of analysts and programmers in the USA (Blomqvist *et al.* 1994).

These figures indicate that the proliferation since the early 1980s of national policy programmes aimed at getting women into IT courses and jobs have not overcome the gender imbalance in this labour market. The division may even have worsened since the early days of computing, when women often worked in operator/programming jobs (Faulkner and Arnold 1985). Given this imbalance, men inevitably take the majority of senior ICT jobs.

In contrast to this situation in design and development work, women form the majority in ICT manufacturing and assembly jobs. These are not only functionally separate from design work but are also often geographically segregated. The bulk of ICT assembly operations—and therefore the places where women are most represented in ICT work—are in developing countries and free-trade zones which rely mainly on cheap female labour.

For example, of 4,500 people employed by Motorola in Malaysia, 3,500 were women, mainly in assembly and clerical work; the 1,000 men were employed as technicians, engineers, materials handlers, or administrators (Women Working Worldwide 1991). Parallels in the sexual and international divisions of labour in ICT work are illustrated by a company with plants in France and Brazil. In this firm, 76 per cent of unskilled assembly workers are Brazilian women, while just one French male is employed in such jobs. On the other hand, only 7 per cent of supervisors and technical posts in the two countries are filled by Brazilian women; 56 per cent are filled by French men (Hirata 1989).

The functions in the ICT industry with the highest proportion of women are therefore those characterized by repetitive and routine work. Opportunities to develop higher levels of skill and technological know-how are limited in these jobs. The organization of work in the offshore ICT industry is sometimes argued to be an improvement on the near-slavery conditions associated with the production of traditional goods in some developing countries. Nevertheless, the work involved in modern ICT factories can be highly monotonous, painstaking, and also hazardous, allowing employees little control over the labour process (Webster 1996).

This ICT work is also subject to the sort of machine pacing which has been common for automobile and other manufacturing assembly lines, in which fast workers are used to set standard times for the execution of tasks as a benchmark against which the rest of the workforce is evaluated. In some factories, talking on the assembly line is forbidden in case it disrupts productivity (Women Working Worldwide 1991: 107).

Even in industrialized countries, the majority of women in ICT design and development jobs are involved in routine programming and operating work. Although there is a world of difference between the working conditions of these jobs and those on manufacturing assembly lines, they are part of an overall pattern in ICT labour markets in which women are either a minority or grouped into women-only workplaces. In both environments, women's job and career opportunities have been severely limited. Opening up new opportunities for women's access to more interesting and influential ICT-related jobs is, therefore, an important step to ensuring there is greater equity in the distribution of benefits from ICT innovations.

References

Blomqvist, M., Mackinnon, A., and Vehvilainen, M (1994), 'Exploring the Gender and Technology Boundaries: An International Perspective', in Eberhart and Wachter (1994), 101–8.

Committee on Women in Science, Engineering and Technology (1994), *The Rising Tide* (London: HMSO).

Eberhart, T., and Wachter, C. (1994) (eds.), 'Proceedings of the 2nd European Feminist Research Conference on Feminist Perspectives on Technology, Work and Ecology', Graz, Austria, 5–9 July. Unpublished.

Elson, D., and Pearson, R. (1989) (eds.), *Women's Employment and Multinationals in Europe* (London: Macmillan).

Eriksson, I. V., Kitchenham, B. A., and Tijdens, K. G. (1991) (eds.), *Women, Work and Computerization: Understanding and Overcoming Bias in Work and Education* (Amsterdam: North Holland).

Faulkner, W., and Arnold, E. (1985) (eds.), *Smothered by Invention: Technology in Women's Lives* (London: Pluto).

Hirata, H. (1989), 'Production Relocation: An Electronics Multinational in France and Brazil', in Elson and Pearson (1989), 129–143.

Stobart, J. (1992), 'Women in New Information Technology Businesses in South Australia', unpublished postgraduate diploma project, University of Adelaide.

Tijdens, K. (1991), 'Women in EDP Departments', in Eriksson *et al.* (1991), 377–90.

Webster, J. (1996), *Shaping Women's Work: Gender, Employment, and Information Technology* (London: Longman).

Women Working Worldwide (1991), *Common Interests: Women Organising in Global Electronics* (London: Women Working Worldwide).

Part IV

Public Access in Politics, Governance, and Education

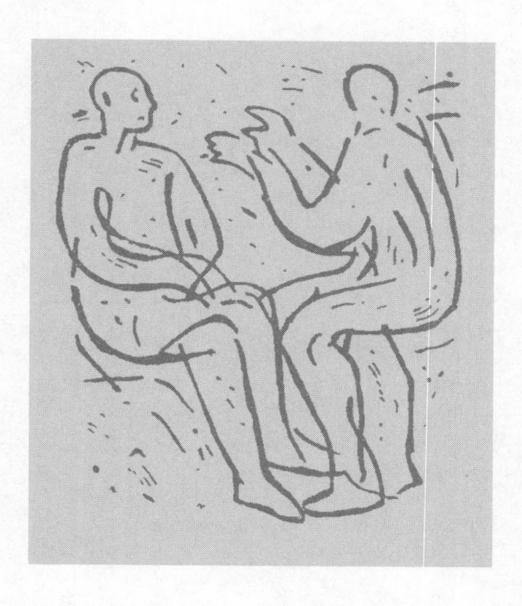

7 Digital Democracy: Electronic Access to Politics and Services

Electronic Democracy: The Centrality of Access

Early models of political influence assumed that politicians would vote in accordance with the interests of the pressure groups that contributed most heavily to their campaigns. However, researchers found that the opening of a channel of communication with a politician is the primary objective of lobbying—not direct political influence. This access enables lobbyists to become a source of information, which can eventually shape the politician's agenda, and views on key policy issues (Milbrath 1963; Bauer *et al.* 1972).

In such ways, access has long been a key resource in politics. This is one reason why tele-democracy proposals, such as the idea of a 'public information utility', have been promoted for decades (Sackman and Nie 1970; Sackman and Boehm 1972). It is also why these proposals are so controversial. For instance, ICTs raise concerns that sophisticated media consultants could manipulate a largely uninformed and disinterested public (Laudon 1977), and more effectively exploit electronic access to politicians (*The Economist* 1995).

ICTs provide a means both for delivering new forms of 'electronic public services' and for extending 'electronic citizen access' to public information and decision-making, in similar ways to earlier extensions of the right to vote (Barber 1984; Becker and Scarce 1986; Grossman 1995). This opens up many new opportunities for changing who gets access to politicians and governments—as well as who politicians and governments can reach with their own messages. The UK minister responsible for using ICTs to transform the provision of public services, David Clark, described his government's vision as one of 'more services being brought directly into the home, accessible 24 hours a day and offering instant or near-instant response. With the progress and processes of new technology at our fingertips, we have the tremendous and radical potential to bring government closer to the people' (Clark 1997).

The profound implications for democratic institutions and processes of 'putting government at your fingertips' have moved up national and international policy agendas following a long period when the technology for facilitating access to public information and decision-making was generally available, but ignored by most governments and politicians (Guthrie and Dutton 1992). This chapter describes some prominent examples of ICT innovations in politics and government, including a discussion of some of the key social factors that are limiting and shaping the adoption and outcomes of ICTs in this arena. It also examines the policies needed to ensure that ICTs complement democratic structures and processes, rather than further distancing citizens from politicians and their governments.[1]

The Information Polity: Institutional Brakes on Digital Government

Every new ICT innovation, from the Internet and the Web to multimedia kiosks and video conferencing, seems to garner proposals for applications in politics and governance. As in the private sector, support for governmental applications of ICTs is often driven by pressures to reduce spending by exploiting the continuing fall in costs and increase in capabilities of ICTs. This has led a growing number of public agencies around the world to pilot and introduce a wide range of innovations in electronic service delivery. Similarly, the widespread enthusiasm for 're-engineering firms' (see Chapter 5) has stimulated attempts to rethink governmental institutions and processes. A key difference between these trends in private and public spheres is that many democratic processes are inherently inefficient, while commercial companies can contemplate the elimination of inefficient ways of doing most of what they do.

Historically, there have been many attempts to modernize governments and make them more 'businesslike', such as America's municipal reform movement (Campbell and Birkhead 1976). Nevertheless, governments at all levels have proven to be quite resistant to structural and technological reform. As John Taylor argues (see Essay 7.1), institutions of governance have evolved through decades of negotiation and bargaining among conflicting groups and interests to balance a variety of relationships within and between governments and citizens. These social institutions are unlikely to be susceptible to radical change. It is in this spirit that many people from across most strands of political opinion regard evolving democratic institutions and processes as too sacred to put at risk over issues tied to technological change, such as in the outlandish utopian images of 'push-button' (Laudon 1977), or, in the digital age, 'point-and-click' democracy.

Despite many brakes on institutional change, a growing number of government agencies are employing ICTs to change the way services are delivered and citizens interact with government and each other. Policy must anticipate and keep up with trends if the protection of freedom of expression, privacy, and equitable access to public services is to be achieved.

Many believe ICTs can also play an important role in creating a more politically informed and active public. A 1996 US district court opinion, for example, argued that the Internet has had a 'democratizing effect' on speech, saying:

> It is no exaggeration to conclude that the Internet has achieved, and continues to achieve, the most participatory marketplace of mass speech that this country—and indeed the world—has yet seen. . . . [I]ndividual citizens of limited means can speak to a worldwide audience on issues of concern to them. (*ACLU* v. *Reno*, No. 96–963 ED Pa., 11 June 1996) (1996 WL 3118)

The US Supreme Court has also judged the Internet and the Web as among the most democratic technologies that exist.

In an era of declining public trust and confidence in democratic institutions and processes, the promise of reform should not be too readily dismissed (Ranney 1983: 11). However, the power shifts associated with ICTs remain a matter of controversy (Danziger *et al.* 1982; H. I. Schiller 1996). Technological change in government could deepen inequalities in access, and further distance the public from government and politicians, rather than enhance democratic control (see Box 7.1).

Box 7.1. The Politics of ICTs: Competing Perspectives

- *Democratic technology*: ICTs seen as inherently democratic in ways that will undermine hierarchy and centralized control (de Sola Pool 1983*a*; Cleveland 1985; Grossman 1995).
- *Technocratic élite*: view that the increasing centrality of advanced ICTs advantages technical experts (the 'cybercrats'), who will exercise greater autonomy and control over decisions (McDermott 1969; Ronfeldt 1992).
- *Economic élite*: ICTs regarded as being driven by military and industrial needs, with public preferences manipulated by marketing and experts accountable to large, mulitnational corporations (H. I. Schiller 1996).
- *Reinforcement*: ICTs seen as malleable resources that are most often controlled by the dominant coalition of interests within an organization or society, reinforcing the prevailing power structure of a political system (Danziger *et al.* 1982; Dutton and Kraemer 1985).

Source: Dutton (1990: 175–81).

Initiatives to Catch Up in Politics, Government, and Public Affairs

Through the 1960s, the public sector often led private enterprise in the application of ICTs. Leadership since then has changed hands, as highlighted by the view of the US White House Office of the Vice-President that the federal government is 'woefully behind the times, unable to use even the most basic technology to conduct its business' (Gore 1994). Likewise, the Labour Party's Shadow Minister for Science observed—before Labour became the party of the government—that the UK had a 'long way to go' to give real meaning to 'open government' (Battle 1995). Something similar could be said of central and local government in most other countries.

As discussed in Chapters 5 and 6, one trend in business since the 1960s has been a fundamental move towards using ICTs to improve the quality of front-line services to customers and away from the focus on 'backroom' administrative support services, such as budgeting, accounting, and payroll. Governments have taken a long time to move in the same direction, although they are increasingly using ICTs for direct support of services to the public.

As major users of ICTs, governments can make a substantial impact over time. For example, the US General Accounting Office reported in 1997 that the federal government spent $350 million on Internet-related activities over a period of three years. The US Telecommunications and Information Infrastructure Assistance Program (TIIAP) of the National Telecommunications and Information Administration (NTIA) provided $79 million in federal funds to support 277 projects over three years after its inception in 1994, including start-up funds for innovative ways of 'advanced telecommunications and information technologies to provide better services, to strengthen community ties, and to provide increased access to information' (NTIA 1996: I). Members of the US National Science and Technology Council advocate 'migrating Federal Information Services from legacy systems, through the interoperable systems of the Internet, and toward more advanced integrated global systems' to create a 'Digital Government for the citizens of the 21st Century' (Schorr and Stolfo 1997: p. vi). The European Commission has also emphasized innovative uses of ICTs in public services, as have successive UK governments in the 1990s (CCTA 1994; Hopkins 1995; Bellamy and Taylor 1998).

Catching up with the times in the use of ICTs in governance is important if the public is to realize the opportunities afforded by tele-access in enhancing democratic relationships among all actors within society. However, as noted by John Taylor (Essay 7.1), technological innovation in governance is difficult and entails serious risks, which is why it needs to be accompanied by careful and open consideration of the full range of opportunities and problems in providing electronic citizen access and digital government.

Electronic Citizen Access

ICTs in Political Campaigns and Elections

Public recognition of the changing role of electronic media in democracies is usually tied to campaigns and elections. Television has for a long time been recognized as central to election campaigns in every advanced industrial nation. More recently, advances in ICTs have held out the promise of more actively engaging the public, better informing voters, and providing access to candidates and causes that get little, if any, exposure on TV.

For example, Speaker of the US House of Representatives Newt Gingrich advocated 'national town hall meetings' during his second term as a Congressman, well before H. Ross Perot promoted interactive TV as a means for discovering and marshalling public opinion in the 1992 presidential election (Gingrich and Gingrich 1981). Perot's campaign renewed US debate over electronic plebiscites, and, by the time of the 1996 US presidential primaries and elections, most major candidates appeared in televised 'town hall meetings' in which they interacted with the audience. Many other uses of ICTs have become elements of campaigns at all levels (see Box 7.2).

Box 7.2. Application of ICTs in Campaigns and Elections

- *Direct mail*: enabling more personalized and targeted correspondence and other telemarketing of candidates and issues over phone, fax, or e-mail.
- *Opinion polling*: computer-assisted dialling, data entry, and analysis for improving the sophistication of public-opinion polling and enhancing a campaign's effectiveness in marketing its candidate or issues.
- *Database systems*: improved techniques for collecting, storing, retrieving, and organizing information about individual voters and contributors, including data matching and profiling to locate individuals for targeted campaigns.
- *Publicly accessible computer networks*: systems like the Minitel in France, and the Internet and the Web for reaching prospective voters, distributing campaign information about candidates and issues, and obtaining feedback.
- *E-mail and conferencing systems*: using ICTs for managing a disbursed field staff, speech writers, and media consultants, and reaching voters more directly, through by-passing media gatekeepers.
- *Desktop computing*: enabling even local campaign managers to handle their own budgeting, accounting, and payroll applications.

Sources: Meadow (1985); Armstrong (1988); Dutton (1990).

Voting and Polling

ICTs like interactive cable TV and the Internet could enable citizens to vote and be polled on matters of public interest from their homes.[2] Warner's QUBE system in Columbus, Ohio, experimented successfully with a variety of town-planning meetings in the early 1980s using interactive polling to tap the responses of citizens to alternative plans (Davidge 1987). However, few experiments with the use of ICTs for voting or polling have been held, because the approach clashes with important aspects of dominant paradigms of the democratic process that place more emphasis on communitarian and representative democracy, as opposed to more direct forms of democratic control. For example, advocates of representative democracy place more faith in the role of politicians, political parties, unions, interest groups, and the press to represent and mediate between government and different segments of the public at large.

ICTs have left these traditional information brokers 'less valuable as repositories of historical and current information and less necessary as interpersonal or interorganizational transmitters of information' (Danziger 1986). Emerging ICTs such as the Internet provide mechanisms for governments to reach citizens more directly, undermining the traditional role of the press, political parties, or other gatekeepers.[3]

Tele-democracy might clash with traditional paradigms of representative democracy, but paradigms can change. Also new forms of tele-access are argued to be in line with a history of democratic reforms, which has consistently expanded access to political participation (Becker and Scarce 1986). The supporters of one new democratic paradigm, labelled a 'Californian Ideology', exhibit 'an impeccably libertarian form of politics—they want information technologies to be used to create a new "Jeffersonian democracy" where all individuals will be able to express themselves freely in cyberspace' (Barbrook and Cameron 1996). New technological paradigms, such as those nurtured by e-mail and the Internet, change the way people think about political communications and access to information and people (Guthrie 1991; Guthrie and Dutton 1992). Technological change revises public models of tele-access, how to get access to public information, politicians, services, and governments. Moreover, the evolving use of ICT networks in the public sector is creating the technological infrastructure for electronic democracy, which means electronic voting and polling is more than ever 'waiting to be plugged in' when society chooses to do so (T. Becker 1981: 7; Tsagarousianou *et al.* 1998).

The geography of ICT networks also challenges conventional paradigms of politics, which tend to have a local basis. The new political map created by tele-access no longer matches well with the boundaries of government jurisdictions. This has been recognized with the formation of single-issue groups that span national and international boundaries, and which could not be sustained on a purely local level.

Economic Constraints on Political Uses of ICTs

One major problem with a greater reliance on ICTs such as TV in campaigns and elections is the degree to which it advantages campaigns with greater economic resources. The costs of high-tech support can be substantial. Instead of giving minor candidates access to voters, ICTs can therefore reinforce the position of candidates with the greatest financial resources, who can use advances such as telemarketing and Webcasting to extend their existing campaign apparatus (Armstrong 1988).

An effort to counter this advantage and to better inform citizens has been to provide public support for independent coverage of candidates and politicians. This includes efforts to obtain or guard existing free TV coverage of candidates, such as is provided by public service broadcasting in Europe and through free 'party political broadcasts' on public-service and commercial TV channels in the UK. C-SPAN provides access to live coverage of Congress for nearly half of all TV households in the USA. ICT-based services for citizens range from traditional TV news coverage to interactive electronic services (see Box 7.3). For example, the Democracy Network offers on-line Web access to information about candidates for office, including their positions on issues and selected text and video clips of their speeches, position statements, and media coverage. It also enables voters to get involved in debate, post their opinions, or volunteer for work on campaigns (Docter 1997).

Conceptions of Voters as Users

Many proponents of using ICTs to improve campaigns believe that the mass media is a significant cause of citizen apathy (T. Becker 1981; Ranney 1983). They and other proponents of new media argue that, if ICTs could be used better to inform voters about the issues at stake, and provide them with a means of engaging more actively in the political process, they might well become more interested in public affairs (*The Economist* 1995). But, even if ICTs cannot overcome widespread apathy, they might well nurture those who are interested in politics, and facilitate their deeper involvement. For example, research suggests that the people most interested

Box 7.3. Citizen and Voter Information and Services

Traditional broadcast TV coverage of campaigns and public affairs
- paid political advertising
- free coverage

Voting and polling systems
- electronic town meetings on QUBE in Columbus, Ohio (Box 4.2)
- polling via mail, telephone, or e-mail, such as Televote experiment in New Zealand to involve the public in long-range planning (T. Becker 1981)

TV, cable, and satellite coverage of public affairs
- cablecasts of local government and other public meetings
- US Cable-Satellite Public Affairs Network (C-SPAN) coverage of the House of Representatives since 1979, and US Senate since 1986
- California-Satellite Public Affairs Network (Cal-SPAN)
- coverage of debate in the House of Commons in the UK

Non-profit and public-interest group Internet sites
- USENET groups discussing public issues over the Internet
- Internet voter guides such as the Democracy Network (Centre for Governmental Studies, Los Angeles), and interactive multimedia information tested on the Full Service Network in Orlando [http://www.democracynet.org]
- Interlinc (Lincoln, Nebraska) permitting Lancaster County residents to register to vote over two dozen public Internet sites [http://interlinc.ci.lincoln.ne.us]

in politics are those most likely to get on-line as one additional means for staying informed and engaged (Guthrie *et al.* 1990). At the margins, new channels of communication can also involve some individuals who are interested but might not otherwise be able to participate in public affairs, such as a lone parent who can use the network to become more directly involved in politics from home.

A study comparing the use of Minitel in France with political campaigns in the USA using the Internet found that these new media played a largely symbolic role in helping to identify the candidate as up to date and technologically youthful (Lytel 1997). One reason was that use of these networks is highly concentrated within a small proportion of the public.[4] Campaigns could develop some presence on the new media at a relatively low cost, compared to TV, but the major campaigns did not place a real priority on the new media, largely because of its limited reach. This means that, in the near term, the new media will be used as a complement, but not a substitute, for the mass media in campaigns and elections.

Digital Government

Trends in computing and telecommunications are extending and transforming the ways in which services are delivered to citizens. In many respects, these moves towards a digital government are more significant in defining the relationships between citizens and governments than are applications tied to campaigns and elections. The following sections describe several broad areas which provide a sense of the ways in which this new spectrum of ICT-based tele-access opportunities could change the relationships between citizens and governments.

Public Information

Government can get closer to its citizens by the provision of public information in electronic form. Such information ranges from library catalogues to welfare benefits, and from information about pending and recent legislation to speeches by politicians. Telephone, e-mail, Web sites, and multimedia kiosks are some of the ICTs being used to support these kinds of services (see Box 7.4).

In 1986 ReferencePoint (RP) was set up in New York as a non-profit foundation to address its founder's concern that these kinds of issues would disadvantage non-profit and public-interest organizations in the information age. RP provided a computer-based network for public access to electronic government and other public information. In 1994, using contributions from a consortium of non-profit organizations, it developed a Public Information Exchange (PIE). This enabled individuals, organizations, and schools to gain electronic access to information provided by governments, non-profits, and voluntary organizations distributed by the PIE through libraries, commercial on-line services, community bulletin boards, and electronic publishers.

Ironically, the very success of the Internet undermined the mission of RP. In 1997 its founder informed his board of directors of the need to close down RP, saying: 'the cold fact is that we and our central theme have been engulfed by the Internet. Every organization, and indeed every individual, can now do the "self-publishing" that RP was trying to organize' (Westin 1997).

Box 7.4. Electronic Access to Government and Public Information

- Community projects: for example, OXCIS (Oxfordshire County Council, UK) providing a free electronic public information service including 'community information points', and Internet sites like Charlotte's Web [http://www.charweb.org/].
- Ask Congress (US House of Representatives): multimedia kiosks for answering common questions and obtaining public-opinion feedback.
- INFOCID (Portugal): access to information from many government departments, including from multimedia kiosks.
- Overheidslokey 2000 (The Netherlands): ICT-supported one-stop shops in various categories of service, such as for elderly and disabled, and real estate services (Lips and Frissen 1997: 76–80).
- Government Web sites: thousands of sites on the Web for all levels of government, such sites as the US White House [www.whitehouse.gov], British Prime Minister [www.number-10.gov.uk], the British Government's Information Service (GIS), and the European Commission [www.europa.eu.int].
- Private intranets, Web sites, and list serves: developed by special and public-interest groups to disseminate government information relevant to legislation, court rulings, and reports, of interest to their subscribers.
- Toll-free telephone access to agencies and services (Fountain *et al.* 1992).

Sources: Dutton *et al.* (1994*b*); NTIA (1996); Taylor *et al.* (1996); Bellamy and Taylor (1998).

The Internet has evolved to contribute to the wider availability of public electronic information. But it is insufficient just to have the technological mechanisms to access and deliver information. 'Open government' also depends on the degree to which public agencies wish—or are obliged—to make information in electronic form available to citizens at a price that approaches the marginal cost of its distribution. This was an objective of the Clinton Administration. Freedom of information legislation might therefore remain a crucial factor in determining the availability of electronic information, particularly in nations without a tradition of open government.

Commercial as well as political considerations threaten access to public information. There are obvious financial incentives to shift valuable information off publicly accessible sites on the Internet, which is becoming increasingly used for commercially based offerings of information and services. As more people send e-mail and broadcast messages over electronic networks, ownership of information and ideas will become increasingly problematic to determine and regulate. Already, copyright restrictions are one of the major costs and barriers to the effective utilization of electronic devices for storing and distributing information. New conceptions and rules for governing intellectual property rights might be essential if the public is to make full use of these new infrastructures for public information (Bekkers *et al.* 1996).

New Channels for Communicating with Citizens

Tele-access can be used to improve the responsiveness of government to its citizens—for example, by opening new channels for the public to request services, voice complaints, and obtain information. Such facilities can go beyond the distribution of information to support the public's use of e-mail for communicating directly with elected representatives, as in the e-mail facilities available to the US President and the British Prime Minister (see Box 7.4). The Public Electronic Network (PEN) in Santa Monica, California, is an example of the potential value of tele-access at the local level (see Box 7.5). When it was used to support interaction between the government and citizens, personnel in the city believed the facility improved the government's responsiveness to the public (Dutton *et al.* 1993). However, this utilization declined as it became more oriented towards the Web in the mid-1990s, and PEN, along with most local government Web sites, became more focused on simply broadcasting information to the public (Hale 1997; Docter and Dutton 1998).

Many citizens are interested enough in politics to desire these new channels for communication, but not most. The proponents of teledemocracy often harbor unrealistic views of the public's interest in politics, and experience with early systems reinforce the degree to which most members of the public are not politically active and attentive to politics. In Santa Monica, for instance, users were more likely than other residents to be active and interested in local politics. Together with the use of public terminals, the important role played by an interest in politics helped the city to create a critical mass of users that was more diverse than the population of home computer users. For example, PENners included the unemployed as well as managers and professionals, the homeless as well as home owners and renters, and it had a larger proportion of women than expected on the basis of computer ownership—although most users were males (Guthrie *et al.* 1990). PENners core user participation may have rivaled attendance at city council meetings, but it fell far short of the expectations of proponents. This is similar to the findings sited earlier on

Box 7.5. Public Electronic Network (PEN) in Santa Monica, California

The PEN 'electronic city hall' was launched in 1986 as a municipally owned e-mail and computer conferencing system operated, and mainly developed, by Santa Monica's Information Systems Department. The city's residents could use a home computer or one of twenty terminals in sixteen public locations to register for PEN and undertake activities on it, such as: retrieving free information about city services; completing some transactions with the city government; sending e-mail to city departments, elected officials, or other PEN users; and participating in numerous computer conferences on topics of local concern. City authorities guaranteed a response within twenty-four hours to complaints and requests made on PEN.

PEN had 4,505 registered public users by 1992, about 5 per cent of Santa Monica's residents. An average of about 400 to 600 individuals used PEN every month. Nearly half of their accesses were to about a dozen PEN computer conferences on local and national issues, such as the homeless. PEN was of value in stimulating discussion, communicating with key opinion leaders, involving people who might otherwise shy away from public participation, and offering an opportunity for a new set of people to become involved in local government. However, participation in e-mail and conferencing declined in the face of controversy over the civility of discussions and with the migration of the PEN system towards a Web-based source of information.

Sources: Guthrie *et al.* (1990); Dutton and Guthrie (1991); Dutton *et al.* (1993).

the use of Minitel in French elections (Lytel 1997). This means that policy should balance the value of connecting the politically active citizens, while being mindful of the many who may never get plugged into an electronic democracy.

Networks for Political Dialogue among the Public

Face-to-face confrontation, as in town-hall meetings (Abramson *et al.* 1988), is another aspect of traditional democratic paradigms and practices that is central to ensuring social accountability. Tele-access challenges this paradigm by enabling citizens to communicate with one another electronically to form new 'virtual' pressure groups and communities of interest linked through ICT networks. Many systems have created opportunities for public dialogue among citizens, with sponsorship from individuals, private, and non-profit organizations as well as government (see Box 7.6). These help to organize and form opinion by supporting horizontal networks

Box 7.6. Systems Supporting Political Debate and Dialogue

Publicly owned or financed

- Santa Monica's e-mail and conferencing system on PEN (see Box 7.5)
- HOST: provision of on-line bulletin boards for public discussion in Manchester, UK
- city-talks and Digital City (City of Amsterdam): used for interactive debates on local government issues and electronic citizen consultation

Private and not-for-profit groups and organizations outside government

- electronic communities accessible over the Internet, such as the WELL (Whole Earth 'Lectronic Link), one of the most widely known e-mail and conferencing systems launched in the 1980s (Rheingold 1994)
- USENET groups on the Internet that provide opportunities for discussion of a vast number of topics across the political spectrum

Sources: Dutton *et al.* (1994*b*); Schalken and Tops (1995); Raab *et al.* (1996); Tsagarousianou (1998).

of communication between citizens, so they have, therefore, been advocated by those critical of networks which primarily support vertical linkages between citizens and government (Laudon 1977).

However, it is not only the structure of networks—horizontal versus vertical—that is expected to reshape citizen access. The move away from face-to-face confrontation can eliminate many social context cues and fundamentally change the nature of interpersonal interaction and political decision-making.

The impact of computer-mediated communication systems on interpersonal and group communications has been studied since the late 1960s (Hiltz and Turoff 1978). Many researchers have attributed a disinhibiting effect to computer-mediated communication, like e-mail. Early evidence of this effect was the frequency of 'flaming'—the rapid escalation of terse remarks or insults in an electronic interchange. This phenomenon is usually suppressed or avoided in face-to-face conversations, but in interpersonal communication via a computer network is likely to be more open and uninhibited. Electronic communication such as e-mail tends to be more spontaneous than other forms of written communication, including many of the grammatical errors and spelling mistakes that come with the spontaneity of a conversation.

This disinhibiting effect is likely to be caused by the way computer-mediated communication tends to eliminate the context cues which add social presence to face-to-face communication (Kiesler *et al.* 1984; Sproull and Kiesler 1991). This reduces the constraints created by such interpersonal factors as the unequal status of the individuals engaged in a dialogue.

Proponents of electronic networking in politics have argued that disinhibition will have a positive impact on democratic dialogue by empowering individuals that might not otherwise be effective in the atmosphere of a town-hall meeting (Hiltz and Turoff 1978). However, open communication can also contribute to a lack of civility and decorum, that led many opinion leaders and politicians to stop using PEN (Dutton 1996*b*).

Establishing Effective Electronic Public Forums

There are two major difficulties in establishing policies and practices to enable public forums to function more effectively:

1. The degree to which any regulation of a public electronic forum would be viewed as an infringement of free speech, which is protected by law in some countries, such as in the US first amendment (Docter and Dutton 1998). This has had the effect of forcing governments to avoid the establishment of public forums, leaving them to private and non-profit organizations that can censor users without the same legal restraints on violating a citizen's free speech.
2. The extent to which participants in electronic communities do not agree among themselves on the norms that should govern dialogue about public affairs. My own research has identified five distinguishable viewpoints, defined by the value judgements they emphasize in discussing the rights and responsibilities of participants within an electronic community (Dutton 1996*b*). For simplicity, I have labelled them: civil libertarians, communitarians, formalists, property-rights advocates, and balancers (see Box 7.7).

The existence of agreed norms are important factors in shaping communication on networks (Kiesler *et al.* 1984; Foulger 1990; and Collins-Jarvis 1992). However, the novelty of the technology has meant that there is no consensus on what norms, etiquette, or rules apply to the new medium. E-mail, for example, is not exactly analogous to a letter, a telephone call, or a conversation. The developing set of conventions and practices—a so-called Netiquette—among experienced Internet users indicates such

Box 7.7. Perspectives on Rights and Responsibilities on Electronic Networks

- *Civil libertarians* stress their status as first-amendment speakers and advocate the rights of the user as a speaker, sender, or user of services. Regulation can be accomplished only informally by the community of users.
- *Communitarians* put privacy concerns over their right to free expression. They are sensitive to the rights of the user as a viewer, receiver, or subject of messages and define limits on speech when harmful to the audience.
- *Formalists* view rights and responsibilities as settled by existing codes, formal policies, and laws—if they are enforced.
- *Property-rights advocates* focus on the rights of those who own the network, forum, or bulletin board, often the system operators, in contrast to the rights and responsibilities of the users.
- *Balancers* hold conflicting rights and responsibilities, often without acknowledging or recognizing the contradictions in their views.

Source: Dutton (1996*b*).

norms could eventually be developed, at least in certain cultural settings. Working against such a consensus is the expanding population of Internet users, which will make the culture of the Internet more heterogeneous than ever.

The role of cultural factors in shaping the actual responses of users can be seen in some corporate settings, where the culture of the organization countered any disinhibiting effect of the technology. For example, a study of mature students taking a correspondence course within Britain's Open University found that their use of computer-mediated communication seemed to have inhibited communication, rather than stimulated discussion (Grint 1992). Older students in the British context might take their written work more seriously than younger students in the US setting. In communities or nations with a more consensual culture surrounding interpersonal communications, public electronic networks might fare better in the short term. However, in the absence of some shared cultural practices, it will be difficult to establish public forums. This means that debates about public issues which use electronic media could be fragmented across various private and non-profit systems, thereby rendering them marginal to debates in real public forums. Private electronic forums have thrived, especially in the USA, but they are subject to strict censorship and control over access.

Public-Service Delivery

ICTs can improve the technical speed and efficiency of delivering many public services. Many of these gains can be realized simply by emulating techniques that have been well proven in the private sector for a wide range of government services, such as allowing motorists to change the address on their driver's licence at a kiosk, fill out an initial application form for welfare benefits at a touch-screen terminal, and process routine, high-volume transactions. Box 7.8 illustrates some electronic services that have been undertaken by public agencies. Not all such initiatives succeed. For example, Info/California was stopped by budget cuts, and, in 1996, the IRS stopped Cyberfile, a plan to permit tax filings over the Internet.

Box 7.8. Examples of Public Electronic Service Delivery

- *Hawaii Access (Hawaii State) and Info/California*: early uses of multimedia, multilingual kiosks for accessing information and carrying out transactions, like an 'automated labour exchange' in Hawaii and driving licence renewals in California (Info/California discontinued by legislative budget-cuts in 1995).
- *Tulare Touch (Tulare County, California)*: multilingual multimedia kiosks advising citizens on eligibility for welfare.
- *SafetyNet (New Hampshire)*: one site on the Internet, operated through the Children's Alliance of New Hampshire, for residents to determine their eligibility for services from over fifty organizations (NTIA 1996: 14).
- *Probation monitoring (Minnesota)*: multimedia kiosk including fingerprint verification and alcohol breath analyser.
- *Singapore Post (Singapore)*: automated post office for purchasing goods, weighing letters, and obtaining information.
- *WINGS (Web Interactive Network of Government Services)*: 1996 pilot project of the US Post Office to provide services via kiosks that cut across government jurisdictions.
- *US Internal Revenue Service (IRS)*: programme launched in 1995 to permit low-income taxpayers to file returns by touchtone telephone.

Sources: Dutton (1994); Taylor *et al.* (1996); Bellamy and Taylor (1998).

Unlike the private sector, concerns over equity create major constraints on the role of ICTs in the public sector. Shifts in the technology of political, or governmental, communication threaten to disenfranchise those who cannot enjoy access to ICTs because of their income, location, physical handicaps, or language skills. Many can still not climb the steps necessary to get into a digital government, unless governments take initiatives to provide public access. The need to maintain a balance between the delivery of more efficient public services using ICTs and the maintenance of democratic equity and fairness was emphasized in a study by the former Congressional Office of Technology Assessment (OTA 1993), which viewed one of the greatest risks from public-sector electronic service delivery to be the potential to widen 'the gap between the advantages that educated, technically proficient citizens have over those less so'.

Equity considerations require public agencies to ensure that the benefits achieved are available to all citizens. In many respects, electronic services are quite well suited to addressing such concerns. For example, well-designed user interfaces and multilingual systems can make information services more readily accessible to different language communities and all levels of skill.

The development of new telecommunications infrastructures—the so-called information superhighways—can also redistribute access to public services across the geographies of urban and rural areas. This could increase inequalities unless proactive efforts are made to develop facilities on a nationwide basis—for instance, by investing in special public facilities and training at locations where access to information technology is more problematic, such as rural areas (Allen and Dillman 1994), and the distressed neighbourhoods of urban regions (Dutton 1993; Goddard and Cornford 1994). About 20 per cent of access to Santa Monica's PEN service came from numerous terminals in public locations, where free training was also provided (Dutton *et al.* 1993). Many pilot TIIAP projects supported by the US Federal government are aimed at supporting public access (NTIA 1996). Box 7.9 provides examples of public-information and infrastructure projects with similar aims.

Valid concerns over privacy and data protection can also constrain the development of ICT-based delivery of public services. Fears in the 1960s of the growth of 'Big Brother' surveillance through the creation of centralized computer databanks eventually led to legislation, such as the US 1984 Data Protection Act. In the 1990s, the migration of ever more information into electronic forms, that can be more easily retrieved and distributed across networked systems, has led to renewed calls for privacy and data protection legislation (see Essay 7.2).

Individuals often choose to trade off their personal privacy for a variety of other benefits, such as public safety or personal convenience (Dutton and Meadow 1987). However, fairly balancing these trade-offs and protecting legal rights to privacy are critical, not only to the democratic process, but also for building public support and confidence in public-sector

Box 7.9. Public Information and Telecommunications Infrastructure Projects

Public access facilities, such as:

- Electronic Village Halls and Host on-line communications and information system (Manchester, UK): supporting local community groups, small businesses, and others through networking, training, and other capabilities
- TeleVillage in Compton, a suburb of Los Angeles: established by the Metropolitan Transportation Authority to provide access to a computer lab, training, teleconferencing equipment, and telework centre
- Plugged In Enterprizes, Inc.: a centre in a low-income urban neighbourhood of East Palo Alto, California, for young people to work in a multimedia lab, create Internet pages, and conference [www.pluggedin.org]

Public telecommunications infrastructures, such as:

- Broadband telecommunications infrastructure set up by the Highland Regional Council, Scotland: promotes innovative applications in a rural area
- Public telecommunication infrastructure projects to serve government and provide advanced telecommunications access to rural and distressed urban areas, such as municipal projects in Los Angeles, California, and Austin, Texas, and regional projects like The Tri-State Network of Mississippi's Department of Economic and Community Development (NTIA 1996: 17)
- Community Area Network (CAN) Forum, Center for Global Communications: promotes development of community-based computer networks in Japan

approaches to digital government. It is, therefore, important that political decision-makers identify fundamental principles of personal privacy and data protection that must not be overridden by efficiency and other concerns less central to the democratic process (Raab *et al.* 1996).

Decision Support in the Public Arena

Advances in ICTs are being used to support public planning and decision-making in ways that create other opportunities and threats to democratic control. For example, computer-based models have long been used to forecast the consequences of alternative decisions on such vital outcomes as public revenues and expenditures, energy, and the environment (Greenberger 1983; Dutton and Kraemer 1985). Increasingly realistic models, and multimedia simulations are being used to support planners in helping the public and decision-makers to visualize the consequences of urban and regional land use and zoning policies, such as showing the visual effect of constructing a new building (Droege 1997*a*).

It has already become necessary for the public to gain an awareness of models, and other virtual representations of reality, in order effectively to recognize and counter biased presentations of public policy options. President Ronald Reagan's former Director of the Office of Management and Budget (OMB) David Stockman created a major embarrassment for the President when he confided to a reporter that he altered OMB projections to support the President's budget (Greider 1981). As modelling and simulations and other decision support technologies become more central to public decision-making on transportation, energy, the environment, and many other activities, the public must find the expertise and resources to assess existing models independently and, if necessary, to provide counter-models and analyses, or be left on the sidelines of debates.

Fostering Digital Democracies: Cultural and Policy Issues

Technological change raises major cultural and policy issues related to enduring values of democratic institutions, including issues of equity, access to information, freedom of expression, privacy, and data protection. As Lawrence Grossman (1995: 7) has concluded, 'in the coming era, the qualities of citizenship will be at least as important as those of political leadership.'

In the area of political dialogue, for example, explicit policy provisions and procedures for editing and controlling of communication over public electronic networks are needed. However, effective policies are not readily at hand to achieve this, as participants in electronic communities generally disagree among themselves on appropriate remedies, nations are imposing their own policies and regulations on this new medium, and international standards appear more problematic as the digital media span increasingly diverse cultural traditions across the globe. Moreover, in the USA, the courts have argued that the Internet is too democratic to permit public regulation (e.g. *ACLU* v. *Reno* 1996). It may well be that users must rely on gradually developing a set of norms that respect the rights and responsibilities of users in ways that allow effective use of networks for political dialogue (Denning and Lin 1994).

There are also conflicting values and standards within and across nations about the role of ICTs in privacy and data protection, which could limit the likely effectiveness of public policy responses. Elizabeth

France, the UK's Data Protection Registrar, has argued the need for a 'culture of privacy' because the force of public policy will be very limited unless individual users understand and value the need for protecting the privacy of individuals—self-regulating tele-access.[5]

The critical realization required to approach each of the issues is that ICTs are biased, but malleable (see Chapter 3). They do not determine outcomes. Governments can meet real needs for service, or waste money and technological potential. Digital government can erode or enhance democratic processes, public responsiveness, equity, access to information, privacy, and national and ethnic cultural traditions. The outcome will be determined by the interaction of policy choices, management strategies, and cultural responses—not by advanced technology alone. That is why it is so important that the discussion highlighted in this chapter moves beyond pro- versus anti-tele-democracy responses that are based on overly deterministic views of ICTs. The debate over appropriate policies for guiding the application of ICTs in politics and governance needs to begin in earnest—before the public sector catches up with the times.

Notes

1. This chapter draws on the author's research on electronic service delivery in the USA (Dutton 1992*c*, 1994), which was further developed through collaborative work on electronic service delivery in Europe that was launched at a PICT Policy Research Forum entitled 'Electronic Service Delivery' (Dutton *et al.* 1994*b*; Raab *et al.* 1996; Taylor *et al.* 1996; Dutton 1997*c*).
2. Audience response systems developed in the 1970s permitted individuals at home to respond to questions and instantly to see the tally of all responses. These systems were used primarily for audience research at a single site, such as a movie theatre, but easily adapted to opinion polling over interactive cable TV experiments (Dutton *et al.* 1987*a*).
3. One assumption of 'democratic élitism' is that democracies depend heavily on the intelligence, values, and attitudes of élites (Bachrach 1967). The public often holds undemocratic opinions and beliefs, which leads many to fear that tele-democracy opens a path to a 'tyranny' of the majority (Grossman 1995).
4. For example, Lytel (1997) cites 3% of French households being responsible for one-third of all Minitel traffic.
5. Elizabeth France developed this argument in remarks made at PICT's International Conference on the Social and Economic Implications of Information and Communication Technologies, the Queen Elizabeth II Conference Centre, Westminster, London, 10–12 May 1995.

Essays

7.1. The Information Polity

John Taylor

Electronic forms of communication and information-gathering and access have become interwoven in the relationships between government and citizens and among citizens themselves. The implications of this for democracy and efficient and effective government are assessed by John Taylor.

7.2. Protecting Privacy

Charles Raab

Citizens and consumers can benefit greatly from the way data about themselves are gathered, stored, and transmitted using ICTs. Charles Raab discusses ways in which personal privacy can be protected during these processes, while still allowing the benefits to be garnered.

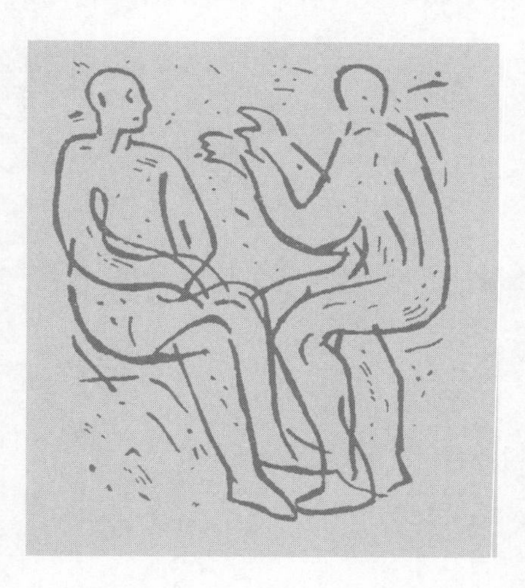

7.1. The Information Polity

John Taylor

The dominant rhetoric of the information age depicts ICTs as bringing profound changes to those aspects of organized society into which they are introduced; yet an examination of the complex relationship between new technologies and society suggests otherwise. As the governments of many countries come to redefine their governance processes through the intensive application of ICTs, so we are provided with opportunities to explore the characteristics of the emergent 'information polity', examining as we do so the technologically mediated relationships which it embodies.

Five sets of relationships lie at the heart of the information polity (Bellamy and Taylor 1997):

1. internal relationships in the machinery of government;
2. the relationships of government organizations to the consumers of their services;
3. the relationship of government to citizens of the state;
4. the relationships between governments and the providers of ICT infrastructure, equipment, and services;
5. the relationship between existing information systems, patterns of communication, and technical infrastructures to the polity's 'appreciative system'—the value system within the polity which attaches importance to some practices and agendas for change rather than others (Vickers 1965).

Research findings suggest that the forces of continuity within each of these core relationships of the information polity are powerful. Theoretically, new computer networks pose challenges to long-standing organizational arrangements within the machinery of government, by providing electronic links between separate units and tiers of government. As well as posing threats to the established domains of the relating organizations, these links offer many potential benefits to them, such as reduced costs of provision and enhanced quality of service. Organizational arrangements establish powerful interests which both shape and resist change, however, and outcomes are shaped from the mutual adjustments made within them (Taylor, Zuurmond, and Snellen 1997).

Similarly the delivery of services by electronic means—at face value one of the most exciting application areas of new ICTs in government—is constrained by the organizational forces within which it is enmeshed (Taylor *et al.* 1996). For example, the delivery of some public services *via* ICTs, such as welfare payments, or housing transactions, is taking place in human settings which shape, and restrain, the specific innovations in ways that raise questions about whether it is providers rather than consumers who continue to be the ultimate beneficiaries.

ICT applications also increasingly pervade the relationship of citizens to government. For example, citizens can now engage interactively in public debates using ICTs. Experiences in countries such as the Netherlands, where forms of tele-democracy have developed rapidly, have raised serious concerns about whether such activities are more likely to lead to an Orwellian nightmare of social control than an Athenian democratic paradise (Van de Donk and Tops 1992). Whilst it is entirely appropriate that concerns such as these are raised, recent work on tele-democracy in the UK suggests that there is little evidence that electronically mediated democratic relationships are fundamentally challenging the settled ordering of the democratic process (Taylor *et al.* 1995).

Private suppliers of equipment, infrastructure, and a variety of information services for governments are also at the heart of the emergent information polity. Research findings show, for example, that the liberalized telecommunications regime in the UK is creating severe discrepancies in the geographical provision of modernized telecommunications, giving rise to profound problems for public organizations wishing to exploit telecommunications in the universal delivery of service (Taylor and Williams 1995).

The fifth set of core relationships in the information polity is that between existing information resources and the state institutions within which they are used. Existing information systems are legacies from previous administrations and, the longer they have been in existence, the more encrusted they will be with pre-existing ways of thinking and assumed values. Dismantling and rebuilding such information systems is both costly and organizationally difficult. Governments are users of data and information on a massive scale. The systems legacies which contemporary government organizations have inherited can be a major obstacle to the delivery of programmes aimed at reinventing and re-engineering government (Bellamy and Taylor 1997).

Analyses of such relationships helps to illuminate interactions between political, social, economic, and technological forces. Moreover, they illustrate the explanatory power of the term 'information polity', for these analyses convey simultaneously the revolutionary qualities popularly associated with the engulfing wave of ICTs and the evolutionary nature of a political system, or polity.

The information polity is concerned as much with those networks comprised of information flows and organizational relationships as it is with the computer networks in and around government. The information polity should be understood neither as a Utopian nor as a dystopian project but as new forms of relationships shaped by a complex combination of technical, political, and social phenomena.

References

Bellamy, C., and Taylor, J. A. (1997), *Governing in the Information Age* (Milton Keynes: Open University Press).

Dutton, W. H. (1996) (ed.), with Malcolm Peltu, *Information and Communication Technologies—Visions and Realities* (Oxford: Oxford University Press).

Taylor, J. A., and Williams, H. (1995), 'Superhighways and Superlow-Ways: Universal Service and Electronic Innovations in the Public Sector', *Flux*, Jan.–Mar., 45–54.

—— Bardzki, B., and Wilson, W. (1995), 'Laying Down the Infrastructure for Innovations in Teledemocracy: The Case of Scotland', in Van de Donk *et al.* (1995), 61–78.

—— Bellamy, C., Raab, C., Dutton, W. H., and Peltu, M. (1996), 'Innovation in Public Service Delivery', in Dutton (1996), 265–82.

—— Zuurmond, A., and Snellen, I. (1997) (eds.), *BPR in Public Administration: Institutional Transformation in the Information Age* (Amsterdam: IOS Press).

Van de Donk, W., and Tops, P. (1992), 'Informatization and Democracy: Orwell or Athens?', *Informatization and the Public Sector*, 2: 169–96.
—— Tops, P., and Snellen, I. (1995) (eds.), *Concurring Revolutions: ICT and Democracy* (Amsterdam: IOS Press).
Vickers, G. (1965), *The Art of Judgment: A Study of Policymaking* (London: Chapman Hall).

7.2. Protecting Privacy

Charles Raab

Information on individuals is collected, processed, and communicated via ICTs as an integral part of the provision of goods and services in the home, workplaces, shops, governmental, and public-service institutions, the virtual space of the information superhighway, and many other activities. This can put an individual's privacy at risk, because personal data are captured in such transactions, often without the knowledge or consent of those concerned.

Data are also obtained through statutory requirements upon individuals to disclose their personal details—for example, as part of contracts between a customer and provider. The circulation and merging of information among organizations is another important data-gathering mechanism.

Important issues of transparency and equity in these transactions need to be addressed—questions about how individuals can know about and influence what happens to their information, and about the imbalances of power between individuals and data-using organizations. Finding answers involves debating the rules and practices that might be instituted for controlling flows of personal data and for arbitrating disputes between individuals and organizations.

Many of the benefits and conveniences which ICTs bring to daily activities —such as shopping, banking, driving, and entertainment—would be impossible without the transfer of personal data from individuals to organizations, or between organizations themselves. Likewise, the provision of public services such as health, social welfare, policing, and education require the use of large quantities of often sensitive information about members of the public. This is necessary whether the activity is performed by electronic service delivery or more conventional methods.

These data are dealt with domestically or globally in ways that the individual can scarcely know or control. One of the main challenges to the information society is to reconcile privacy, and its protection, with these myriad uses of information.

Concern for privacy existed long before the development of modern ICTs, but ICTs have changed the conditions under which privacy can be maintained by widening access to information. Privacy has increasingly become the subject of public policy and law as new technologies and their wide-ranging use have posed new surveillance risks (Lyon and Zureik 1996). Although these risks are difficult to specify and measure, exposure to them appears to be unequally distributed across categories of class, race, gender, and spatial location.

Individuals who are data subjects and organizations that are data users can be involved in conveying or using personal information within and across countries. Privacy protection therefore requires a multi-level and multifaceted strategy which encompasses the interrelation of actions taken in several domains. These include national and international legal systems, self-regulation by data users, and steps taken by data subjects themselves. This involves complex patterns of rights, relationships, and responsibilities in an environment of rapid technological change in which there is considerable uncertainty about the means and ends of any controls adopted (Raab 1997).

Since the 1970s, many countries have enacted national privacy or data-protection laws which incorporate important and widely- agreed principles of fair information practice (Flaherty 1989; Bennett 1992). In the 1990s, the growing globalization of economic and other processes using advanced ICTs stimulated renewed efforts to establish binding regulations at the international level.

Privacy legislation typically establishes an official agency to enforce and supervise the law, to encourage good practice, and to publicize the importance of data protection and the availability of rights of redress. These regulators can become involved in disputes over interpretations of the law, over the meaning of 'privacy', and over privacy's weight when balanced against the objectives of governments, businesses, and individuals who are involved in data-using activities. Regulatory agencies are also concerned with inserting privacy considerations into public policy-making about new technological applications—for example, national identity cards, police computer systems, and road-pricing schemes—and with encouraging data users to improve any practices which have privacy implications.

Much also depends upon the policies and practices devised by organizations to abide by laws safeguarding the data they use in their daily business. This may involve the formulation of a code of practice, often in consultation with the regulatory agency. However, the implementation of such codes is frequently ineffective, because good practice costs money, requires a degree of technical expertise, involves staff training, and may require new ICT systems.

Some data users may also consider the results of good practice as intangibles whose importance is not easily appreciated in relation to the organization's overall goals. But good data protection depends on more than legislation and the actions of data users. Individuals themselves need to act to protect their own privacy by becoming informed of the risks they face and gaining knowledge of their rights and how to exercise them. This knowledge may enable persons to bring more effective pressure to bear on organizations to improve their practices.

ICT-related developments such as the identification of incoming telephone calls, electronic cash payments, direct marketing using specialized mailing lists, smartcards, and video surveillance are usually presented as overwhelmingly beneficial to individuals or the general public. However, their negative implications for privacy can also be highlighted to help people understand the full implications, take precautions where necessary, seek redress if harmed, and pursue political actions relating to the role of ICT in social and economic processes. Regulatory agencies and privacy or civil-rights pressure groups play an important part in assisting individuals in these ways.

Although it is difficult to evaluate the quality of its results (Raab and Bennett 1996), privacy protection is therefore an interactive process among many participants, whether as partners or as adversaries. Beyond that, the development of systems of encryption and their availability to individuals holds great potential for enabling persons to protect their own privacy and thus regain some control over their information (Chaum 1992).

However, privacy protection should not be viewed as just a matter for specific technological fixes for data subjects or data users. The broader social, political, and legal issues relevant to privacy require debate and policy-making on a broader canvas. In particular, privacy protection depends to a considerable extent on individuals, organizations, and governments learning about ICTs, their effects on society, and the opportunities for, and limits upon, the instruments of regulation in the midst of such rapid technological change.

References

Bennett, C. (1992), *Regulating Privacy* (Ithaca, NY: Cornell University Press).

Chaum, D. (1992), 'Achieving Electronic Privacy', *Scientific American*, 267: 96–101.

Flaherty, D. (1989), *Protecting Privacy in Surveillance Societies* (Chapel Hill, NC: University of North Carolina Press).

Lyon, D., and Zureik, E. (1996) (eds.), *Computers, Surveillance, and Privacy* (Minneapolis, MN: University of Minnesota Press).

Raab, C. (1997), 'Co-Producing Data Protection', *International Review of Law, Computers & Technology*, 11/1: 11–24.

—— and Bennett, C. (1996), 'Taking the Measure of Privacy: Can Data Protection Be Evaluated?', *International Review of Administrative Sciences*, 62/4: 535–56.

8 Knowledge Gatekeepers: ICTs in Learning and Education

Knowledge Access in Society

Education and training are widely regarded as being the key to social and economic development in the future, as indicated by the invention of terms such as the 'learning society' and 'knowledge society' (Drucker 1993: 17–42; Freeman 1996*a*), and the identification of education, science, and technology as the basis for a sound industrial policy (Nonaka and Takeuchi 1995; Melody 1996). These views are allied to a belief that investment in ICTs can nurture a virtuous cycle in which education supports innovations in the technologies, which in turn improve learning and education (Noll and Mays 1971: 2; Castells 1996: 68–9; Freeman 1996*a*).

For example, universities helped to develop research and educational computing networks, like the Internet, that have helped to enhance research and education. The United States' leading proponent of information infrastructures, Vice-President Al Gore, initially focused his efforts on building information infrastructures for education through the National Research and Education Network (NREN). The success of networking for research, particularly of the Internet, led the Clinton–Gore Administration to promote a new generation of high-performance computing initiatives, along with the end-of-the millennium goal of connecting all classrooms to the National Information Infrastructure (NII) by the year 2000. The US Secretary of Education, Richard W. Riley, even claimed that 'Computers are the "new basic" of American education, and the Internet is the blackboard of the future' (Cimons 1996). This focus on education is evident in many other advanced economies, such as in Britain, where education was a central aspect of ICT policy during the early 1980s (ITAP 1983). In 1997, the newly elected British Prime Minister Tony Blair promised that his government's three main priorities would be 'Education. Education. Education.' He also emphasized that the use of ICTs in education would be a key priority for the UK, which has among the highest proportions of schools in the world with access to computers (Carvel 1997).

Knowledge is a key resource. But it—like information—is not a new one, as has been acknowledged even by a leading popularizer of the concept of a knowledge society, Peter Drucker (1993: 41). The concept of a knowledge society itself has several weaknesses (Robins and Webster 1989*a*, *b*; H. I. Schiller 1996). One might expect, for example, that a knowledge society would be ruled by knowledge élites, although those in positions of power and authority have long been able to obtain expertise.

What is changing is how people—you, politicians, experts—gain access to knowledge, information, or expertise. The idea of a knowledge society could even be misleading if policy-makers conclude that ICTs create knowledge, rather than shape access to knowledge and expertise. It is more likely that ICT initiatives will be balanced with other educational priorities if educators and politicians focus on the value of ICTs as carriers which support tele-access, rather than as creators of knowledge (Kay 1991: 100), as discussed in Chapter 2.

Tele-access is why ICTs are central to shaping the future of education, research, and the sciences. Applications such as e-mail, multimedia communication, and the Internet—like the printing press before—are not just better black or white boards. They are changing the way we do things. They can undermine or support the role of traditional gatekeepers in education, such as teachers, but also foster new gatekeepers. ICTs shape who gets access to which knowledge producers and users—and blur the accepted distinctions among them. To develop these points more concretely, this chapter focuses on the opportunities and risks of tele-access in the classroom, in higher education, and in learning and education more generally. I then discuss some of the major factors that are shaping tele-access in education, which map well into the factors shaping tele-access across all sectors of society (see Box 1.1).

Reshaping Tele-Access in Education and Learning

Tele-access helps clarify the multifaceted role of ICTs in education, where they are more than just instructional technologies. The technologies cut across the different specialized institutions, skills, and people involved in each educational sector, such as what is called Kindergarten through to twelfth grade (K–12) and high school in the USA, or Nursery School to year 13 in the UK, as well as college and university education. ICTs are obviously important to the storage and distribution of information. However, ICTs are not simply a new multimedia encyclopaedia. Tele-access to people, services, and technology—not just information—is also of broad importance to education (Box 8.1), such as in helping to teach individuals how to learn (Tehranian 1996).

Box 8.1. The Interrelated Roles ICTs Can Play in Learning and Education

Information access

- searching, screening, and obtaining multimedia information
- drill and practice with immediate, personalized feedback
- visualizing and learning by doing

People

- networking with students, teachers, experts
- institutional networking among administrators

Services

- facilitating routine transactions
- packaging and distribution of educational services
- breaking down distinctions between producers and users

Technology

- learning about ICTs through routine exposure and use
- using ICTs to improve learning and education

Access to Multimedia Information

The Internet has highlighted the role ICTs can play in searching, screening, and obtaining multimedia information and in supplementing other educational media, such as textbooks. For example, in the weeks following the 1997 landing of the Pathfinder mission on Mars, images relayed from the Sojourn rover to the Jet Propulsion Laboratory (JPL) in California were seen by millions round the world over the Internet.[1] Similarly, video of a university lecture can be put up on the Web—Webcast—to enable it to be seen and heard anywhere in the world as it occurs.

Video technology, of course, also enables such events and other multimedia material to be recorded and stored for future viewing and use. For instance, schools in Cerritos, California, experimented with the provision of educational video on demand (VOD) for classrooms, delivered over a fibre-based telecommunications network. Video cassette tapes remain far more cost-effective. However, visionaries of broadband networks see a day when schools can search and retrieve a far wider array of high-quality multimedia materials, at lower costs, through electronic networks, creating an educational video jukebox for households and schools.

Digital imaging can do more than just allow more students to access multimedia documents and video from anywhere in the world. High-resolution cameras and digital enhancements can provide access to new information, such as better satellite images of the earth. One of the 'Initiatives for Access' undertaken by the British Library, for instance, was the digitization of the earliest known manuscript of *Beowulf*, which included images from passages of the poem that were hidden by damage that occurred in a fire in 1731 (Kenny 1994). Digital imaging provided access to more information about the manuscript, while also making it accessible worldwide.

A powerful means of coping with the huge and growing volumes of electronic information has been the development of hypertext versions of course material on CD-ROMs, such as one for PICT research and publications (Dutton *et al.* 1997), and on the Web. Researchers, teachers, and students can benefit from these tools for locating information relevant to their studies. A rapidly accumulating corpus of electronic news coverage, trade and professional magazines, scholarly journals, and books has increased the importance of tools for searching the Internet and the Web (see Box 8.2).

Box 8.2. ICTs for Searching Electronic Files

- On-line library catalogues, such as the British Library's Online Public Access Catalogue (OPAC), that enables searches by keyword, author, and title.
- Software that searches across a network of computers and databases for specified keywords supplied by a user and electronic documents that contain these words in their contents as well as their titles. Web Crawlers search the Web for specified keywords and provide feedback on their likely relevance.
- Menu-based search tools, such as Gopher software—developed at the University of Minnesota, where the gopher was the university's mascot—which is available on Internet servers, from where it will 'go fer' the specific information located on a menu-driven search path.
- Hypertext Web browsers, such as Mosaic, Netscape's Navigator, and Microsoft's Explorer, that help a user moving across linked Web sites and indices to locate information.
- Web pages, developed by any individual or organization, which embed hypertext links with other Web pages. Many Web services began as an individual's Web page.

Learning by Drill and Simulation

Early uses of computers in education focused on the development of instructional software that emphasized drill and practice, particularly in managing repetitive exercises in the basics, such as arithmetic, spelling, and grammar. Early drill and practice software generally failed to utilize the value of computers as a means for encouraging students to discover facts and ideas on their own (Miles *et al.* 1988: 205–12). This criticism has been addressed in the best drill-and-practice modern multimedia software by incorporating many game-like features and colourful elements of 'discovery software'. Modern multimedia ICTs permit immediate and personalized feedback that can be far more imaginative, flexible, and engaging than earlier generations of computer-aided instruction. These techniques are also of value to adults, as in grammar and style checks in some word-processing software.

However, educators have increasingly emphasized the value of ICTs in helping students and researchers at all levels to visualize and learn by doing. One of the leading exponents of computers in the schools (Papert 1980: p. viii) argued long ago that simulation is the primary advantage of the computer: 'The computer is the Proteus of machines. Its essence is its universality, its power to simulate. Because it can take on a thousand forms and can serve a thousand functions, it can appeal to a thousand tastes.'

ICTs now support an enormous variety of models, simulations, and games that permit students to discover how a process works, instead of only reading about it (Gell and Cochrane 1996; Schroeder 1996). For example, flight simulators have been used in 'virtual-reality' environments to train pilots (see Box 8.3). Hughes Training Incorporated in Texas has adapted such simulations to training railroad engineers, permitting an engineer to take a realistic drive on a route many times in varying weather conditions before actually driving a real cargo.[2]

Box 8.3. Virtual Reality (VR)

The term 'virtual reality' was coined in the late 1980s by John Lanier (Schroeder 1995: 387), a pioneer in human-computer interface design. He was describing the sensation of being in the middle of (as opposed to looking at) an artificial three-dimensional (3-D) world. A virtual reality is created when a user looks at a computer-generated image through 'eyephones'—a headset that positions computer displays very close to each eye—rather than viewing an image framed on a computer screen. Users control movement in this 3-D space with devices like a 'data-glove'.

Work on head-mounted human-computer interface designs was pioneered by Ivan Sutherland in the 1960s at MIT and Harvard University, funded by the US Defense Department's ARPA agency. Researchers at AT&T also worked on telecommunications systems using head-mounted audio and video systems, as well as force feedback, to create a sense of 'telepresence' (Noll 1972, 1976).

Early R&D of VR was supported by military and defence needs, such as for training pilots, and for the remote control of robotic systems to undertake tasks such as repairing the outside of a space vehicle or handling hazardous materials. Since the 1980s, efforts to commercialize VR have led to applications in other areas of training simulations and in entertainment, such as video arcades (Schroeder 1996). Realistic VR environments require a level of computing resources that has kept the cost out of reach for many potential educational and training applications.

There is an exploding variety of other models and simulations that can be applied to education and learning. Off-the-shelf software can be used by young children to design 'totally trendy' doll clothes (Digital Domain 1996), or by older children to simulate the development of a city, or navigate through the anatomy of the human body. PC pioneer Alan Kay (1991) has become a major advocate for using simulation as a tool for learning, for example, by working with students to construct and use a simulation of ocean ecology to teach principles of marine life and science.

Networking People

ICTs such as voice mail and the Internet can be used to support parent–teacher communication, student–student dialogue, and collaboration among researchers across space and time. For instance, teachers in a number of US schools can leave voice-mail messages for parents about the day's lessons and homework assignments. Parents can call in the evening to hear what their children covered during the day and leave questions or comments for the teacher.

Networking over e-mail and the creation of Web pages are promising ways to disseminate research, encourage children to write, and perform many other educational tasks. An early use of e-mail in one US school district created an opportunity for inner-city and suburban school children to overcome some barriers to communicating across the ethnic, socio-economic, and geographical divisions of the metropolitan area. Administrators in this district used a computer program to filter out profanity—but they quickly learned that many students could not spell even four-letter words!

E-mail, video conferencing, and other ICTs can be used to link students and specialists. Apple Computer, for instance, experimented quite early with the use of desk-top video communications for schoolchildren to ask questions of experts, who could respond to them when they were available. This idea is reflected routinely on computer networks when individuals post questions such as: 'Does anybody know about . . . ?'

This communication support is also of value to institutional networking. Teachers, students, and administrators can use e-mail or other ICTs to network with other schools for sharing ideas and coordinating joint events, projects, speakers, or other resources. In working with the head of an inner-city school within a distressed area of Los Angeles, I was impressed by how important she felt it would be to have e-mail as a tool to support such resource-sharing (Dutton 1993). For example, neighbouring schools could access each others' calendars of events so that children could participate in another school's concert, play, or exhibit. This potential was instrumental to the launch in 1988 of 'Big Sky Telegraph' in rural Montana, which foresaw the benefits now available more widely through the Internet (see Box 8.4).

Box 8.4. Big Sky Telegraph (BST): Networking Rural Schools

The sparsely populated Rocky Mountain state of Montana on the Great Plains in the north-west USA has less than 1 million people in an area of 147,138 square miles (381,087 sq. km.). It has continued to rely on one-room schoolhouses in rural areas.

In the early 1980s, many of these one-room schools had a PC, but they were used infrequently and for very limited purposes. This underutilization helped to inspire Montana educator Frank Odasz to propose a network that would use this equipment to link schools throughout the region. Teachers and students could then share resources, access libraries and information providers, and communicate with one another. In collaboration with a computer-conferencing proponent, David Hughes, and the financial support of several foundations, Odasz launched BST in January 1988 as a dial-up electronic bulletin board, providing facilities for e-mail, conferencing, information retrieval, and on-line training in the use of the system.

About thirty schools became part of the BST within two years, representing about a quarter of the targeted schools. In the years since its launch, the Internet and the Web have provided a means for these schools to access similar resources. Ironically, BST's rather homogeneous community of users might be dispersed over the Internet.

Source: Uncapher (1996, forthcoming).

Re-Engineering the Management of Educational Services

ICTs can facilitate routine transactions and services in education, just as they can in government and business. For example, students at many universities can register for courses over any push-button phone rather than having to wait in line during the days of regular registration. Such electronic services may not be as interesting and widely discussed as those tied to the provision of teaching, but they can be important both to the efficiency of services and in shaping the choices of students. For example, a student browsing the Web for courses, or to decide which university to attend, might make different selections—better or worse—than those students sitting down with their parents, friends, or teachers. This is one reason why teachers market themselves, and their courses on the Web, and universities seek to have a greater presence on the Web.

Packaging information in electronic forms can also create new ways to distribute and sell educational services. Parents can buy CD-ROMs to help their pre-schooler learn how to spell. Professors can subscribe to on-line services, like LEXUS-NEXUS, to facilitate their access to information.

Schools can obtain 'free' services like Channel One in the USA (see Box 8.5). Electronic services such as these, particularly those with interactive capabilities, are changing the producers of information, at the same time as they are helping to break down distinctions between the producers and users of information. For example, the users of an electronic forum also contribute to the creation of its content.

Box 8.5. Whittle Educational Network (WEN): Channel One

Chris Whittle developed a media business that thrived for a time by providing targeted audiences to advertisers. For example, Whittle Communications L.P. would provide a free satellite dish, related television monitors, and specialized programming to the waiting room of a dentist's office. Whittle could attract advertising by using this access to a captive audience with a known profile.

He applied this same targeted market model to American secondary schools with WEN. This was developed in 1989 to provide free equipment to schools in return for a guarantee that teenagers would watch twelve minutes of Whittle programming each day, during which time they would see news and ads targeted to their age group. Since schools could use the equipment for a variety of other purposes, this seemed like a fair trade to many schools. WEN invested more than $300 million in video equipment and programming, and was able to attract major advertisers such as MacDonald's and Pepsi. Many schools had previously allowed ads on their premises, such as corporate logos on facilities donated to the school, so they viewed this trade as an extension of the practice.

Whittle had grander ambitions to provide low-cost, private educational alternatives to students. However, in 1994 he sold Channel One to K-III Communications Corporation, which also publishes the *Weekly Reader* for students. Although Channel One was resisted by state boards of education, and generated controversy, it received numerous awards and has been adopted by thousands of US schools, reaching over 8 million teenagers by 1997.

Source: Woo (1997).

Access to Technologies: Learning about ICTs

At the K–12 levels, the routine use and integration of ICTs in the school curriculum enable teachers and students better to cope with and exploit PCs and other ICT tools in their everyday life. I am not suggesting, as some do, that teachers can simply put children in front of a computer and expect them to learn. However, guided exposure can enhance a student's awareness of ICTs, which is one important step towards learning how to understand and exploit tele-access. A leading politician and advocate for minority communities in Los Angeles, Mark Ridley Thomas, maintains that computers are one aspect of a contemporary definition of literacy.

A variety of initiatives are seeking to improve 'computer literacy' —proficiency in their use (see Box 8.6). The most widely publicized are initiatives to get computers, educational software, and broadband communications into the schools to support students' access to such services as multimedia CD-ROMs and the Web. In the UK, private firms, such as CRT Group plc, and the telecommunications companies, as well as the public-financed British Broadcasting Corporation (for example, through its BBC microcomputer), and various forms of parent and government support have sought to provide greater access to computers in the schools.

Box 8.6. Computer Literacy

Policy-makers have sought to encourage education about computers for decades, often using the term 'computer literacy' to underscore the degree to which knowledge about computers will be increasingly important within modern societies.

For example, advisers within the US Office of Science and Technology set up under President Richard Nixon urged: 'If a nation as a whole is to benefit from the efficient use of computers, then a population conversant and knowledgeable about computers and computing power—in effect, a computer literacy—will be required' (Noll and Mays 1971: 1).

As simple as this concept may seem, it has created much controversy along two related themes:

1. an inability for experts to agree on what people should know to be computer literate; for instance, must they be able to program a computer, or only know enough to use and consume computer services produced by others?
2. concern over the appropriation of the concept of literacy, which is almost universally valued, by a field that might divert educational resources away from reading and writing to the study of technology.

Given these definitional and symbolic issues, many prefer to speak of teaching a 'proficiency' in the use of ICTs for particular tasks, rather than 'computer literacy'.

Source: Dutton and Anderson (1989).

Within colleges and universities, there are parallel efforts to involve students and teachers with the development and utilization of advanced technologies. In Europe, for example, BT's broadband Joint Academic Network (SuperJANET) links higher educational institutions in the UK. In the USA, the National Science Foundation (NSF) has supported a variety of high-performance computing initiatives, such as the Consortium for Education Network Initiatives in California (CENIC) to design and deploy CalREN-2, an advanced network for linking state educational institutions to each other and the national high-speed information superhighway.[3] The success of the universities in nurturing the Internet and mining its capabilities has reinforced investment in advanced network initiatives, such as the Internet2 and SuperJANET.

Social Factors Shaping the Value of ICTs in Learning and Education

Despite these opportunities for tele-access, and the momentum behind ICT initiatives in education, there is a countervailing view that devoting scarce resources to ICTs will seriously undermine the quality of education (Robins and Webster 1989*a*; H. I. Schiller 1996: 27–41), turning many universities into 'digital diploma mills' (Noble 1998). In California alone a 1996 state commission argued that $11 billion should be budgeted to give students from kindergarten through to twelfth grade access to computers and the Internet by the year 2000 (Woo 1996). Concern about this level of investment has been a source of growing opposition to pushing high-tech on schools at all levels (Hecht 1997). As one critic put it, 'the Clinton Administration has embraced the goal of "computers in every classroom" with credulous and costly enthusiasm' (Oppenheimer 1997: 45). A Stanford education professor, Larry Cuban (1997), concludes that the push to wire schools has 'enriched high-tech companies but produced underwhelming results for students'.

Many critics and proponents of ICTs in education, like other fields, seem to be inhabitants of the 'certainty trough' (see Essay 2.3). This recognizes that the implications of investments in ICTs for learning and education are more uncertain than the confidence conveyed by the technologies' most avid promoters and critics, as the role of ICTs is not predetermined by the nature of the technology, but will depend on social and institutional responses to innovations in tele-access.

Limited Technological and Educational Paradigms

As in business and government (Chapters 5–7), new paradigms are needed to make effective use of ICTs in learning and education. The film projector, programmed texts, computers, radio and television, the video cassette recorder, PCs, and, more recently, the Internet have yet to revolutionize learning and education in ways that proponents had forecast (Robins and Webster 1989*a*). Instead, each technology has been adapted to play relatively marginal roles within traditional paradigms of the educational process. Traditional educational paradigms are under challenge, but no potential replacements have been widely accepted.

Without a new paradigm, educators are likely to use ICTs to do things the way they have always been done, but with new and more expensive equipment. For example, field research on how computers were being used in classrooms in the Manchester, UK, area in the early 1990s found that schools often sidelined PCs against the walls of the classroom, where their use primarily for copy typing failed to teach students even how to exploit word-processing software to compose and revise their work (Croft 1994). Similarly, many lecturers use a PC with software like PowerPoint in the classroom as nothing more than a high-powered overhead or 35mm slide projector.

The realization that emulating traditional practices with new equipment contributed to the productivity paradox was one impetus behind BPR in the business setting (see Chapter 5). Educators may also need to redesign how they do their work in order to achieve the potential benefits of ICTs. However, in the case of education, the new paradigms lack credibility and need to be better developed.

One paradigm is technology-driven, and has no clear foundation in an approach to education and learning. Many simply urge schools at all levels to modernize their technology, which often means making them ubiquitous. Universities compete to be on the list of most 'wired' campuses. One university advocates the installation of one computer port per pillow (PPP) in their dormitories.[4] A problem with a technology-driven strategy is the risk of 'overscaling', which could be as inefficient as 'underscaling' and which few educational institutions can afford to do (Schramm 1977).

A more educationally anchored paradigm regards ICTs as 'teaching machines'—substitutes for teachers—as opposed to tools for teachers and students to use in some of the ways outlined above (see Box 8.1). This belief originated with optimistic views about how TV would revolutionize education. For instance, when I was a college student in 1966 I took an introductory course in psychology taught by a professor who had died several years before. I joined about 200 other students each morning in a darkened auditorium to watch his recorded lectures, which we could discuss later with the teaching assistants. Although the recorded lectures were well prepared, organized, and presented, they did not succeed for me and many others of my fellow students. I cannot even recall his name now—and I never took another psychology course at that university.

Today there is a similar level of optimism placed on the computer and the Internet as teaching machines. The interactive nature of computers give them a 'holding power' and enable them to be more self-directed than many other ICTs, such as TV (Papert 1980). Some teachers have found that challenged students have often made miraculous progress on a computer. However, teachers generally find that computers do not improve the performance of most students (Cuban 1997), who usually also need other forms of human direction. Educators have labelled this the 'fingertip effect'—the false hope that you can sit a person in front of a computer and walk away (Perkins 1990). Over twenty years ago, Wilbur Schramm (1977: 13–14) reminded those who looked at TV as a teaching machine that 'almost all teaching is multimedia . . . with an active teacher plus textbook plus much more'.

A more contemporary paradigm accepts the need for teachers to use ICT support to enable them to switch more towards a role of facilitator, or coach—rather than lecturer, or authority (Gell and Cochrane 1996). As information and expertise is increasingly accessible over electronic networks, it is argued from this paradigm, the teacher's prime function will be to help students navigate through and interpret that wealth of knowledge. In many respects, this extends a traditional role of education in helping students 'learn how to learn' (Tehranian 1996). But at the same time, this perspective seems to dismiss the degree to which teachers have still had to teach despite the availability of growing numbers of books, libraries, films, and other instructional technologies. However, this paradigm has started to shape policy and practice in education. For example, one professor at my university assigns no 'reading' for his doctoral students, but encourages them to pursue their interests over the Internet.

One risk of this approach is that teachers are encouraged to move their vital gatekeeping role to others. In a world exploding with electronic and published information, a positive gatekeeping role will become more critical than ever in determining what educational material is best for students. ICTs can shift this educational gatekeeping role in two very different ways.

On the one hand, some emerging educational technologies can give more authority to top administrators. An example of this shift is provided by Whittle Communications' Channel One (see Box 8.5). By broadcasting to the nation, Whittle could generate the advertising revenue to put substantial investment into high-quality production to create more profssional television productions than local schools could afford to do on their own. This creates an incentive for ever larger economies of scale, which means the gatekeeping function could become more concentrated, and further removed from the teachers.[5]

In contrast, the Internet could shift gatekeeping in a quite different way, while just as surely diminishing the teacher's role in the choice of content. Using the Web and hypertext search tools, students can follow their interests wherever they may lead, escaping the designs of a teacher as well as any author. One reason why people speak of Web 'browsers' and 'surfing' the Internet is that many users do not read anything. Instead, they scan and download images as they click from one hypertext link to another. However, logical arguments and complex problems are often best approached through sustained attention along a linear path, as is encouraged by a book or even sitting down with a blank piece of paper.

A focus on tele-access—as one aspect of any paradigm for the use of ICTs in education—can alert educators to the basic point that more information *per se* is not necessarily adding value to education. All educators need to be alert to how ICTs reshape tele-access, including who plays the role of gatekeeper to educational material. Schools should try to influence access to the outside world in ways that will enhance education. That might mean shutting the windows on some occasions and opening them on others. Schools, libraries, museums, and universities need walls as well as windows —and responsible gatekeepers, who are accountable for their choices.

The Geography of Learning and Education: The Virtual Campus

Any new paradigm of learning and education needs to incorporate the changing boundaries of educational institutions made possible by ICTs which make learning and education more independent of time and place. Online uses of ICTs have begun to raise more fully the promise of distance education, which since the late 1800s has sought to provide access to classrooms for rural, working, handicapped, mature students, and others who found it difficult to be educated at a particular location at certain times (see Box 8.7).

The geography of education and learning can change through such innovations as:

- using Webcasts in the same way videotapes have been used to make some of the most gifted teachers accessible to more students;
- making published work more widely available by putting it into electronic as well as print forms;

Box 8.7. Distance Education

Since the late-1800s, colleges and universities have offered opportunities for students to take courses from a distance, using the mail, telephone, radio, TV, and, more recently, networks, PCs, and CD-ROMs. Some examples include:

- The UK's Open University (OU), one of the world's leading organizations offering opportunities for mature students to study and take degree-level courses from their homes. Based at a campus in Milton Keynes, north of London, the OU delivers its study material using multiple media, from the mail to TV, and computer conferencing. in conjunction with attendance at residential summer schools (McLain 1994).
- The University of Wisconsin pioneered distance educational opportunities in the USA. The state's large rural population created incentives for providing continuing education courses, many of which employed audio conferencing systems over the telephone to facilitate interaction between instructors and groups of students at remote sites.
- The Instructional Television Network (ITN) in the School of Engineering at the University of Southern California has offered one-way video, two-way audio, interactive courses for engineering students since the early 1970s. Originally delivered by microwave to students working in companies in the LA metropolitan area, ITN has expanded via cable and satellite provision of its offerings (Nilles *et al.* 1976).

- employing computer-conferencing software and e-mail to enable students to communicate about specific topics with other students and instructors between classes and without being in the same place.

The notion of a 'virtual' university or classroom is based essentially on this idea of ICTs eliminating the necessity for students to be physically present on a campus or in a classroom (see Box 8.8). At the extreme, some virtual-university proponents believe that media like the Internet can serve the functions of the traditional university by substituting for face-to-face education and learning (Noam 1995; Denning 1996). However, as in discussions of the virtual organization and the changing geography of the firm (Chapters 5 and 6), this perspective often fails to recognize the importance of tacit knowledge and values that cannot be codified well and distributed over the Internet (Faulkner and Senker 1995: 200–12; Lamberton 1997: 74–5). It also underestimates the degree to which learning can depend on factors that can be undermined by mediated communication, such as inspiration, socialization, a sense of personal obligation, and accountability (Tehranian 1996). For instance, the OU finds it critical that students attend

Box 8.8. On-Line Educational Initiatives in Higher Education

- Globalwide Network Academy (GNA) claims to be the first on-line university. Incorporated in Texas, GNA has no physical centre (McLain 1994).
- The UK's Open University has used computer networking and multimedia CD-ROMs to complement distance education offerings (Box 8.7).
- University of Phoenix's private, San Francisco-based 'on-line university' began in 1988, and graduated its first students in 1992 (Garson 1996: 404–5).
- In 1994 the University of Southampton began to offer the first B.Sc. degree in Information Engineering in Britain to be conducted over the Internet.
- The Center for Innovation in Engineering Education, established in 1995 at Vanderbilt University, set up the Asynchronous Learning Network (ALN), with corporate sponsorship, to create Web-based tools to support university engineering courses, such as at the University of Illinois, Urbana-Champaign.
- Governors of eleven states in the western USA developed a consortium in 1995 to offer a regional on-line college. In 1996 California's Governor Pete Wilson launched a separate Californian consortium combining existing offerings from the state's colleges and universities (Wallace 1996).
- In 1998, Oxford University launched an on-line degree programme at postgraduate level.

intensive sessions on its Milton Keynes campus to support distance education. ICTs could even make the geography of education and learning more important, because the need for face-to-face communication, rather than access to information, should become even more central to the choice of an educational institution.

Regardless of the fate of on-line universities, the geography of education is already changing. For example, the diffusion of ICTs among households has made the home an increasingly important locus for learning about ICTs and using educational software. As a complement to a teacher or a parent, some CD-ROMs can be of value to a child learning to read, write, add, and subtract, thereby offering educational advantages to families with computer facilities in their homes.

Economic Resources and Constraints: Knowledge Gaps

The educational value of having a computer in the home illustrates how the cost of access to such ICT capabilities could widen, rather than narrow, knowledge gaps in society (Miles *et al.* 1988: 205–12). Disparities can grow even if an ICT is universally available. There are many homes without books. Similarly, there is educational material on non-subscription television in many countries, but it can still widen disparities. For instance, studies in the USA have found that those households with a college education were twice as likely to watch public television once a week—and therefore be exposed to educational versus commercial programming—than were families whose formal education stopped in grade school (Comstock *et al.* 1978: 119). Educational TV programming will not only be watched more often by children in households with more educated parents, but also be more beneficial to children who already have greater verbal and analytical skills, and are higher on a learning curve.

In the case of emerging ICTs like the multimedia PC, access remains far from universal. In distressed neighbourhoods of many urban areas, for example, there are children who are never exposed to ICTs like a PC, unless it is at school (Dutton 1993). Therefore, addressing the issue of access to technology underlying knowledge gaps is of crucial social significance. However, as the Californian 'NetDay' events (see Box 8.9) have indicated, even when government and private industry promote initiatives to build

Box 8.9. NetDays: Wiring America's Schools

Promoted as a 'high-tech barn-raising', 9 March 1996 was designated California's NetDay [http://www.netwday96.com]. Inspired by President Clinton's goal of connecting all classrooms to the Internet, this effort enlisted nearly 20,000 volunteers—including President Clinton—to install the interior wiring necessary for about 3,500 California schools to gain access to the Internet (*Los Angeles Times*, 2 June 1996).

NetDay also sought the commitments of telephone companies to provide schools with free dial-up Internet access. Sponsors, ranging from major corporations to individual parents, donated funds for NetDay 'kits' that included wire, jacks, patch panels, and other equipment essential for installing a local area network that connects classrooms and the school library for Internet access. Interactive Web sites posted volunteer information, using a clickable map of the state to permit individuals to discover sites where they could volunteer their time. The Web was also used for on-line sign-up and to organize face-to-face meetings of the volunteers, with toll-free telephone numbers used to recruit volunteers. A second effort—NetDay2—followed in October 1996, wiring an additional 200 schools.

Although used as a model by Vice-President Gore and others, NetDay raised concerns over the degree to which disparities in access were widened by the greater volume of volunteer efforts in wealthier school districts.

ICT infrastructures, like wiring schools, prevailing disparities can be reinforced because volunteers come forward for the wealthier schools.

More targeted public initiatives are therefore necessary to redress this imbalance, such as setting up technology centres to make PCs and other ICTs more accessible to teachers, parents, and students in rural and distressed areas of inner cities. In Los Angeles, for example, the TeleVillage (see Box 7.9) provides access to PCs and the Internet as well as training.[6] Joseph Loeb, President of Break Away Technologies, a computer training centre launched in the wake of the 1992 disturbances in Los Angeles, thinks of computers as 'a great equalizer' (Gold 1997). But this is impossible without access to the technology.

Conceptions and Responses of Educational Users

PICT researchers Geoff Cooper and Steve Woolgar (1993) show how conceptions of the audience can have a major impact on the design of ICTs. In education, this process is made more difficult because it is often driven by weak conceptions of the user in two major ways.

First, students are often stereotyped as ready-made users of ICTs because they grew up surrounded by the technology, while teachers are perceived as being caught in the old, linear print generation (Tapscott 1997). This generational stereotyping stems, in part, from studies based on observations of the interactions with computers by exceptionally bright children of well-educated parents, such as the 'child programmers' of MIT faculty and staff (Papert 1980; Turkle 1984). It is also anchored in anecdotal observations by engineers and scientists whose kids grow up with computers as a part of their everyday life. However, computer proficiencies and attitudes are far more varied among the young and old than portrayed by these generational stereotypes.

The way many children initially associate computers with TV might help to account for children's attraction to ICTs (Greenfield 1984: 127–54). Multimedia presentations such as IBM's *Ulysses* have demonstrated how hypermedia could be used to help jump start a child's interest in literature or music that they might otherwise never explore.[7] What has been called 'edutainment' or 'infotainment' can take advantage of popular interest in video by embedding lessons in anything from soap operas to interactive computer games. However, this approach can underestimate the sophistication of children and adults as consumers of video programming. People accustomed to video arcades and flipping through dozens of TV channels—watching TV rather than a programme—are not automatically captivated by 'infotainment' unless it is outstanding.

A second way in which ICTs are challenging conceptions of the user is by the way they can 'reconfigure' relations between users and producers (Woolgar 1996). For example, I worked on a university committee that sought to provide our alumni—graduates of our university—with better electronic access to information, such as through use of the Internet. It became immediately apparent that ICTs permitted the alumni to be not only a consumer audience, but also producers of information for others, contributing knowledge in their disciplines and locales to enrich a university's pool of expertise and experience. This indicates the potential obsolescence of the old paradigm, which viewed the faculty as the producers of information and students as an audience, not active participants in education. This is a gross misconception of the way in which knowledge is distributed among faculty, students, and alumni.

Educational Institutions and Policies

Leading advocates of educational innovation have criticized schooling at all levels as a 'holdover' of the industrial age (Toffler 1970; Williams 1982: 215–26). Institutional change in this area is a major challenge because educational institutions, from kindergarten to advanced research centres, are generally very conservative. Lecture and discussion in many contemporary classrooms still follow a very traditional chalk–talk format. Many law professors, for example, pride themselves on their ability to employ a Socratic dialogue—a technique for posing questions and pursuing the implications of answers in order for students to discover, rather than being told, the answer.

There are a variety of reasons why education has been slow to re-engineer itself. One is the degree to which excellence in teaching remains an art that cannot be simulated by machines. People still speak of 'gifted' teachers because no one knows how to teach others how to emulate their performance. Some of the most creative developers of educational software have spent a great deal of time observing gifted teachers in the classroom setting.[8] More incentives should be provided for gifted teachers to work with software developers.

Another major challenge is the degree to which ICTs can be used to 'reduce educational costs per credit hour' (Garson 1996: 403) by cutting staff numbers and improving efficiency. Many administrators see the potential for ICTs to be used to allow fewer staff to teach more students using business management paradigms such as the virtual organization and outsourcing (see Chapter 5). However, critical analysis of the assumptions underlying many proposals for downsizing, such as through on-line universities, suggests that the proponents are also underestimating the resources necessary to produce high-quality educational materials using qualified teachers and multimedia producers (Garson 1996; Noble 1998).

That said, there are many institutional arrangements and policies in education that constrain more innovative uses of ICTs. For example:

- The incentive structures in most higher educational institutions continue to reward narrow definitions of scholarship, thereby failing to encourage work with new media—such as in authoring or collaborating on the development of educational software—or in reaching new audiences.
- Copyright and Intellectual Property Rights are a barrier to the dissemination of educational material. British universities, for example, have often waived their rights to teaching materials produced by their faculty, but the prospect of expanded access to virtual universities might lead many universities in North America and Europe as well as Britain to assert such rights (Weston 1994).
- Concerns over maintaining the security of computer networks and systems prevent many educational institutions from providing more open access.
- Many educational institutions depend heavily on the provision of software, databases, and services provided by private industry under licence agreements that restrict their use to regularly registered students and staff.
- Educational networks and institutions have a need to guard the rights of their members, such as in protecting their personal privacy (Katz and Graveman 1991).

In some cases, a lack of competition might also be a brake on innovation. However, in opening more opportunities for tele-access, ICTs have created more options and competition among schools, including from outside the traditional education sector. Computer manufacturers, software companies, and media businesses like Whittle Communications see market possibilities in the provision of multimedia education. Private educational offerings, such as Phoenix's on-line university (see Box 8.8), create a long-term threat to traditional institutions. The provision of distance education by traditional universities will put them into more competition with one another for new markets. And all kinds of private industries are undertaking their own education and training programmes which increasingly compete with the continuing education ambitions of technical institutes, colleges, and universities, such as Motorola University in the USA. BT in the UK, for example, has created its own 'virtual university' (Gell and Cochrane 1996: 260).

The Broader Ecology of Knowledge Access

ICTs can 'open doors to a learning society'. Students can become their own gatekeepers on the Internet or in the library. But, in the classroom and in their courses, teachers need to select which doors should be opened and closed to achieve their ends. Students and educators do not need more of everything, but control over access to the best information, services, and people. If educators and politicians understand ICTs as tools for tele-access, as opposed to teaching machines, they might balance their priorities better.

Another theme underlying this chapter is that the role of ICTs in learning and education will be highly dependent on the course of development in other sectors of society, such as government, private industry, and the household. The boundaries of education are being eroded by new ICTs, new paradigms for education, a changing geography, and revised conceptions of knowledge producers, users, and gatekeepers. For example, the role of ICTs in education will depend heavily on the degree to which households provide more or less equitable opportunities for tele-access, an arena of activity that is the focus of the next chapter.

Notes

1. JPL recorded an average of 40 million–45 million 'hits' each day, with a peak of 80 million hits on 7 July 1997 (Yates 1997). A 'hit' is a record of someone accessing a Web page over the Internet. A single access to one page can be recorded as multiple 'hits' if the page embeds more than one image, so hits are not an accurate reflection of either the number of users, the number of pages accessed, or whether the information is even viewed for any length of time.
2. The engineer of a train must often begin braking miles before an upcoming curve is visible. Simulations provide a type of apprenticeship that permit the drivers to memorize a route before they actually take control of a real train.
3. This consortium was formed by the University of California, the California Institute of Technology, the California State University, Stanford University, and the University of Southern California.
4. This acronym is a play on point-to-point protocol (PPP), which is one software scheme for creating a connection between two computers, such as using a modem and phone line to dial in and connect to a computer at another location.
5. There are many other and more subtle commercial transactions that could influence the choice of content, such as professors receiving a free 'desk copy'

of books that they require for a course. However, the expense of big media, such as television and multimedia productions, could more seriously undermine the gatekeeping role of teachers whose choices are dictated by major state and national choices of content.

6. Also the County of Los Angeles, with gifts from AT&T, has funded the development of computer learning centres. By early 1997, ten of twenty five planned learning centres had been established (Gonzalez 1997).
7. IBM funded the production of educational hypertext treatments of Tennyson's poem *Ulysses* and the discovery of America by Columbus.
8. For example, Apple Fellow Alan Kay has been studying gifted teachers as an approach to the development of educational software.

Part V

The Virtual City: Shaping Access in Everyday Life

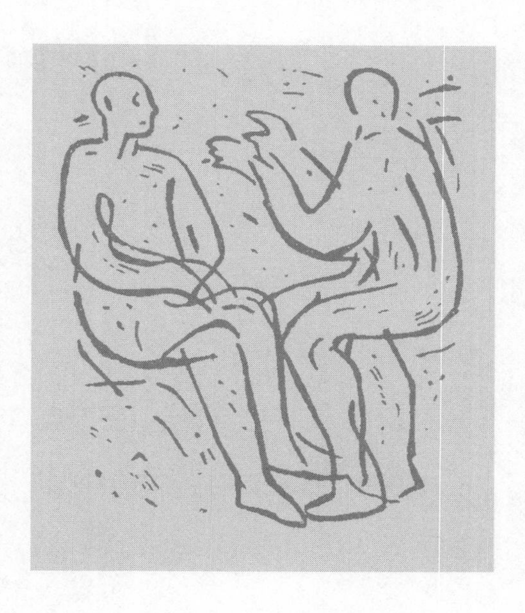

9 The Intelligent Household: For Richer or Poorer

Technological change evokes dramatic responses from proponents and critics alike because it 'entails a change in a network of social relations' (Woolgar 1997: 7). This is illustrated by examples in this book of the way tele-access intertwines with changing social relationships in a variety of settings. However, the implications of ICTs are possibly nowhere more threatening or promising than in households and communities, where relationships defining day-to-day life are at stake.

This chapter looks at the adoption and 'domestication' of ICTs among households and the public at large, establishing a foundation for discussion of how the media seek to wire households for access (see Chapter 10). It shows how choices in the home are shaping access, and how social and economic constraints structure these choices.

The ICT Revolution Hits Home

Most households in the USA are on-line, at least in the sense that over half have used a computer at home, work, or school.[1] Triggered in large part by the popularity of the Internet and the Web, the number of Americans joining an on-line service grew from '5 million in the winter of 1994 to nearly 12 million by June' 1995 (Times Mirror 1995: 1). This near doubling of Internet use continued from 1995 to 1996 (see Table 9.1). By 1996, almost ten percent of Americans over 17 years old had accessed the Internet (Mediamark Research 1996). Such growth in the popularity of the Internet has been mirrored in other nations, bringing estimates of the number of Internet users worldwide to 35 million.[2]

Table 9.1. North American Access to the Internet and the Web, 1995–1996

Percentage of respondents	Aug. 1995	Mar.–Apr. 1996
With access to the Internet	14	22
Who used Internet in last 3 months	9	15
Who used Web in last 3 months	6	12

Source: Nielson (1997), stratified random telephone survey of US and Canadian households. 4,200 interviews were conducted in 1995, with two-thirds of this panel reinterviewed in 1996.

Before the mid-1990s there was a comparatively gradual adoption in households of computers and on-line services, such as Compuserve and America Online (AOL). Early adoption rates of home computers paralleled the diffusion of cable TV in its early years, in contrast to the much faster diffusion of black and white, colour TV, or the VCR (see Figures 9.1 and 9.2). Public interest in the Internet and the Web led to a substantial increase in the rate of computer adoption among households, much like

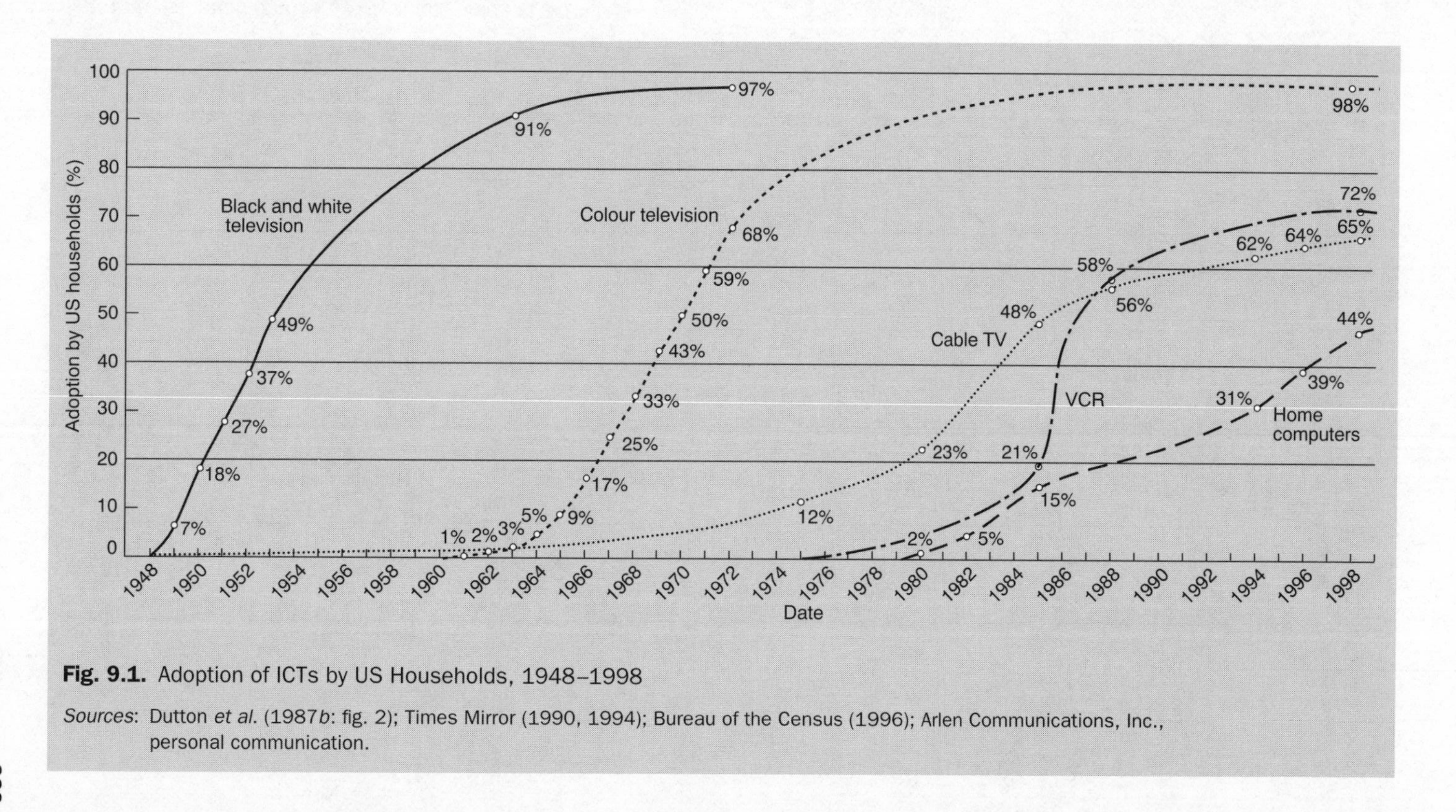

Fig. 9.1. Adoption of ICTs by US Households, 1948–1998

Sources: Dutton *et al.* (1987*b*: fig. 2); Times Mirror (1990, 1994); Bureau of the Census (1996); Arlen Communications, Inc., personal communication.

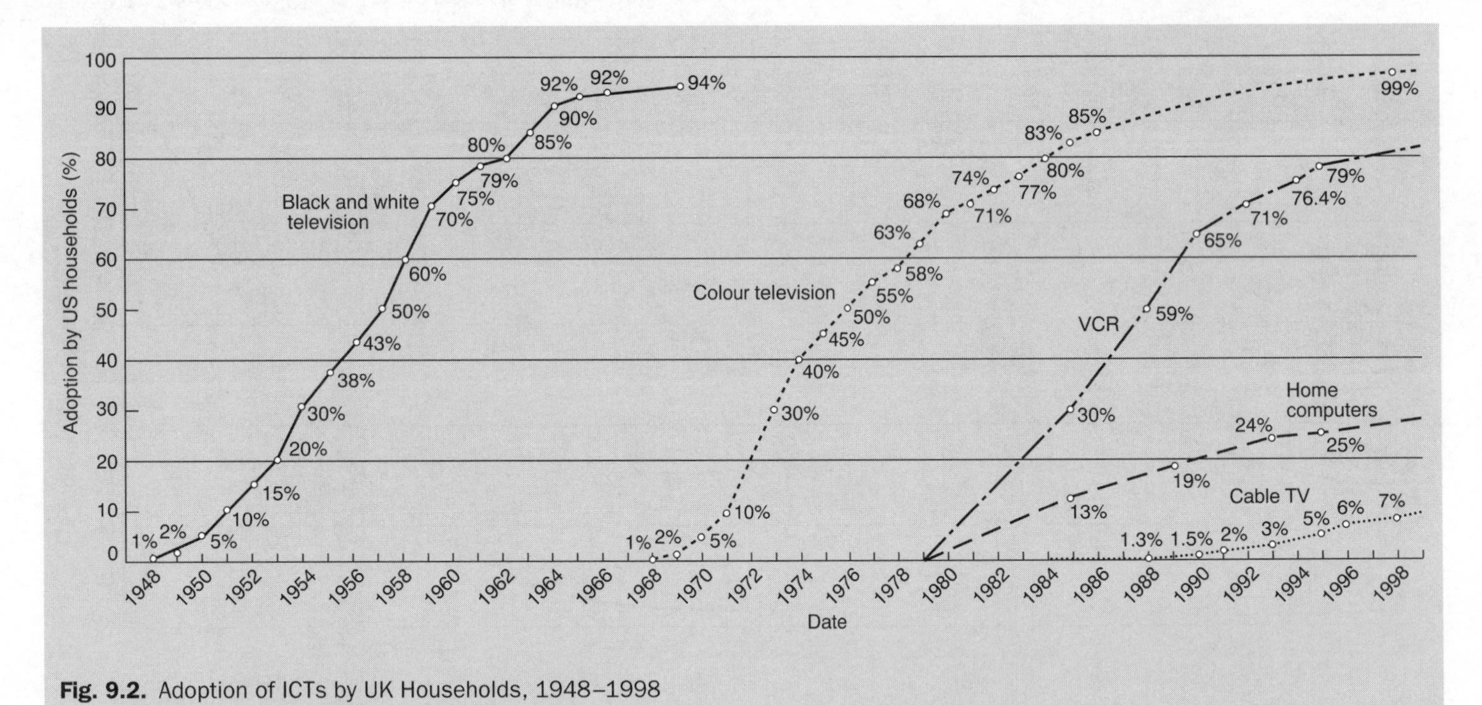

Fig. 9.2. Adoption of ICTs by UK Households, 1948–1998

Sources: Dutton *et al.* (1987*b*: fig. 3); OPCS (1993); Central Statistical Office (1996); ONS (1996, 1997); *Screen Digest* (1996–7); European Audiovisual Observatory (1997).

the increase in cable TV adoption among US households achieved once satellite-delivered programming was ushered in during the late-1970s.

Despite an increased rate of PC adoption, a substantial proportion of households do not have a computer in their home. An even smaller proportion are able to access on-line services from their home. The social implications of this division among households is related both to the characteristics that differentiate adopters from non-adopters, and how households with similar social and economic profiles vary in the ways they fit ICTs into their day-to-day lives.

Domesticating ICTs

Conventional discussions of the diffusion of ICTs often neglect the degree to which households and the public at large are themselves 'domesticating' ICTs to serve their own particular values and interests through their everyday choices in consuming ICTs (see Essay 9.1). Social research on TV audiences and PC users has underscored the active role that consumers take in deciding how to use these technologies, and how to interpret the messages they convey (Silverstone and Hirsch 1992; Cawson *et al.* 1995). In doing so, the public actively shapes tele-access, sometimes by design, but often inadvertently in the pursuit of other values and interests.

The negotiation of tele-access is an important factor in many households, as demonstrated in the 'politics of the remote control' as children, teens, and parents seek to watch their favourite programmes. The average American household has more than two TVs, which is one strategy for resolving these conflicts. However, the addition of VCRs and multiple TVs to the mix has also expanded the range of issues tied to access in the household. Increasingly, families must negotiate when and where TV is watched, as well as what programming is suitable for children. Cable 'lock-boxes' can be used to keep children from watching an adult channel, and the V-chip (see Box 3.6) to help parents regulate access to violent or sexually explicit programming.

Lock boxes and V-chips also raise new issues in the household over whether and how these ICTs should be employed. Many households consciously regulate access over and through other ICTs other than TV, such as books, answering machines, services that screen and return telephone calls, software allowing children to use their parents' PCs while preventing them from destroying electronic files or accessing pornographic sites on the Web, and subscriptions to cable or pay TV services which are suitable for children.

Household ICT Paradigms

Decisions about where to put TVs, telephones, and PCs can influence how and when they are used. The television can be a centrepiece that the family gathers around to watch, or a bedroom fixture. The single telephone in the hallway has given way to multiple fixed and cordless phones that permit calls from anywhere around the household.

These choices are structured by culturally accepted beliefs and attitudes about the organization of the household that are comparable to management and ICT paradigms in business organizations. There are also explicit efforts to promote new ICT paradigms for the household. For example, proponents of the 'intelligent home' have offered a variety of very different visions of how ICTs should be deployed and interconnected within the household (see Box 9.1). Acceptance of one paradigm could lead a family onto the information superhighway, while another could put them on an information cul-de-sac. However, visions of the intelligent household do not come out of thin air, but are most often based on prevailing technological paradigms. The mainframe era spawned images of a centralized

Box 9.1. The Intelligent Home of the Future: Alternative ICT Paradigms

1. *Centralized intelligence.* The centralized control paradigm of mainframe computers is reflected in early visions tying a wide range of electronic devices throughout the home to a central computer that would enable the automated control and monitoring of all household functions.
2. *Distributed intelligent appliances.* The PC explosion was extended to all sorts of independently intelligent devices—PCs, TVs, telephones, fax machines, displays—being used throughout the 'smart' home.
3. *Electronic cottage.* Use of ICTs for telework rekindled romantic images of a return to lifestyles that blend work and family life in the household.
4. *Networked cyberHome.* The Internet has supported designs for plugging into networks from anywhere within the home over upgraded internal wiring that interconnects all kinds of electronic devices to information superhighways. Householders can also control lights and other appliances remotely, such as over a PC at the office, or a cell phone in the car.
5. *Information fortress.* One reaction to the information explosion has been the use of ICTs to build an electronic castle, buffering the household from unwanted interruptions, and channelling electronic access to areas such as the home office that do not interfere with family activities.

Sources: Miles (1988); Forester (1989*a*); Negroponte (1995); Gates (1995).

information utility for the home, while the Internet encourages many to conceive of a cyberhome being networked to the digital world.

There is an economic rationality driving new paradigms of the 'intelligent home'. For example, during the construction of a new home, the low marginal cost of wiring additional rooms for telephone and cable TV services makes it economical to wire every room, thus encouraging the use of multiple telephones and TV sets. Many families will put much thought into where to place a telephone or cable outlet in a room, because it will influence how the furniture will be arranged, but neglect to put as much thought into the decision of whether or not to wire a room at all.

Domesticating Space

Decisions about the allocation of space in a household can influence the use of ICTs. A decision to dedicate a room as the husband's office, but provide no private room for the wife, will make it more likely that the PC is placed in the husband's space and constrain its use by his spouse (Haddon and Skinner 1991). Such 'gendering of space' is but one aspect of how the use of space influences the way men and women, and girls and boys, use ICTs, whether it is in the home, classroom, or video arcade (see Essay 9.2).

While the arrangement of the household can influence the use of ICTs, the opposite is also occurring. The use of ICTs is shaping the design of homes. Just as households sought to accommodate print technologies, such as by installing bookshelves, more homes are being built to accommodate PCs and other ICT equipment.

The design of homes and their interior decorating can also be assisted by ICTs such as computer-aided design (CAD) and modelling. Off-the-shelf software permits homeowners to: 'Turn your 2-D sketch into a 3-D design you can walk through' (Softkey International 1997). This new technology does not only change the way households design their home. It can change the way consumers interact with architects and designers, as well as the final choices they make.

These are some of the many ways in which households domesticate ICTs and govern their impacts on society. The choices of households are not open-ended. For example, they are constrained by prevailing attitudes to the arrangement of the home, which varies across cultures, socio-economic groups, and time. In North America and West Europe, it used to be considered rude to use an answering machine. In the 1990s it is more likely that you would be criticized for not having one. The choices of consumers are also constrained by the influence of producers as well as by the larger social, economic, and political ecology outside the household.

Configuring the Consumer: The Influence of Producers

Generally, the producers of ICTs have more power to create and manage the use of ICTs than do individual consumers (Mansell and Silverstone 1996). However, the rejection of many ICTs, such as the video phone, demonstrates that consumers can also play a pivotal role in ICT developments (see Chapter 4). As Steve Woolgar explains, the degree of influence exerted by producers depends on their ability to 'configure the user' (Essay 12.2). For better or worse, producers can position ICTs in relation to users long before they reach the market place.

For politicians to represent the public's interest well, they need to understand all sides of issues important to business and industry. This was made clear by venture capitalist John Doerr in commenting on the value of Vice-President Al Gore's involvement in a monthly meeting of Silicon Valley entrepreneurs: 'if we're not in an ongoing conversation with the elected officials, we'll get . . . totally blindsided' (Shogren 1997). Consumers can also be blindsided if they and their advocates are not wired into early discussions of emerging technology and policy, as indicated by developments in Caller ID and digital TV.

Laying the Groundwork for Caller ID

Telephone companies had been developing the technology that underpinned call line identification for twenty years before introducing it to the public (see Box 3.5). Nevertheless, caller ID's arrival generated battles with consumer advocates over the propriety of transmitting a phone number with a call. These discussions were primarily centred on issues of tele-access, such as whether the right of the called party to be left alone should take precedence over the right of the caller to prevent the unauthorized disclosure of personal information (see Box 9.2).

The telecommunications industry was able to win this debate in the US courts and with regulators in Britain, for example, by introducing options for consumers to block their number from being transmitted with a particular call, or on a particular line, called per-call, or per-line blocking, respectively. It was also successful in defining Caller ID as a general way for consumers to protect themselves, as highlighted by one Caller ID ad campaign claim: 'Now you can stop talking to strangers' and have 'a safe way to answer your phone' (GTE 1996). This illustrates how producers can frame a product for the public years before most consumers learn of its existence.

Box 9.2. Consumer Tele-Access Issues Raised by Caller ID

1. *Privacy.* This ICT protects a person's right to be left alone, but only by infringing the caller's right not to disclose personal information (a phone number).
2. *Freedom of speech.* Access to the caller's phone number would prevent anonymity and stifle the use of phones for political speech, such as in expressing unpopular political views.
3. *Safety.* Caller ID would help identify the source of 'unwanted' or obscene calls, but also give a person's phone number to anyone called.
4. *Equity.* The cost is within reach of most households, but wealthier households can afford multiple telephone lines, caller display devices, and answering machines useful to protect privacy.
5. *Etiquette.* Caller ID simply enforces conventions of etiquette for callers to identify themselves, but etiquette can be exercised only if there is choice.

Sources: J. E. Katz (1990); Dutton (1992*b*).

Establishing a Case for Digital TV

Another example of the producer's role can be seen in the way the computer industry and broadcasters configured the TV consumer for the introduction of digital TV. While American broadcasters, politicians, and the computer industry compromise over the details of a transition to digital TV, it appears that the outcome of two decades of struggle over TV standards is likely to create a further divide between an A team and B team of TV viewers, with and without digital TVs.[3] Of course, politicians —and the regulators they appoint—are elected to represent the public's interest in such negotiations. Competitive markets should also push producers to further the consumer's interests. Yet the evolving debate over TV standards shows how producers can structure consumer choices that are further downstream, wielding influence and enjoying levels of access to politicians and regulators that are seldom matched by consumer groups.

Generally, producers orchestrated a national commitment to a new generation of advanced digital TVs well before market forces could exert influence on its course of development. Consumer interests could have played a role at several points in the establishment of digital TV standards. For example, the Federal Communications Commission (FCC) set up an Advisory Committee on Advanced Television Service in 1987 to ensure discussion that would include consideration of consumer interests as well as the views of industry. In fact, representatives of American industry used consumer protection as a justification for objecting to Japanese and European standards for analogue 'high definition television' (HDTV). They argued that the USA needed an HDTV standard that was 'downwardly' compatible to guarantee access to content by consumers without a HDTV set. Another consumer-based argument against HDTV was that early trials suggested that viewers did not really notice an improvement in the resolution of HDTV as compared with standard National Television Systems Committee (NTSC) colour TV in the USA (see Boxes 9.3 and 9.4).

Box 9.3. Standards for Conventional Analogue Colour TV

The USA developed the first nationwide standard in 1946, but committed the country to standards with lower resolution than nations that adopted later standards, such as throughout Europe.

- National Television Systems Committee (NTSC) standard of 525 lines per screen was adopted by the USA, Canada, Japan, and most of Latin America.
- Phase Alternate Line (PAL) standard from Germany of 625 lines per screen was adopted by most nations of Western Europe and Anglophone Africa.
- Sequential Colour and Memory (SECAM) system standard of 819 lines per screen developed in France was used in France and Francophone Africa.

Seemingly a technical matter, the adoption of analogue TV standards was infused with the high-stakes politics of international trade and industry in the production and distribution of consumer electronics (Crane 1979).

Ironically, given such consumer-oriented arguments against HDTV, the FCC committed the USA to advanced TV standards for digital systems that would be incompatible with TV sets in every American home. Many argued that TV sets must be brought up to date with new technology in any case, given the emergence of digital satellite channels competing with conventional over-the-air and cable TV systems. Inexpensive converter boxes—about $US 200—would be available also, so that consumers could keep their existing TV sets. Moreover, advances in digital compression technologies could permit the same frequency spectrum used to transmit one analogue colour TV signal over the air to carry up to four compressed digital TV channels to households—creating the potential to quadruple consumer choice.[4]

In the USA, digital TV standards were also believed to be in the interest of American industries, and therefore of value to the economy as a whole.

Box 9.4. High Definition and Digital Television Standards

Analogue high definition television (HDTV). Since 1972, Japanese industry saw higher resolution TV sets replacing existing colour TVs and developed the first set of standards for 'Hi-Vision' analogue transmission of TV signals, followed by Europe:

- Muse HDTV, Japan's standard for a more rectangular screen, such as in a movie theatre, was composed of 1,125 lines as proposed in 1981.
- HD-MAC was a European proposed alternative HDTV standard.

Digital television. In the HDTV debate, industry convinced the FCC to reject new analogue standards in favour of digital standards for 'advanced television' (ATV) that would permit a conversion to a new generation of TVs that could receive and display signals recorded by a digital TV camera, or digitally coded, compressed, and transmitted to homes. Digital compression standards have been promoted by the Motion Pictures Expert Group (MPEG). Digital signals could be displayed on multimedia PCs, on an analogue TV through a set-top box converter, or some new hybrid TV/PC, such as Gilder's (1994) 'teleputer' as well as on a digital TV, fuelling competition for creating the box that will entertain households.

Sources: Neuman *et al.* (1997); Noll (1998).

The USA had lost its base of TV-set manufacturers, but American semiconductor, computer, and software manufacturers could benefit from extending microcomputer technology to the TV set, creating a far greater mass market for computer equipment.

Despite public-interest arguments, the shift in standards has placed universal access to TV at risk. Those who can afford new TVs or set-top boxes to convert digital signals for display on an analogue TV set will have access to a different mix of programming than those who cannot afford these ICTs. Moreover, as the TV becomes more like a computer, it is likely to develop at a pace not dissimilar to the PC. The life expectancy of a new model of a PC is approaching six months before it is outdated by new technology. Over time, TV households will be distributed across a spectrum of low- to high-end users, who have the latest technology and features for interactivity, Web browsing, higher resolution, and screening content, creating a TV underclass.

To transition consumers to a new generation of TVs, the FCC agreed to give broadcasters additional spectrum to enable them to transmit two sets of signals, analogue and digital, until the public had made the conversion to digital TVS. As one analysis of this policy process concluded:

'Inadvertently . . . Americans may be required to replace 270 million analog TV sets by 2006. . . . All this is a result of the attraction of free spectrum, not advanced television technology' (Neuman *et al.* 1997: 223).

This conversion could take decades, if broadcasters ever do make this transition. This has led some to call this allocation a $70 billion windfall for broadcasters (Ratnesar 1997; P. Taylor 1997*b*). Broadcasters could use this spectrum in ways that do not provide better pictures, or multiply programming choices, but instead offer new revenue, such as for paging services, or subscription TV.

However, the impact could be mixed even for broadcasters. Some TV stations are concerned that the allocation of digital TV assignments will restrict the reach of their over-the-air broadcasts, and therefore threaten their share of the TV audience (Shiver 1997*a*). Broadcasters are also concerned that the computer industry will exercise undue influence within the administration's advisory committees and thereby on the public interest requirements placed on broadcasters in return for this free spectrum, such as requiring more children's programming, or free air time for political candidates (Shiver 1997*b*).

Social and Economic Constraints on the Consumption of ICTs

Threats to equity, such as that posed by digital TV, are among the societal risks most frequently raised by critics of the information society (H. I. Schiller 1981, 1996; Hamelink 1988; Mosco and Wasco 1988). A key concern is that new technology erects barriers for disadvantaged groups to gain access to the information and communication resources available to others. In this way, technological change will not only perpetuate inequalities, but widen gaps in tele-access between the haves and the have-nots.

Socio-Economic Status and Access to Computers

Research on the adoption and diffusion of ICTs among the public at large highlights widening gaps in society between the information rich from the information poor, such as between 'high-tech teleworker' and 'lone-parent' households (Silverstone 1996). Many factors are associated with the adoption of computers in the household, such as age, race and ethnicity, gender, the presence of children, and the use of ICTs at work (Dutton *et al.* 1987*b*; Vitalari and Venkatesh 1987; Wilhelm 1997). Nevertheless, the consumption of ICTs has been marked by inequalities (Mansell and Silverstone 1996).

The most consistently dominant set of factors associated with the early use of ICTs in the household has been tied to the socio-economic status of the household—the income, occupation, and education of the head of household (Dutton *et al.* 1987*b*). For example, the heads of households with a computer in the home are more likely to have a higher income, be employed in a managerial or professional occupation, and have a college education than those in households without a computer.

Education, occupation, and income are interrelated, so it is difficult to determine the precise nature of the relationship. Clarifying the role of each factor would help to assess alternative explanations for widening gaps (see Box 9.5).[5]

Box 9.5. Why Households Adopt Computers

Four alternative explanations have been offered for the relationship between socio-economic status and computer adoption:

1. *Media style*. Education, in particular, cultivates more active and selective use of media, and the adoption of active innovations such as PCs, video games, and on-line services as opposed to more passive activities, like watching TV.
2. *Learning culture*. Education might foster an interest in learning generally, and learning about ICTs in particular.
3. *Instrumental reason*. Occupation can drive the adoption of ICTs because certain groups, such as managers and professionals, have specific instrumental reasons for looking to ICTs to facilitate work at home.
4. *Economic inequality*. ICTs are general purpose machines, supporting work as well as entertainment, but the cost limits access by lower-income households.

Source: Dutton *et al.* (1989).

Cumulative Inequalities, Persisting over Time

Data from the late 1980s indicated that income had a more direct influence on early computer adoption than did either education or occupation (Dutton *et al.* 1989). Education and occupation were also important factors structuring the adoption of computing, but primarily because of the degree they contribute to the income of the household. This finding supports the model illustrated in Fig. 9.3, which lends credibility to the 'economic-inequality' explanation in Box 9.5 for the uneven adoption of computers in the home.

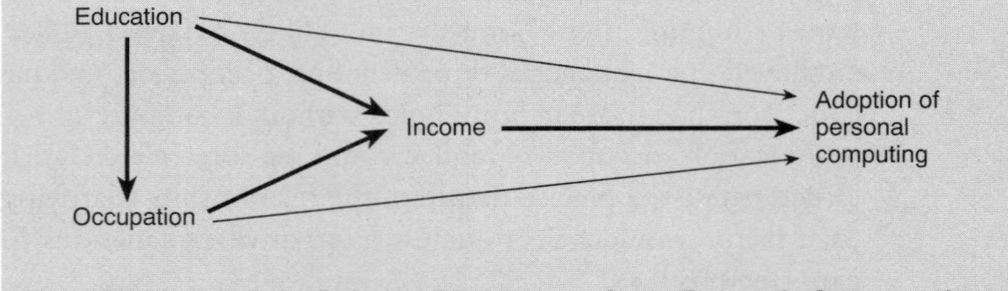

Fig. 9.3. Model of the Relationship between Socio-Economic Status and the Adoption of Home Computing

Source: Dutton *et al.* (1989: fig. 1).

Findings like this were dismissed by many in the mid-1980s as being limited to the initial years of computer adoption. Optimists saw a continuing decline in the price of ICTs that would diminish, if not erase, differential access based on income—a world in which the 'teleputer' replaces the TV (Gilder 1994). However, there is evidence that the role of economic inequality has persisted in the multimedia world of PCs, and is cumulative across a number of ICTs (Anderson *et al.* 1995).

In 1984, for instance, less than one-fifth of UK and American households had a PC in their household (see Figs. 9.1 and 9.2). This proportion has increased, but by 1997 a majority of households, even in the advanced industrial nations of the USA and UK, still did not have a PC. Moreover, the gap between rich and poor households increased—rather than decreased—between 1984 and 1994, with 10 per cent of households with incomes under $25,000 having a computer, compared with 54 per cent of households with yearly incomes over $50,000 per year (Wilhelm 1997; see also Katz with Aspden 1997; and Kraut *et al.* 1996).[6]

A similar association of access to the Internet with income, education, and occupation seems likely to persist despite its growing popularity. A Rand study found that socio-economic gaps in access to network services increased in the USA (Anderson *et al.* 1995). Another study of Internet 'drop-outs' found that those who stopped using the Internet after a short time tended to be 'less affluent' and 'less highly educated' (Katz and Aspden forthcoming). Patterns of rejection and adoption reinforce gaps between the information haves and have-nots.

Such disparities persist and are becoming cumulative across a growing array of ICTs in most wired nations. The numbers, speed, and capacity of computers, CD-ROM drives, and modems create a wide range of disparities even among homes that have access to one or more computers. Moreover, the PC is only one of a much wider assortment of ICTs—cell phones, answering machines, and TVs—consumed by households (see Table 9.2). The use of one ICT tends to be associated with the use of others. You need a PC to access the Internet. And access to all these ICTs are related positively to income and education. For example, even the telephone, which is available in 93 per cent of American households, is not accessible to nearly a quarter of low-income households in the USA, which are phoneless.[7]

Table 9.2. American Households Using ICTs, by Income, 1995

Percentage of households	Family income				
	$50K+	**$30–49K**	**$20–29K**	**Less than $20K**	**All**
with a computer	57	37	23	12	32
with a CD-ROM	29	18	9	4	15
with an on-line information service	14	6	4	1	6
using e-mail regularly	13	7	5	2	7
using the Web	5	4	2	1	3

Source: Times Mirror (1995: 8), a national sample of 4,005 US households.

In all advanced economies, TV is an exception to the trends shown in Table 9.2. TV is so widely available, for instance in over 98 per cent of American homes, that income does not discriminate among households (Bureau of the Census 1996: 561). However, access to pay-TV services is related to income. Also, as discussed above, the TV is soon to become a more differentiated commodity that will be unevenly distributed across income groups.

Cultural Conceptions and Responses of Users

Cultural differences can also facilitate or constrain tele-access, and often reinforce economic disparities. For example, the dominance of the English language on the Internet is an advantage to English-speaking countries and English-speaking groups in multilingual nations, as well as to anyone able to read and write in English in many scientific and technical professions.

However, ICTs can also be used to overcome language barriers. For example, bilingual information business and public information kiosks have been valued by consumers and citizens (Dutton 1993). In California, for example, over a third of residents are Hispanic or Asian, and over 7 million individuals depend on a language other than English. In Los Angeles, the County Library offers services to over forty-five different language communities. Telephone translation services set up by telephone companies, like AT&T, to facilitate international teleconferencing have supported local communication between minority-language communities and public agencies.

Assumptions about print literacy can lead ICT designers to have misconceptions about the users of their systems. For instance, the National Center for Education Statistics (NCES 1993) reported that 47 per cent of US adults are below the level of literacy required to read and interpret a bus schedule. Yet many suppliers of ICTs assume their software and documentation is written for consumers in the USA with at least a high-school education, knowledge of the English language, and a background within the mainstream culture.

More ICT suppliers should follow the lead of those in the mass media who have addressed the needs of minority communities in the provision of news and entertainment. For instance, the language, illustrations, metaphors, and cultural references used in computer software and documentation needs to be aimed at minority languages and cultures and to the poorly educated in mainstream English-speaking communities. The development of multimedia might help reach many who are unable to read print sources, but basic literacy—education and training in reading and writing—remains one of the most critical barriers to tele-access.

The Geography of Household Access

There is a geography of tele-access that is shaped by the socio-economic segregation of households. This can reinforce socio-economic disparities, but also help identify strategies for creating more equitable access (see Essay 9.3).

Distressed Areas within Nations

Issues of tele-access are particularly acute in distressed areas of inner cities and of rural communities. For example, on-line users in the USA are not only more affluent and better educated; they are also more likely to live in the suburbs (Pew Research Center 1996*a*: 5). As discussed in Chapter 8, ICTs can be invisible to many within the inner city, putting a premium on exposure to ICTs that children might gain at schools, libraries, and museums. As one local government official said: whole communities are 'out of the loop' and 'distanced from technology' (Dutton 1993).

Managers and professionals who are overloaded with information might find it hard to imagine a true lack of essential data. In many large urban areas, individuals go without food and shelter simply because of their lack of access to information on the availability of programmes designed to deliver emergency food and shelter (see Box 9.6). Thus, when focusing on households without computers, it is important to bear in mind that a sizeable number of individuals have no household and are dependent on public phones and public information for their day-to-day subsistence.

Box 9.6. INFO LINE in Los Angeles: Putting Food and Shelter on Hold

INFO LINE is a non-profit agency offering public information and referral services over the telephone to anyone calling within the County of Los Angeles. It receives hundreds of thousands of calls per year. Financial restrictions mean they cannot increase staff numbers to enable more calls to be handled. Information specialists refer callers to agencies, depending on the caller's location and the type of service required. However, there has been an average wait of seven minutes per call before a specialist can speak with the client. There is no way of tracking how many people hang up before their call is answered, but INFO LINE estimates that it is a high number. The two most frequent calls to the agency are for emergency food and shelter, so a failure to obtain information could deprive some individuals of a meal or a place to sleep.

The agency has tried to speed services by creating more neighbourhood locations and computerizing its database of 5,000 social-service providers in the County.

Source: Dutton (1993).

International Disparities

The divides across households within different nations and regions are even greater. Four-fifths of the world's population has no telephone. Half of the world's population has never used a telephone. The sixty-seven highest income countries account for only 15 per cent of the world's population, but they are served by 71 per cent of the world's main telephone lines. The lowest-income countries account for 59 per cent of the world's population, but only 4 per cent of main telephone lines (ITU 1994).[8]

There are also major divides within the most advanced industrial nations. For example, depending on definitions of 'on-line' households, there are orders of magnitude differences across nations in the use of on-line services such as the Internet. The USA alone had close to 15 million on-line households by 1996, compared to 2 million in Germany, and less than one million in any other nation (Jupiter Communications 1996). The scale of US participation on the Internet advantages its users and producers as well as other English-speaking nations, for similar reasons to those which have made English a 'language of advantage' for the provision of satellite TV (Collins 1989). More than 80 per cent of Web users, for instance, reside in the USA, compared with about 10 per cent from Europe, and 6 per cent from Canada and Mexico (GVU 1995). Likewise, within regions, such as Europe, there are major cross-national differences in access to ICTs like telephone services (Gillespie and Cornford 1996: 339). The use of on-line media like the Internet also varies within Europe with use concentrated in nations of northern Europe, such as Finland and Denmark, as compared to Italy and Spain.[9]

The Role of Tele-Access in the Household and Community

This chapter has shown that the public—as consumers, citizens, audiences, and users—limit, open, and otherwise shape tele-access. Households domesticate ICTs in far more ways than technological determinists acknowledge. But, as the growing gap between information haves and have-nots suggests, the public varies in ability to do so effectively. As a consequence, the adoption and diffusion of ICTs have widened disparities between the information rich and the information poor. Narrowing this gap needs to become a major priority of policy and practice at all levels and in all arenas if positive visions of an information rich society are ever to be realized.

Computerization can 'advance the interests of the richer groups in any society because of the relatively high costs of developing, using and maintaining computer-based technologies' (Kling and Iacono 1990: 72). Nevertheless, socio-economic disparities do not provide grounds for rejecting innovations in ICTs. On the contrary, recognition of the priority that must be given to closing this gap can guide users, producers, and public policy interventions. There are important roles that consumers and their representatives can play if they are conscious of the significance of tele-access. Suppliers and public agencies can also address disparities, such as by providing training, education, and ICT infrastructures in distressed areas of the inner city, rural communities, and developing nations. Public interventions, such as subsidizing the cost or purchase of computers in ways analogous to the subsidization of telephone services, might be effective. Disparities are not being eroded by technological advances and market forces alone.

Virtual Communities or Cities?

Inequality of access is a central problem for claims that the Internet and information superhighway, more generally, are tools for enhancing community (Gore 1991; Rheingold 1994; US Advisory Council 1996*b*: 9). This is reminiscent of the hopes surrounding earlier innovations in broadcasting, cable TV (Dutton *et al.* 1987*a*), and computer conferencing (Hiltz and Turoff 1978), which were also expected to enhance community. The growing disparities of ICTs for richer and poorer suggest that a more realistic expectation might be captured by Paul Resnick's (1997) concept of a 'virtual city', in which ICTs are used to build centres of learning, museums, and amusement parks, but also gated communities, and red-light districts.

The telephone, in conjunction with highway development, has been tied both to urban sprawl, encouraging movement to the suburbs, as well as to the growth of major world cities (Castells 1989; Graham and Marvin 1996). This apparently 'dual effect' (de Sola Pool 1977) can be explained by the degree to which telecommunications reduce constraints on tele-access, permitting factors outside communication to drive locational decisions. But the telephone has also been linked to changes in the 'psychological community' in that it provides a feeling of connection with those at a distance, while reducing unmediated interpersonal communication with nearby friends and neighbours, and thus contributing, according to Wurtzel and Turner (1977: 256), to the 'malaise of urban depersonalization'.

Advances in ICTs, such as satellite broadcasting and the Internet, have extended communication over time as well as space (de Sola Pool 1990). Electronic discussion groups—such as 'chat groups' on the Internet have been credited with creating electronic social relationships that do not depend on face-to-face interaction (Hiltz and Turoff 1978; Shields 1996; McLaughlin *et al.* 1997; Smith and Kollock forthcoming). Others are looking at ways in which social relationships over computer-mediated networks can undermine or support geographical communities, such as a neighbourhood.

The long-term social implications of ICTs for community are great. At the same time, these long-range issues are not necessarily the most meaningful focus for research and policy intervention. For example, cable TV experiments of the 1970s were also promoted as a means for creating a stronger sense of community (Dutton *et al.* 1987*a*). Survey research in Japan focused on this issue, but failed to discern a significant impact of cable TV viewing on the public's orientation to the local community (Ito and Oishi 1987: 206–10). One reason was that relationships between cable viewing and one's sense of community were overwhelmed by such factors as the educational background, occupation, and local-versus-cosmopolitan orientation of the viewers.

Such findings do not provide sufficient evidence to dismiss the role of ICTs in shaping communities, either real or psychological. An explosion of locally based Web sites, for example, attests to the positive role that organizers believe the Internet can play in supporting access to people nearby, as well as to those at a distance. However, the outcome will depend on the day-to-day choices made by countless numbers of individuals about how to manage tele-access. These choices will be shaped by their beliefs and values surrounding community—ICT paradigms and practices—as well as by the capabilities of ICTs. The structure of communities can be too far removed in time and in scale to be demonstrably affected by the use of ICTs in the short term (McQuail 1987). But, as a UK broadcasting enquiry once argued, many major social implications of ICTs are analogous to 'water dripping on a stone' (Annan Committee 1977: 26).

Notes

1. By 1995, 54% of US respondents over 16 years of age said they had used a computer at home, work, or school (Times Mirror 1995: 6).
2. This estimate was based on a study by Morgan Stanley reported by Denton (1997).

3. European debates over access to ICTs refer to the creation of an A team and a B team.
4. Digital transmission of all the information contained in an analogue colour TV signal would take more bandwidth than the analogue signal. For an excellent discussion of the trade-offs between analogue and digital TV, see Noll (1998).
5. The literature and arguments in support of these alternative explanations are elaborated by Dutton *et al.* (1987*b*: 260–2).
6. Wilhelm (1997) based his analysis on a November 1994 Current Population Survey, conducted by the US Bureau of the Census, which added questions on computer access beginning in 1984.
7. 76% of households in the USA with annual incomes under $5,000 had telephone service (Bureau of the Census 1996: 564, table 883).
8. Low-income nations include 54 countries with gross domestic product per capita (GDP) below $US 600 per year, while high-income countries include 67 with GDPs above $US 10,000 per year (ITU 1994).
9. One indicator of Internet use is the number of hosts per country, which is traced by Internet domain surveys, such as that conducted by Network Wizards (1997).

Essays

9.1. Domesticating ICTs

Roger Silverstone

The household is a complex social, economic, and political space that powerfully affects both the way technologies are used and their significance. Roger Silverstone evaluates how the household environment influences the diffusion and use of information and communication technologies.

9.2. Gender and the Domestication of the Home Computer: A Look Back

Leslie Haddon

The image of the home PC as a masculine technology first emerged in the home computer boom of the 1980s. Leslie Haddon highlights the aspects of household life that reinforce gender differences in relation to a technology like the home computer.

9.3. The Geography of Network Access

James Cornford and Andrew Gillespie

The concept of an information society presupposes a diffusion of networks beyond larger companies into smaller enterprises and the domestic household. James Cornford and Andrew Gillespie look at patterns of inequality in the competitive provision of telecommunications services.

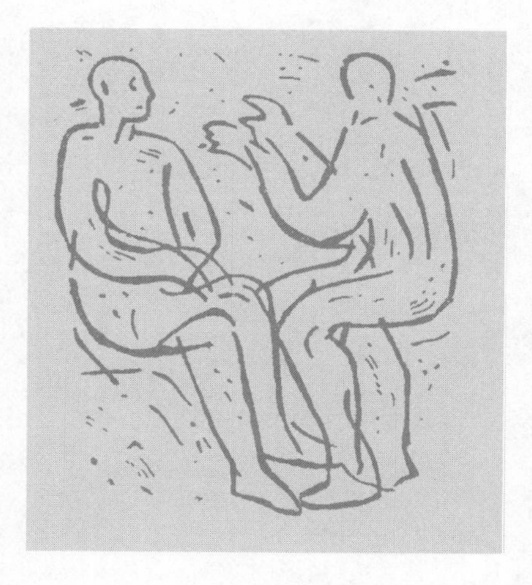

9.1. Domesticating ICTs

Roger Silverstone

Innovation is neither linear nor does it stop once an object—be it a new computer, software program, or interactive service—leaves the literal or metaphorical shelves of the supplier. For example, ICTs and the information they transmit exist in public spaces and have meanings defined by the combined efforts of designers, regulators, and all those who create the market for new technologies and services. Once the innovations move inside the household, however, consumers and users seek to gain control over their newly acquired objects.

Domestication involves fitting and fixing the new into the familiar and the secure, moulding its novelty to the needs, desires, and culture of the family or household (Silverstone and Hirsch 1992; Silverstone 1994; Silverstone and Haddon 1996). It is, therefore, essentially a conservative process in which consumers try to incorporate new technologies and services into the patterns of their everyday lives in such a way as to maintain both the structure of those lives and their capacity to control that structure.

Domestication is, therefore, a struggle between the familiar and the new, the social and the technical; between the revolutionary potential of the machine and the evolutionary demands of family and household. It is also a struggle within households—for instance, as parents and children, male and female, the computer expert and the computer illiterate or semi-literate seek to manage space, time, and technologies without losing position and identity in the complex and uncertain politics of age and gender in the home.

Our domesticity can be described as a 'moral economy'—a distinct social and cultural space in which the evaluation of individuals, objects, and processes forming the currency of public life is transformed and transcended once the move is made into private life (Silverstone *et al.* 1992). In our domestic spaces we are more or less free, depending on available material and symbolic resources, to define our own relationships to each other and to the objects and meanings, the mediations, communications, and information that cross our thresholds.

Households can, therefore, be seen as both economic and cultural units within which their members can—and do—define for themselves a private, personal, and more or less distinctive way of life. The materials and resources they have at their disposal both come from the inner world of family values, as well as being provided by the public world of commodities and objects. This moral economy is constantly changing, affected by the relentlessness of the human life cycle, as well as by the buffetings of everyday life, historically specific yet uneven in their consequences.

The dynamics of the moral economy are defined through an eternal cycle of consumption and appropriation in which the pasteurized commodity is accommodated to the spaces, times, and functional requirements of the home. New hardware and software technologies and services are bought home to be placed and displayed, to be incorporated into the rituals of domestic and daily life, to enhance efficiency or increase pleasure.

The new skills that may be developed, the meanings that are generated, the new conflicts that are, or are not, resolved are all expressions of the constant tensions between technological and social change within the household. The novelty and the achievements, the significances created and sustained in the ownership and use of the new machines or access to new computer software or television channels, then become part of the currency of everyday discourse—discussed, displayed, and shared in the endless gossip and talk of neighbourhood, school, and workplace.

Computers, information, and other media enter this complex and dynamic space once they enter the home. They become articulated into domestic life, shifting, puncturing, and redefining the boundary between public and private spheres of life.

For example, the combined effects of market forces and government policies in some countries created a boom in low-power home computers in the early 1980s. These were often bought for their educational potential, but were quickly converted to games-playing by teenage boys—or increasingly left to languish in dusty corners and rarely, if ever, used. At this stage, neither family culture nor connections to an emerging commitment to computing at school were able to support and sustain the kinds of culture required to turn a general promise into domestic reality (Haddon 1988).

The more recent history of home computing indicates that individuals in the household construct and affirm their own identities through their appropriation of the machine via processes of acceptance, resistance, and negotiation. What individuals do, and how they do it, depends on both cultural and material resources. For many, of course, neither is sufficient to enable them to participate in a world of networked information services (Silverstone 1996). The choices made first at the time of purchase, and consistently and insistently thereafter, are the product of a complex and contradictory situated rationality of desire and discrimination, rather than of a simpler monochromatic rationality of efficiency and need.

This rationality has various expressions in the broad range of households in different national or local cultures—such as urban or rural, wealthy or impoverished, nuclear or lone-parent, elderly or teleworking. To make sense of the situated rationality in these varied contexts, we have to make better sense of the process of domestication that underlies and underlines it.

References

Dutton, W. H. (1996) (ed.), with Malcolm Peltu, *Information and Communication Technologies—Visions and Realities* (Oxford: Oxford University Press).

Haddon, L. (1988), 'The Home Computer: The Making of a Consumer Electronic', *Science as Culture*, 2: 7–51.

Mansell, R., and Silverstone, R. (1996) (eds.), *Communication by Design: The Politics of Information and Communication Technologies* (Oxford: Oxford University Press).

Silverstone, R. (1994), *Television and Everyday Life* (London: Routledge).

—— (1996) 'Future Imperfect: Information and Communication Technologies in Everyday Life', in Dutton (1996), 217–31.

—— and Haddon, L. (1996), 'Design and the Domestication of Information and Communication Technologies: Technical Change and Everyday Life', in Mansell and Silverstone (1996), 44–74.

—— and Hirsch, E. (1992) (eds.), *Consuming Technologies: Media and Information in Domestic Spaces* (London: Routledge).

—— Hirsch, E., and Morley, D. (1992), 'Information and Communication Technologies and the Moral Economy of the Household', in Silverstone and Hirsch (1992), 15–32.

9.2. Gender and the Domestication of the Home Computer: A Look Back

Leslie Haddon

Many potential industrial players who could have developed a home computer initially saw no use for small computers at all and, later, no use for such a machine in the home (Haddon 1988*a*). It was thought at one point that a possible computer could develop as a crossover product which originated from the office. In that case, it may well have been promoted for word processing. Given the association of typing with female clerical labour, such a machine could have had very different early gender connotations from the more masculine orientation which eventually emerged.

The reality was, therefore, more complex than that claimed by the suggestion that home computers were made by males for males. In practice, it was indeed male electronic and computer hobbyists who took the lead in developing the microcomputer and they certainly did produce early systems mainly for other male hobbyists.

British high-tech entrepreneur Sir Clive Sinclair launched one of the first really low-cost machines in the early 1980s packaged to create a mass computer market. Although, in part at least, he too appealed to existing male hobbyists, he also invited a wider audience to explore the world of computing—to play with the machines rather than use them for particular purposes.

Apart from a masculinity implicit or explicit in the 'techie' adverts and manuals of many computers of that era, the hobbyist appeal to explore a technology implied free disposable leisure, which many women felt they did not have. Even in those cases where there was some interest, the time just to play could not so easily be justified.

However, there were countervailing influences in those early years. These computers also featured in government policies in many countries. In the UK, for example, the BBC, the publically funded broadcaster, campaigned for computer literacy through TV series and publications (Haddon and Skinner 1991). In this climate, many families bought machines as a stake in the future, especially for their children—and that meant for both boys and girls.

The story of how the early home PC was gendered must also take account of the history of interactive games. Games had originally been developed to show the potential of computers, a use later taken up by computer hobbyists (Haddon 1988*b*). Yet those games also migrated to arcade and video-game machines. Once a predominantly male audience for games had emerged, games software became geared to its tastes.

However, the social environment in which games are played may be a more important factor than the content of games in shaping male interest. Action games based on fast reflexes were originally developed to mimic pinball. In practice they replaced pinball machines in male-dominated arcades. It was in these public spaces that the practices of games-playing were established. For many males this involved more than the moment in front of the screen—it also involved watching others, learning tips, talking about games.

Later, when computers became the main vehicle for games playing—and before the rise of Nintendo and other portable hand-held games consoles—these practices were transferred to the home machine. In addition, home computers produced their own set of social interactions, such as copying and swapping software and exchanging the information read in computer-games magazines.

It is a myth that girls do not play games. It may be true that fewer play games and do so less intensively than boys. But the distinctive feature is that it is solely a domestic experience for girls, who do not discuss it so much outside the home or read about it so much in magazines. Boys who take an interest in games have an additional culture outside the home which supports and reinforces their interest (Haddon 1992). Hence they are more likely to appropriate the machine, to create demand for computers within the home, and to influence the choice of games software entering the home.

Hardware suppliers were actually ambivalent about games. While games offered an important application area, the manufacturers were also threatened by the possibility that the PC could have been viewed as just a down-market games-machine product. That would have limited its spread to other potentially lucrative applications. And in the early years, manufacturers did lose some control over the product's identity as games became the most significant application, with the male games-player seen as the main user—visible both in public spaces and to other family members in the home.

This perspective on the history of the home computer's emergence highlights the involvement of multiple actors, including producers and users, and an interweaving with other social contexts, such as gendered game playing. It indicates that the domestication of the PC and its association with one gender, as for any technology, can be understood fully only by considering experiences outside the home as well as within it.

References

Haddon, L. (1988*a*), 'The Home Computer: The Making of a Consumer Electronic', *Science as Culture*, 2: 7–51.

—— (1988*b*), 'Electronic and Computer Games: The History of an Interactive Medium', *Screen*, 29: 52–73.

—— (1992), 'Explaining ICT Consumption: The Case of the Home Computer', in Silverstone and Hirsch (1992), 82–96.

—— and Skinner, D. (1991), 'The Enigma of the Micro: Lessons from the British Home Computer Boom', *Social Science Computer Review*, 9: 435–49.

Silverstone, R., and Hirsch, E. (1992) (eds.), *Consuming Technologies: Media and Information in Domestic Spaces* (London: Routledge).

9.3. The Geography of Network Access

James Cornford and Andrew Gillespie

As their market place has been liberalized, telecommunications suppliers have become more sensitive to the geographical composition of expressed demand, as well as the relative costs of supplying that demand. For example, profits for a telecommunications supplier are high in areas where large numbers of densely located firms generate a high level of effective demand that can be met at relatively low cost, such as in central business districts of major cities. Such locations tend to experience greater telecommunications competition, more investment, faster upgrade of existing services, and earlier introduction of new services in a liberalized environment.

Competition, new investment, and new services may also eventually reach areas of low demand and/or high costs of supply. However, by the time this has occurred, the core locations will have already moved on. The result is a continually shifting, but durable, telecommunications supply gap in terms of the levels of competition, pressures for discounting, and availability of enhanced or advanced services.

We can view this unequal geographical pattern at a range of spatial scales, such as at the level of a whole region, a single nation, or an individual city. At the European level, for instance, there are major disparities between different EU member states in terms of the degrees of access to even the most basic telecommunications services, the quality of service provided, and the prices consumers are expected to pay for access. To take just one example: the average annual bill in the mid-1990s for a standard digital mobile telephone service, including calls, was over 22 per cent of per capita GDP in both Greece and Portugal, compared with less than 8 per cent in the UK and Germany (calculated from Mercer Management Consulting 1994).

The take-up of basic telephony also varies widely. In 1993, for instance, there were more than ten domestic main lines for every ten households in all Scandinavian countries and France—but there were fewer than eight in Ireland and Portugal and fewer than nine in Greece and Spain (ITU 1995). At the same time, 10 per cent or more of households were still not connected to the basic telephone service in EU countries such as Ireland, Portugal, Greece, and Spain.

This diverse pattern of supply is by no means homogenous within each national territory. Rather, there is a kind of 'fractal' effect in which the basic pattern is repeated again and again as we change the geographical scale at which we examine the issue. Similar patterns emerge to those at the wider European level if the focus is narrowed to examine just one country, or an area within a country.

In the UK, for example, effective competition is far from universal—even though liberalization began in the early 1980s. In the City of London, the country's main financial centre, there are a multiplicity of suppliers. Central business districts of most other major cities have a more restricted choice. In the residential suburbs of these cities, competition is usually restricted to BT, the former monopoly state supplier, and, possibly, the local cable company. For most free-standing towns and rural areas, firms and households are confronted by a *de facto* BT monopoly in terms of their access to the Public Switched Telecommunications Network (PSTN).

In short, hot spots of intense competition are surrounded by warm shadows of duopolistic rivalry, which give way in turn to cold shadows of *de facto* monopoly. The same can be seen by zooming down to examine in detail the pattern of network supply in a city such as Newcastle upon Tyne in the north of England.

At the level of local authority wards on which parliamentary and local council elections are based (each with a population of around 5,000), research by the City Council in the mid-1990s revealed that the proportion of households without a telephone varied between 1 per cent and 40 per cent (Newcastle City Council 1997). Even more dramatically, on one deprived inner-city council housing estate, only 26 per cent of households had a telephone (Newcastle University 1992).

What is most stunning about this finding is that the housing estate in question stands next to a newly developed business park catering for sophisticated users of advanced telecommunications services. The business park has telecommunications provision from a number of competing operators who have created a highly robust system, with multiple access points and direct fibre optic links.

The information society has been heralded as freeing us from the limitations imposed by geography, abolishing the distinction between core and periphery, and overcoming the 'friction of distance' through the ubiquity of information. Such claims are fanciful. They ignore issues that go far beyond the availability of telecommunications networks, such as the complex social construction of core and periphery areas.

Even if these kinds of issues are set aside, the geographical ubiquity of information presupposes the ubiquitous presence of the means of disseminating or obtaining it. There is no evidence of such geographical ubiquity. Rather, patterns of access to, and the take-up of, the networks which are expected to underpin the information society are characterized by complex but structured patterns of geographical inequality.

References

ITU (1995): International Telecommunications Union, *World Telecommunications Indicators 1994/5* (Geneva: ITU).

Mercer Management Consulting (1994), *Future Policy for Telecommunications Infrastructure and CATV Networks: A Path Towards Infrastructure Liberalization* (Report to the European Commission, DG XIII; Brussels: Commission of the European Communities).

Newcastle City Council (1997), *City Profiles* (Newcastle upon Tyne: Newcastle City Council).

Newcastle University (1992), *West End City Challenge Monitoring Report* (Newcastle upon Tyne: Department of Town and Country Planning, University of Newcastle upon Tyne).

10 Wiring the Global Village: Shaping Access to Audiences

The most important deterministic claim about new ICTs is that they are 'technologies of freedom' (de Sola Pool 1983*a*) because they promote greater freedom of expression and more democratic modes of communication (see Chapter 3). From this perspective, ICTs will revolutionize tele-access by providing more channels of communication, a greater range and diversity of offerings, easier access to larger volumes of information and entertainment offerings, opportunities for interactive versus only one-way communication, and capabilities for individuals to determine the time at which they wish to be entertained or informed (see Box 10.1). The creators

Box 10.1. The Democratic Bias of the New Media

ICTs like digital media, and compression technologies, combined with new channels of communication such as cable, satellite, and fibre optic transmission, offer:

- *abundance*: the elimination of barriers to the provision of content, making the scarcity of over-the-air spectrum a less critical bottleneck;
- *horizontal linkages*: the facilitation of one–one and many–many forms of communication rather than one–many, vertical communication networks;
- *interactivity*: the ability to provide two-way, rather than only one-way communication;
- *asyncronicity*: the ability for viewers to store or retrieve content at the time and place they choose, using such ICTs as the VCR, PC, or voice mail.

Source: Dutton *et al.* (1987*a*: 13–16).

and users of information are viewed as the chief beneficiaries of this shift. The losers are expected to be those who package and distribute content—the publishers, film distributors, broadcasters, telephone, cable, and satellite companies—as their control over the bottlenecks between creators and users are eroded by ICTs (Gilder 1994).

However, this democratic vision of the future of tele-access ignores the social shaping of ICTs. Just as ICTs can be domesticated by households (see Chapter 9), so the mass-media industries can use ICTs to reinforce their role in tele-access. This chapter looks briefly at some of the strategies of an increasingly competitive array of media—print, cable, satellite, telephone, and emerging media—to gain and retain access to audiences. Consumers may now have more choice than ever before, but the media remain powerful shapers of tele-access. The role of the viewer, as well as the role of public policy and regulation, can be improved by understanding the tele-access strategies of the media.

The Battle for Eyeballs: Conceptions and Responses of the Audience

Andrew Grove, the chairman and CEO of Intel (the world's leading maker of microchips), declared that they were in a 'war for eyeballs' between the PC and TV set (Kehoe 1997; Miller 1997). The battle for eyeballs is nothing new to the major media organizations around the world. One of the primary advantages accruing to the mass media is a historically anchored understanding of their audience. Nevertheless, many proponents of new media, who are critical of the mass media, fail to appreciate the degree to which the mass media know what their audience likes, and how to deliver it. This is true of public-service broadcasters as much as it is of the commercial broadcasters.

Managing the Creation, Distribution, and Rights to Content

In the early decades of radio and TV broadcasting, the mass media could never assume an audience. They were the gatekeepers—arbiters for what the public would listen to on the radio and watch on TV—but they were subject to some competition for the time and attention of their audience and to public criticism if they failed to meet public expectations.

In broadcasting, for example, even in the early 1980s, most nations of Europe had one or two broadcasters (Noam 1991). One was likely to be a public-service broadcasting organization, financed by a licence fee, like the UK's British Broadcasting Corporation (BBC), with a public mission to 'provide entertainment, information, and education', and 'to enlarge people's interests' (Annan Committee 1977: 27). From its early years, the BBC faced competition from foreign broadcasters. Since 1954, private advertisement-funded UK stations licensed by the Independent Broadcasting Authority (IBA)—later replaced by the Independent Television Commission (ITC)—have also become competitors. The BBC and IBA developed sophisticated studies of their audiences and trained highly skilled professional broadcasters.

In the USA, commercial TV stations financed by advertising competed in most major TV markets. Critics of American commercial broadcasting argued that the primary role of broadcasters was to create programming that would deliver an audience to advertisers (Mosco 1989: 31–2). In a commercial system, success was measured by counting the number of eyeballs and dividing by two.

In both public-service and commercial broadcasting regimes, therefore, there were major incentives for broadcasters to develop accurate conceptions of their audience and develop, distribute, and obtain the rights to content that would attract it. Popular TV programmes are obviously a key means for gaining access to viewers. One strategic advantage of established newspapers, broadcasters, cable companies, and satellite firms is accumulated knowledge of their audience, combined with expertise in the development and marketing of programming. This is dramatically illustrated by the large proportion of time that households devote to the mass media.

The New Media Hub

Forecasts of competition between the PC and TV reflect speculation about changes in the media habits, values, and choices of households. The production of any media is based on assumptions about the audience. For example, audience considerations for a multimedia Web page would include such factors as its attention span, interest in colour, and proficiency in using the Internet. Are these assumptions changing? Will the TV, the computer, or some new form of digital TV be the future 'hub of integrated entertainment, information and communication systems in the homes of the next century' (Rawsthorn 1997)?

TV has already had critical implications for how the public gains access to information, such as an increased reliance on the electronic media, particularly TV, for news and entertainment. In asking Americans whether they read a newspaper or watched TV news 'yesterday', the Pew Research Center (1996*b*) found that 59 per cent got their news from TV, compared to 50 per cent from the newspaper. Generally, declining proportions of the American public rely on the newspaper as a source of news. This is one trend in many that supports the growing importance of electronic versus print media.

Americans were traditionally more oriented around TV than Europeans (Becker and Schoenbach 1989). However, TV viewing might have hit a ceiling in the USA, while it continues to grow in many other nations with higher levels of newspaper readership. In Germany, for instance, where 80 per cent of households read a daily newspaper, as opposed to about half of American households, TV viewing has increased with the advent of new cable and satellite channels from nearly two to three hours per day (Neuberger *et al.* 1997). As this time expands, cross-media competition for the public's time is likely to increase.

Since the early 1990s, TV has also begun to decline as a source of news, at least among the American public (see Table 10.1), suggesting a change in media habits beyond a simple substitution of TV for print media. One possibility is that the use of computers and the Internet is increasing in households (see Chapter 9) and drawing from time devoted to other media.

Whether computer users appear to have less time to read a newspaper, or are using on-line services for their news, they have declined in their use of both the newspaper and TV for news (Pew Research Center 1996*b*).

Table 10.1. Americans Who Say They 'Regularly' Watch Certain Programmes or Read Certain Publications, 1993–1997 (%)

Questions asked	Percentage who say 'regularly'		
	1997	1995	1993
Watch the national nightly network news on CBS, ABC, or NBC?	41	48	58
Watch the local news about their viewing area?	72	72	76
Watch Cable News Network (CNN)?	28	30	35
Read a daily newspaper?	56	—	66

Source: Pew Research Center (1997: 33–4).

Even before the widespread use of the Internet, households that were early adopters of computers spent more time studying and reading, less time watching TV, more time indoors, and in isolation, and more time working at home (Dutton *et al.* 1987*b*). This suggests that one of the major impacts of the diffusion of ICTs has been the reallocation of time in the household, reinforcing the belief that ICTs could be associated with changes in the media habits and values of the public at large. Bill Gates's prediction that 'Americans will live a Web lifestyle' embodies what many see as a fundamental shift in how the public will use media (*Los Angeles Times*, 1 Oct. 1997).

Dinosaurs or Chameleons? Responding to New ICT Paradigms and Practices

Forecasts of the demise of established print and electronic media often overestimate the speed at which media habits and values evolve, but they can also underestimate the degree to which established media can exploit ICTs to reach new audiences or provide new services. In 1997, for instance, the revenues of US media, film, television, publishing, and on-line publicly quoted groups rose more than 10 per cent for the third year in a row, and at a rate nearly three times that of the overall gross domestic product of the nation (Parkes 1997). The mass media are not dinosaurs, made obsolete by emerging ICTs. The media are more analogous to chameleons, who can easily adapt to their surroundings.

For example, the media have experimented with emerging ICTs for decades as a means to anticipate and respond to changes in the media habits of their viewers. TV broadcasters have experimented with the provision of new services like paging, cable operators with interactive services, telephone companies with broadcasting, and Internet service providers with Webcasting—a means for using the Web for broadcasting (see Table 10.2). ICT networks are commonly viewed as point to point or broadcast, but many are capable of delivering a wide variety of services—one–one, one–many, and many–many.

Table 10.2. The Flexibility of ICT Networks: Old Media in New Forms

ICT	Traditional design	New media strategies
Newspaper	Local, physical delivery	Electronic delivery via satellite, on-line news
TV	One–many	One–one paging services; two-way shopping
Cable	One–many	One–one telephony; many–many Web access
Telephone	One–one	One–many, as with '900' numbers in the USA
Internet	One–one, many–many	One–many Webcasting

Print publishers, concerned with a decline in readership, have used innovations in ICTs to experiment with providing the news electronically—on fax machines, telephones, TVs, and computer screens (see Box 10.2). The invention of viewdata, later called videotex, at the British Post Office in 1979 created great enthusiasm for the telephone being used to convey text and video images to a mass audience by using the TV set to display information stored on computers and delivered to households over the telephone network. Videotex and teletext provided a means for households to obtain text and graphics over their TV sets, which posed a threat to newspapers for the provision of news, as well as classified advertising. However, leading papers saw videotex as another channel for distributing news and advertising that was already in electronic form and that might reach screens that nearly all households watched for hours every day.

In the USA, newspapers took a more prominent role in efforts to develop videotex delivery systems, but they were only one of many key

Box 10.2. Electronic News: Strategies over the Years

- *Facsimile*: experiments with the provision of news via fax as well as scrolling text over the TV set in early cable trials (Kawahata 1987).
- *Teletext*: a system for broadcasting hundreds of frames of text and graphics over the vertical blanking interval of an analogue TV signal, enabling viewers to select pages for news, weather, sports, and movie listings.
- *Videotex*: the delivery via telephone line or cable of news and information to be displayed as text or images on a TV screen, or specialized videotex terminal, from remote computers on which information is stored as hundreds of thousands of videotex frames, many supplied by newspapers. Major systems have included Prestel (UK), Bildschirmtext (Germany), Captain (Japan), and regional systems, such as Times Mirror's Gateway (USA).
- *Audiotext*: using the telephone to provide prerecorded news and information, such as through interactive voice-mail systems.
- *Information Services*: text-only news and information services provided through such computer-based information service providers as Prodigy, and America Online (AOL), before they became more Web-based services.
- *The Internet*: electronic versions of newspapers accessible over the Web, often embedding links to other on-line newspapers and information.

Sources: Greenberger (1985); Aumente (1987).

actors, along with telephone operators and broadcasters, in the development of videotex systems. As described earlier, videotex failed to gain a substantial market (see Chapter 4), and newspapers folded their commercial videotex ventures. Times Mirror, publisher of the *Los Angeles Times*, and Infomart, a company owned by two Canadian newspaper publishers, Southam and Torstar, stopped their joint videotex venture—Gateway—after investing $15 million and failing to reach more than about 3,000 subscribers. Other videotex ventures, such as Knight-Ridder Newspapers Inc.'s Viewtron experiment, invested even more, without any greater success.[1] Nevertheless, these ventures demonstrated that the industry was ready and able to invest substantial sums to seize whatever opportunity ICTs presented.

In a similar way, the growth of the Internet and the Web has fuelled the rapid development of on-line news services—putting the newspaper on-line so that it is available over the Web. By the late 1990s, thousands of print newspapers established on-line news services. They also use the Web to advertise the print and electronic versions of their products. As in earlier years, this has been a defensive strategy, in part, for many within the newspaper industry, since the Web has created the potential for nearly anyone to create an on-line news service.

However, it is also an opportunity for existing newspapers to reach a target audience of computer users whose time demands and changing reading habits have cut into the readership of the traditional newspaper. In fact, the profile of newspaper subscribers is similar to the profile of early computer users, both of which are biased towards managers and professionals with higher incomes. These advances are not simply threats to print, but give the print industry tools for enhancing tele-access in the face of changing media habits and, very possibly, more ICT-centric lifestyles among an important segment of the newspaper's market. The BBC has followed a similar strategy in its Ceefax teletext service, as well as an on-line news service, both of which complement its televised news programming.

Re-Engineering Media Structures and Processes

Organizationally, the media are also far from dinosaurs. In fact, ICT industries were among the earliest to re-engineer management and business processes to take fuller advantage of ICTs (see Chapter 5). Television networks, cable TV multiple system operators (MSOs), newspaper chains, and other media businesses created networked management structures and business practices well before many other industries.

The print media, for example, have been aggressively using ICTs to re-engineer their business processes, to compete both with other media as well as with their industry rivals. The computer has become the printing press, and a means for extending access to the printed word (Eisenstadt 1968, 1980; de Sola Pool 1983*a*). Book, magazine, and newspaper publishers have long envisioned the use of ICTs to create the 'ultimate system' for the creation, production, and distribution of news (A. Smith 1980: 73–134). Computers have replaced the typewriters of journalists and replaced the printing press of the typesetters. In the 1980s, for example, a small group of journalists in Britain engineered the launch of a new national newspaper—the *Independent*—by outsourcing many aspects of its production. The paper's launch provides a case study of creating a virtual organization.

Despite a protracted battle, and great public controversy, public broadcasters, such as the BBC, have introduced new management practices aimed at making their operations more business-like. They have also employed ICT strategies to create new sources of revenue, such as through the worldwide distribution of news. While many critics have viewed these management strategies as an attack on the institution of public-service broadcasting, many other broadcast organizations have since sought to downsize and streamline their operations (Barnett and Curry 1994), and create new sources of revenue (Blumler and Nossiter 1991).

Virtual organizations have long been the standard approach to organizing film productions, in which each film is treated like a separate business and produced by a temporary network of actors, directors, and other specialists, brought together with film studios. Television stations are locally owned in the USA, but early in their development they formed national 'networks' which also reflect many attributes of the virtual organization that other industries seek to emulate. By staying on the leading edge of new management and business practices, old industries have used ICTs to remain more competitive.

Extending Geographical Reach

Another major organizational strategy of the media has been to take advantage of the distance bias of electronic media to expand the geography of their market. Advances in ICTs, such as the Internet and communication satellites, make it no more costly to provide information to users across the world as across the hall.

However, the cost of building the technologically complex and large-scale networks, such as direct broadcast satellite systems, to provide these services is immense (see Box 10.3). This cost structure creates a major economic incentive to enlarge their markets to metropolitan, national, regional, and global scales (Garnham 1990). The *Financial Times* of London and the *Wall Street Journal* have used satellites to deliver the electronic text for printing in sites distributed across the world's major markets. Likewise, on-line news services provide an even greater reach to large and small newspapers.

The use of ICTs to enlarge the geography of media services can advantage consumers by creating more choice. It also can advantage larger firms, which are better able to acquire content, market services, and manage operations on a national or global scale (Melody 1996). Global markets are not simply a matter of transmission. To cite one example, the worldwide release of a major animated Disney film is accompanied by the availability of sound tracks in over two dozen languages. Those with the resources great enough to manage global operations have a major advantage in using the reach of global media.

Box 10.3. Direct Broadcast Satellite

Early broadcast satellites, positioned in geosynchronous orbits 22,300 miles above the earth, used relatively little power but required large satellite dishes on the earth to receive signals retransmitted from these microwave relays in the sky. Most broadcast satellites, particularly over North America, were used to distribute programming to cable TV operators, creating satellite-linked cable systems. Later generations of satellites have employed higher power transponders and digital technology to enable more channels to be transmitted for reception by much smaller 'dishes', making this technology an increasingly attractive alternative to cable TV. Technically, satellite systems are able to transmit as many as 1,000 television channels to households, making the management of programme packages an increasingly major undertaking. Technologically, operationally, and financially, the scale and complexity of launching and sustaining a satellite programming venture creates major barriers to entering this business with multimedia giants such as Rupert Murdoch's News Corp., Germany's Bertelsmann, the Kirch Group, and Canal Plus, and suppliers like Hughes Electronics Corporation.

Sources: Collins (1992); Kleinsteuber (1997).

The distance bias of ICTs finds support in the repeated failures of local networks. Broadcast licences and regulations on media ownership were set up in the USA to promote localism, but were undermined by national networking. Cable TV was thought to be a local medium because the attenuation of signals limited early community antenna television (CATV) systems to covering small geographical areas. However, terrestrial microwave relays and satellite-linked cable systems enabled cable TV systems to import and export signals within regional and global networks (see Box 10.4). The industry has been transformed since these early years by the growth of MSOs. By purchasing a large number of local systems, often in geographically concentrated areas, an MSO could consolidate some functions such as administration, billing, and legal services to lower the per household cost of local systems. Over the decades, the cable industry not only grew to reach over 60 per cent of television households, but also consolidated to the point that a majority of cabled households in the USA were controlled by just five leading MSOs.

In similar ways, local computer bulletin board systems flourished when individuals relied on directly dialling into a local computer bulletin board. The Internet overcame the costs of connecting with computers at a distance, creating incentives to permit worldwide access to the same information. Advances in ICTs have made it easier for broadcasters, cable

Box 10.4. The Changing Technology of Cable TV over the Decades

- *Community antenna television* (1950s into the 1960s). CATV systems were local, one-way, tree and branch networks, with up to twelve channels, based primarily on the simple relay of over-the-air broadcasts, but later supplemented by signals imported by terrestrial microwave from major TV markets.
- *Interactive cable TV* (early 1970s). Two-way cable systems were local, tree and branch systems capable of two-way communication and up to thirty or more channels, enabling applications such as teleshopping, and pay per view. Few were built due to the cost.
- *Satellite-linked cable systems* (since the 1970s). National networks of local cable systems relied on local, regional, and satellite distributed programming, one-way to the home, but with over fifty channels, including addressable pay TV.
- *Fibre-to-the-home systems* (FTTH). Switched networks for point-to-point or broadcast transmission of hundreds of channels and interactive services have been piloted since the 1980s. Expense and reliability refocused development towards fibre to the curb (FTTC), where fibre runs only to a local hub from which phone and TV services are provided to the home via copper and coaxial cable.

Source: Baldwin *et al.* (1996).

operators, and on-line service providers to reach a global scale—one to millions, and more difficult for any community, nation, or region to retain a captive audience (de Sola Pool 1990).

Economic Resources and Constraints: Why Size Still Matters

ICT innovations in such areas as the Internet and the Web, coupled with new management and business practices, have indeed expanded the capabilities available to individuals and small companies to reach audiences. The Internet is called a democratic technology because it brings the cost of providing electronic information within the reach of more individuals and organizations than ever before possible. However, the costs associated with the production and distribution of media remain great, and greater transmission capacities can provide even more formidable advantages to those with the economic resources to exploit them (Collins *et al.* 1988; Garnham 1990).

The Changing Structure of Costs

In an earlier era, the invention of the easy-to-use and inexpensive cameras, such as the Polaroid in 1947, promised to democratize photography. The declining cost of camera equipment has made it a feature of most households, even a disposable item. Nevertheless, very few photographers have made a business of photography, nor reached a large audience.

One reason is that the cost of transmission, or even production equipment, is not always the major cost of a media enterprise. ICTs change the structure of costs within an industry, potentially putting the largest and dominant players in a better position in relation to would-be competitors. At one level, the telephone is among the most democratic of media, because the users control its content. At another level—that of its provision—it has been highly centralized and removed from public control, provided as a monopoly service until the 1980s. The cost of using a telephone system needs to be disentangled from the cost of designing, building, and maintaining such a large and complex technological system, which remains great.

In the case of newspapers, users of the Internet can get access to thousands of on-line newspapers over the Web. The costs of distributing an on-line newspaper may be low, but the costs of employing journalists, editors, and other specialists are not, which advantages the established papers. For example, the revolution in printing technology during the 1980s created expectations of many new papers destroying the established press. However, as two British researchers argued, 'these predictions were based, as it turned out, on a myth: the widespread belief that overpaid print workers accounted for the major part of newspaper costs. In fact, production wages comprised only 21 percent of Fleet Street costs before new technology was introduced' (Curran and Seaton 1991: 121). To remain competitive, it is critical to reduce costs in an area comprising 20 per cent or more of a newspaper's costs. However, the costs of publishing a newspaper remained high, maintaining very real barriers to entry.

Moreover, the established papers have the branding, reputation, and infrastructure in place to produce on-line news at a marginal cost, and also the economic resources to invest in the R&D and commercial ventures that are critical to learning how to use advances in ICTs to their advantage.

A dramatic illustration of the advantages accruing to established actors can be found in the cinema. Over 90 per cent of box-office receipts in the USA are accounted for by a half dozen major film studios, which produce the lion's share of films and videos distributed worldwide. They have the financial muscle and expertise to develop vertically integrated business plans for marketing and distributing films with top celebrities to cinemas, pay-TV, video rental stores, cable and satellite providers, and broadcasters around the world. The total long-term returns on a major motion picture across all media and merchandise are driven largely by the box-office receipts in the first weeks of release. This gives a decisive financial advantage to those studios with the resources and expertise for worldwide marketing and the distribution of a major motion picture within such a short time frame.

The Capacity to Achieve Economies of Scale and Scope

Advances in ICTs have expanded the number of broadcast channels into the household from a handful in most advanced industrial nations in the 1980s into the hundreds. Yet, the investment needed to be an important player in this game requires firms with the economic resources to package and manage hundreds of channels across nations and regions.

The logic of market competition serving the public interest is to create incentives for competitors to move the prices of services as close as possible to actual costs and to develop innovations in ICTs that lower costs and improve service quality. That said, a key business strategy for reducing the costs of services to individual customers is to achieve economies of scale and scope, both of which favour larger companies with the resources to invest in ICT innovations and business expansion. Companies expand the scale of their business by expanding the geography of their market, and their share of the existing market. Media mergers and acquisitions that seek to achieve these economies have accelerated with the fall of more legal and regulatory barriers, leading to increasing concentration within most ICT industries (Bagdikian 1997; McChesney 1997).

Size also supports efforts to create economies of scope, gained by using the same infrastructure for a wider array of services. For example, telephone companies have long sought to expand the range of services provided over their networks, such as by providing video telephony, or entertainment TV, as with video on demand (VOD). Regulatory restrictions, expense, and limited markets have limited the viability of such services. However, the very likelihood of failure also favours those with deep pockets, capable of sustaining formidable loses.

New Gates and Ladders in the Tele-Access Game

The media can shape and exploit the design of ICTs to advantage themselves in reaching consumers. These strategies include efforts to control networks, establish a gatekeeping role, and in other ways enhance the media's communicative power in relation to the audience.

Network Control: Establishing Designs and Industry Standards

ICT producers prefer network architectures and standards that reinforce their interests (Dutton 1992*a*). For example, broadcasters were supported by the scarcity of broadcast spectrum, which limited access by competing providers. They resisted the introduction of multichannel cable TV because it enabled more competing channels.

In the telecommunications sector, equipment manufacturers, such as the computer industry, prefer telecommunication systems that embed more intelligence in equipment purchased by the customer, such as in PCs. In contrast, telecommunications carriers, like the major telephone companies, prefer networks in which intelligence resides within the central offices of their companies (Sirbu 1992).

An example shaping competition among telephone companies can be found in the design of intelligent networks (Mansell 1993). The telephone industry moved from electromechanical to digital switching systems, permitting companies to offer new services, such as call forwarding, and also make major reductions in personnel. However, as Robin Mansell's research has pointed out, the design of these intelligent networks can favour the established public telecommunications operators and their equipment manufacturers, over their competitors (see Essay 10.1).

Computer hardware and software manufacturers battle among themselves to design and promote ICTs that provide a competitive advantage. For example, Microsoft, the leading developer of operating systems for PCs in the 1990s, has been frequently cited by competitors for creating designs that favoured other Microsoft products and services. Kenneth Flamm (1998) illustrated the significance of computer operating systems to tele-access by comparing them to the railroads of the industrial age:

> There is a large fixed cost in developing railroad infrastructure, creating huge economies of scale; the cost of railroad services declines with traffic as more customers use the system. And there are . . . network externalities—the more that other locations and

customers connect to the same rail network to which you are connected, the more valuable it is to both you and them.

This gave the owners of the railroads monopoly power over local delivery services, leading to the establishment of anti-trust laws to curb their abuses. In many respects, the providers of software for PCs must deliver their services through the operating system in ways analogous to the rail network, which places a company like Microsoft in a position such that it can choose to compete with any provider of software, such as a provider of word-processing software, and use knowledge of its own operating system, and its ability to reshape its design in ways that unfairly advantaged it over competitors.

One case in point was the competition between Microsoft and Netscape for the Internet browser market (see Box 10.5). In such ways, competitors seek to design networks and control tele-access in ways that advantage their products and services. Whether they undermine fair competition is a highly contested matter and needs to be a continuing subject of oversight by the regulatory authorities in a competitive regime.

Box 10.5. The Browser War: Gateway to the Internet

The browser is a computer program that supports the use of a computer for accessing information on the Web. It converts the hypertext markup language (HTML) of the Web into the figures and text with the italics, underlining, and other features that the creator wants the user to see. Browsing (reading) the Web became a 'killer application' that drove the purchase of equipment such as the PC in the 1990s in a way comparable to how word-processing packages and spreadsheet programs drove PC adoptions in the 1980s. An early and major success was the development of Mosaic at the University of Illinois, which spawned the development of a commercial browser by one of the developers of Mosiac—Netscape Communications' Navigator. This software became a standard in the industry before Microsoft acquired a licence to market software developed at the University of Illinois in its Internet Explorer. The browser market created two related threats to Microsoft. First, the browser was central to use of the Internet, which threatened to make an end run around companies like Microsoft, if they had remained focused on the standalone PC applications. Secondly, the browser presented an approach to accessing files and applications that could eventually replace the standard operating system for which Microsoft enjoyed over 85 per cent of the world market. Microsoft developed its own browser and made it an integral part of its operating system. This gave Microsoft an advantage over rival software companies, leading to a challenge by the US Department of Justice over whether this bundling of software violated antitrust rules agreed between Microsoft and Justice.

Electronic Bottlenecks and Intelligent Gatekeepers

The browser is related to a more general tele-access strategy—that of positioning a company as a gatekeeper, or undermining existing gatekeepers. Different search engines on the Internet, for example, lead to different sites (see Box 10.6). If a user employs a search engine over the Internet to look for information related to a general topic, the search engine might identify thousands of potentially relevant sites. Most users will go to those sites recommended by the search engine, such as one of the top ten sites, putting the search engine in a powerful gatekeeping role.

Box 10.6. Search Engines for the Web

Early users of the Web created pages with lists of addresses to interesting sites on the Web organized by topic or other type of category. Daniel Filo and Jerry Yang developed one of the earliest search sites on the Internet—Yahoo! —that has become one of the five most popular indexes or directories for looking up other sites on the Internet. Excite, Lycos, and Alta Vista are among many other popular search engines. Listings are added to these sites in a variety of ways, including submissions from their developers, and sites discovered by the creators of the search engine, such as through the use of automated tools that look for sites with particular words, phrases, or subject matter. Search engines develop reputations based on such characteristics as their speed, user friendliness, number of listings, or how rapidly they are updated. These tools could be major gatekeepers to content on the Web, as they vary widely in the millions of sites they include, and exclude.

One of the chief attractions of new media has been the potential to circumvent the power of media gatekeepers. Prior to multichannel cable and satellite systems, for example, broadcasters controlled a major bottleneck on access to the household—the broadcast frequency spectrum. To Americans, these new channels meant an opportunity for more cultural, educational, and minority programming, breaking the hold of the three commercial networks (Mahoney *et al.* 1980). To Europeans, this same technology created a threat to public-service broadcasting, undermining existing gatekeepers and opening the public to a flood of commercial imports (Dutton *et al.* 1987*a*). In both cases, ICTs undermined the role of dominant gatekeepers and had major implications for access to the public.

In the early decades of TV, broadcasters did have a stronger gatekeeping role. They are still powerful gatekeepers in deciding what gets on a broadcast channel, and also in arranging the time and order of programming. With a limited number of broadcast channels during the early decades of broadcasting, most viewers would watch one channel through an entire evening. Given this pattern of audience flow, a key access strategy of broadcasters was to attract an audience early during prime-time viewing to hold them throughout the evening.

Likewise, US cable TV operators were required to carry local broadcast channels, but they could use channel positioning in ways that would advantage one or another broadcaster, such as by putting a channel further away or closer to the most popular channels. The provision of more channels, and the remote control, has combined to lessen the ability of broadcasters and cable operators to control audience flow with the same ease, and strengthened the role of the viewer as gatekeeper—for better or worse.

The power of some gatekeepers might be declining over time, but other gates are being built for the public at large. The major MSOs and satellite companies are putting together huge packages of entertainment programming—providing a wealth of programming choices for viewers. Consolidation within the cable and satellite areas could increasingly enhance the communicative power of these programme packagers, making them major gatekeepers to the home. Their role could be reinforced or reduced by the design of such ICTs as the set-top box.

For example, a direct satellite broadcaster might supply a set-top box to permit a paying subscriber to descramble and convert a digital satellite channel for display on an analogue TV set. The box could be designed in ways that make it difficult or easy for the subscriber to switch to another service provider. It is in the interest of the consumer to retain easy access to competitors and in the interest of the provider to make this a difficult switch.

New TVs and set-top boxes are being designed to provide WebTV and other TV-computer interfaces that permit households to access entertainment and other electronic services (see Box 10.7). In ways analogous to the set-top box, WebTV systems could advantage some equipment manufacturers, and content providers, by creating the menus from which viewers choose their programmes and services.

Communicative Power of Senders and Receivers

Technologies like the TV remote control and caller ID for the telephone have changed the relative communicative power of senders and receivers. Caller ID shifts more power to receivers by allowing them the opportunity

Box 10.7. WebTV: The New Gateway to a Mass Audience

WebTV is the idea of using the TV set as a multimedia display of Web content. Interactive and broadcast versions of this technology have been promoted since the mid-1990s. Interactive versions, analogous to the earlier technology of videotex, use a remote control and set-top box to connect to the Internet over phone or cable lines for browsing the Web and displaying the images on television. With an optional keyboard attachment, viewers can also receive or send e-mail. Broadcast versions of WebTV, more analogous to the earlier technology of teletext, also employ a remote control and set-top box but to retrieve Web pages stored on hard disc in the set-top box or TV. The pages can be received by a simple antenna and downloaded overnight or continuously.

Great expectations have surrounded the potential for WebTV to create a mass market for two-way, interactive TV services. Major corporate investments were made by Microsoft in acquiring 'WebTV Networks' in 1997. The company envisions the use of versions of its popular Windows operating system and browser to enable TV viewers to surf the Internet via two-way cable, extending the reach of computer firms beyond those homes with PCs.

to know who is calling before they decide whether or not to answer the phone. Knowing who is calling, sending e-mail, or communicating in other ways can extend other advantages to the receiver of a message. Recognition of this potential can be seen in the development of 'push technologies' (see Box 10.8). The sophisticated use of push technologies on the telephone systems, and on the Internet and the Web, has become a powerful tool for enhancing the communicative power of service providers, such as by facilitating telemarketing. For example, the customers of an on-line bookstore can ask the bookstore to send them an e-mail whenever a new book is available by one of the customer's favourite authors. In these and many other ways, ICTs are changing the communicative power of service providers in relation to the household.

ICTs can also empower the consumer as a gatekeeper. The V-chip provides an example of an ICT that viewers can use to define and screen objectionable material. But it is also possible to provide positive gates—letting the most desirable programmes through, rather than blocking the most objectionable. TV guides can serve this function. One example is a Los Angeles newspaper column called 'TV Smarts' that identifies 'programs that contain material included in the public school curriculum and on standardized examinations' (Kahlenberg 1997).[2] Firms with outstanding reputations for quality, such as Encyclopaedia Britannica, have

Box 10.8. Push Technologies: Provider versus User Driven ICTs

Most letters, telephone calls, and e-mail messages are initiated by a user who sends (pushes) messages to a receiver, the audience. The Web created an innovative approach to access, that turned the concept of sender and receiver on its head:

- *Pull technologies.* The Web popularized a system in which audiences would go to a provider for information. For example, by going to a Web page, users locate information on the computer storing the Web page and 'pull' the information to their computer.
- *Push technologies.* The Web has itself been used to push information once again to audiences. This happens in two ways:
  1. *Users initiate.* Visitors to Web sites can purposefully leave information, such as an e-mail address, in order to receive particular information. Users might ask the provider to send information that meets a certain specification, such as news of particular interest to the user.
  2. *Senders initiate.* The information provider collects e-mail addresses and information about individuals to create a targeted database of individuals likely to be interested in particular products or services. An organization, such as a business or university, might collect the e-mail addresses of individuals who look at its Web site, or employ a 'collection tool' with their Web browser that will 'spider' from site to site for hours collecting a targeted e-mail list.

Source: King (1997).

employed editors to identify the most valuable Web sites to link with their own guides to the Web. In specialized areas, such as health and medical information, institutions like the National Library of Medicine offer gateway sites to help users locate accurate and up-to-date information.

The technology behind the V-chip could also be turned to the identification of the best content, based on criteria supplied by the viewer. Whether the chip is lodged in the receiving equipment, or in the network itself, viewers could take an active role in constructing positive filters and indices for identifying the news, entertainment, and other programming of greatest interest (Mulgan 1991). Nicholas Negroponte (1995: 20) has promoted this concept, arguing: 'The answer lies in creating computers to filter, sort, prioritize, and manage multimedia on our behalf—computers that read newspapers and look at television for us, and act as editors when we ask them to do so.' Such positive editing functions can be performed by a guide printed on paper, and it was certainly once perceived to be the role of the news media and broadcaster, but, in the face of a profusion of choice, the computer provides new options to those viewers who have the ICTs, the proficiency, and the awareness to employ them.

The Social Implications of the Battle for Viewers

This chapter has highlighted only a few of the many ways in which key actors seek to use ICTs to improve access to their audiences—wiring businesses and households to their products and services. In the words of a leading communications consultant: 'a good business is one in which you can erect barriers after you enter' (Bortz 1985: 118). Barriers can be legal, such as through exclusive franchises for cable TV, or protected monopolies for voice telephone services. These legal restrictions on competition are being eroded in the 1990s. Advances in ICTs are expected to reduce many technical barriers to more democratic control over the media.

However, the mass media are not passive victims of the ICT revolution, but are actively seeking to exploit advances in ICTs to retain and expand their audience. The social implications of these tele-access strategies are not necessarily benign.

As discussed in Chapter 11, the prevailing policy paradigm sees the public interest served by competition across and within media. Whether consumers choose cable or satellite, for example, should have little social significance. If competition is effective, so this argument goes, all content providers will strive to provide what the public wants. Competition among providers will help lower the costs of all services, move industries towards the most efficient technical solutions, and expand access to all kinds of home entertainment. Government simply needs to ensure fair competition and let market forces drive the developments.

However, this positive vision assumes that advances in ICTs will decentralize control over tele-access, without paying sufficient attention to the strategic use of ICTs by key actors—the media strategists or 'road warriors'—in ways that can distort competition (Burstein and Kline 1995). Ensuring fair competition is not straightforward and must be pursued at all levels, ranging from equipment and software in the household to the design of telecommunications infrastructures.

A focus on tele-access provides a perspective on the ways powerful actors in government, business, and industry are actively seeking to exploit ICTs for tele-access to the public at large. This is how they compete with one another. Society cannot afford to wait and see if the information superhighway brings more or less democracy, privacy, freedom, and sense of community. The public's stake in the outcome of these competitive strategies is enormous—they are not simply an issue for media firms.[3]

The central thesis of the proponents of the information superhighway is that advances in ICTs enable the public to enhance their communicative power (Dyson 1997). However, these same advances can simultaneously enable the providers of ICT services to enhance their communicative power in relation to the public and their competitors. Tele-access strategies appear to be a driving force behind the consolidation and globalization of multimedia firms that can place such valued goals of the media as quality, diversity, and localism at risk (Blumler and Nossiter 1991; Blumler 1992). As Geoff Mulgan (1994: 1) has argued: 'Ensuring that technologies achieve their potential to liberate will not be easy.'

It is critical, therefore, that the public at large better understands how ICTs provide the industry with mechanisms to gain access to information about its audiences and wage the battle for eyeballs. The outcome will depend on regulatory oversight and public policy, to which I turn in the next chapter.

Notes

1. These videotex ventures are described more fully by Aumente (1987: 54–65).
2. This column is compiled by Richard Kahlenberg (1997) in consultation with Crystal Gips for the *Los Angeles Times*.
3. Interest in the role of ICTs in the communications sector takes place on two tracks. One concerns the competitive implications of advances in ICTs, such as in speculating on the implications of convergence for the structure of the communications industry. A separate track follows the impacts of ICTs on society, often dismissing concern over the competitive strategies of communications companies as a diversion from social issues (McQuail 1987). However, social issues involve concerns over the public's access to information, and competition entails access to the public. It is the interactions of producer and consumer strategies over time that shapes the societal implications of ICTs.

Essay

10.1. The Bias of Information Infrastructures

Robin Mansell

Technical decisions can advantage some actors over others. Robin Mansell discusses how the details of the technical designs of network infrastructures can favour dominant telecommunications companies over new entrants.

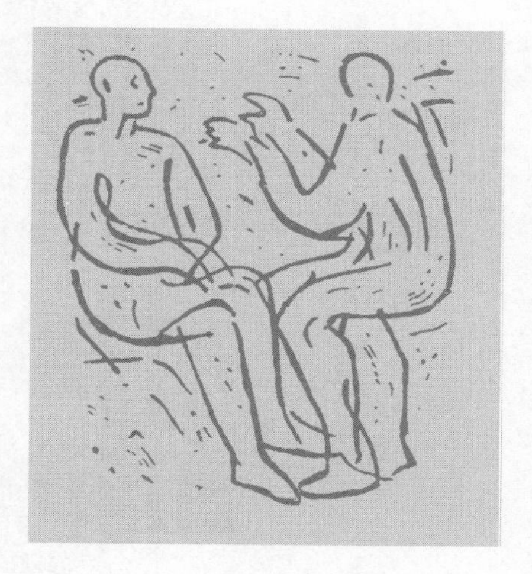

10.1. The Bias of Information Infrastructures

Robin Mansell

An area of technical decision-making which has crucial implications for the information society concerns the design and implementation of the architecture and control features of the underlying information infrastructure. As happens with many complex innovative technical systems, the move towards the implementation of advanced networking has become differentiated along several distinct paths as early visions have been moulded by the strategic interests of the various suppliers of related equipment and services.

A major rethinking of the architecture and functionality of automatically switched networks was initiated by Bellcore in the USA in the mid-1980s. This advanced architecture, called the 'intelligent network', aimed to provide a framework that would enable the major incumbent national public telecommunications network operators to meet new commercial challenges. These were created by rapidly changing technologies and by markets newly characterized by competition rather than the traditional monopoly structures.

The key to this framework lies in its software and computerized switching systems. By introducing decentralized computer-operated network nodes, the architecture achieves the aim of distributing some of the control over network functions throughout the public telecommunications network. The use of computer applications based on the latest software techniques can enable service environments to be reconfigured much more flexibly than with the centralized architectures embedded in existing networks.

This vision of an advanced intelligent network was important in terms of the introduction of specific new services. Facilities such as toll-free freephones, premium-rate services, customized tariffs, voice mail, and many other sophisticated telecommunications services are not dependent on having the original version of an intelligent network as developed by Bellcore. Nevertheless, the growth of these services has been greatly stimulated by the way public telecommunications operators acted to ensure that their networks could support new software-driven applications efficiently and reliably.

The development, diffusion, and use of advanced networking technologies since Bellcore first defined its vision have provided an opportunity to look at the many factors which influence technical issues of design and implementation. For example, the relatively slow introduction of pervasive intelligent networking capabilities was associated with constraints like: the need to negotiate standards for the interfaces within the network and between different components of software; inadequate software performance for meeting the stringent demands of real-time network management and control; the high cost of installing new switching equipment; the uncertainty of market demand; and uneven development of the network infrastructure, especially in Europe.

Studies of the implementation of advanced networks around the world also clearly show that biases reflecting the different strategic interests of suppliers and users become incorporated in new generations of the public telecommunications infrastructure (Mansell 1993). Far from taking the opportunity to design completely open and decentralized networks as originally envisaged, equipment manufacturers and network operators are introducing features into the design and implementation of new systems which tend to favour certain actors.

Some users do benefit from the greater variety and flexibility provided by software-driven services. Yet they do not encounter these new opportunities in a neutral way. The major manufacturers and network operators continue to establish the boundaries of network control. They construct the new advanced networks using proprietary standards, as they have done in the past, and locate interfaces in ways that minimize—rather than maximize—opportunities for competitors to make use of the software functionality. They also establish tariffs for accessing services which bundle functions together, thereby making it difficult for some users and intermediate service suppliers to exploit the separate components of the network.

Many of these technical decisions are taken to protect the integrity and security of the public network. However, others are made to protect market position, as some strategies for network design and implementation jeopardize the dominant market position of the traditional public telecommunications operators more than others.

Most advanced networks have implemented either proprietary standards or standards drawn up by private consortia. In addition, the possibilities for the interoperability of services between network operators in national or international markets are the result of subsequent bilateral arrangements concluded after the initial installation of equipment. The actual outcome is highly differentiated because of the relative strengths of different actors in the market. It is also strongly influenced by the actions of policy-makers, regulators, and new players, such as Internet Service Providers, who want to use the infrastructure in very innovative ways.

Rivalry, monopolization of markets, and the strategies of major players in the telecommunications and computing industries are leading to many changes in the strengths of the players involved in implementing the visions of an advanced network infrastructure. These processes are creating new opportunities for competitors and for all kinds of users of future information highways (Bernard *et al.* 1997). On the other hand, they are also resulting in new forms of technological lock-in which are frequently biased in favour of the dominant firms and larger customers. This means there will be a continuing need for policy-makers and regulators to invest resources in ensuring that the networks of the future are constructed and used in ways that are of collective benefit to the largest possible number of customers, including the new virtual communities.

References

Bernard, J., Cattaneo, G., Mansell, R., Morganti, F., Silverstone, R., and Steinmueller, W. E. (1997), 'The European Information Society at the Crossroads' (European Commission DGXIII, FAIR Project Report, Brussels).

Mansell, R. (1993), *The New Telecommunications: A Political Economy of Network Evolution* (London: Sage Publications).

Part VI

Industrial Strategies and Public Policies

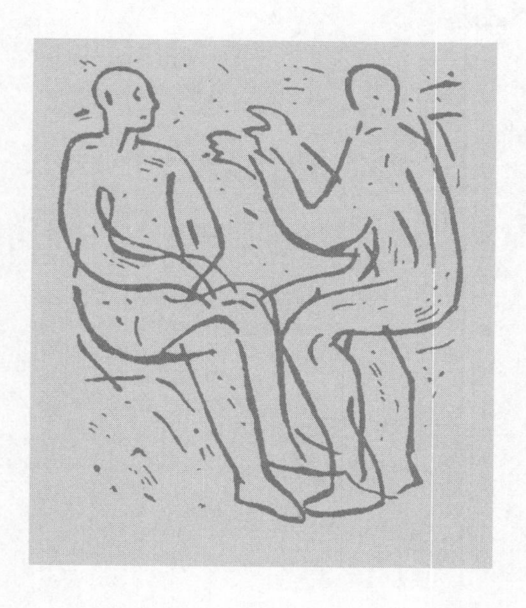

11 Regulating Access: Broadening the Policy Debate

Public policies and regulations are of particular significance in shaping tele-access because they represent purposeful governmental efforts to achieve broad social and economic objectives. The centrality of tele-access to many of these objectives also means that tele-access provides a framework for connecting the public policies and regulations tied to ICTs, as was recognized and promoted by the Clinton Administration's vision of an information superhighway (ISH) in the USA, which has been the subject of national policy initiatives across the globe. But such a vision also needs to be sustained and strengthened internationally.

The Ecology of Policy and Regulatory Games

I introduced the concept of an ecology of games in Chapter 1 to emphasize the degree to which tele-access is the outcome of an unfolding history of choices in many different arenas.[1] In discussing policy, the ecology of games offers a crucial understanding of why tele-access can be a victim of well-intentioned actors seeking to achieve a diverse range of objectives. For instance, access to information might become less equitable, even though policy-makers seek the opposite outcome, unless the range of policy choices shaping tele-access is well understood.

The ICT Policy Process as an Ecology of Games

Interactions across sectors are routine and often account for unanticipated change. A perspective on the policy process that anticipates and incorporates these interactions across policy sectors is the ecology of games. Players—stakeholders—are involved in an ecology of games that has outcomes which are more far-reaching than any single game within it (see Box 11.1).

From the perspective of an ecology of games, no one is playing in a tele-access game. 'Tele-access' is not even a word used by policy makers. It is an analytical concept. Nevertheless, there are many policy issues, as varied as universal service and trade barriers, that are explicitly focused on one or more dimensions of tele-access as an end of policy, but which seldom deal comprehensively with all its dimensions. Moreover, there are many other ICT policy issues, such as competition, privatization, freedom of expression, privacy, and copyright, that shape access to information, people, services, and technology (tele-access). However, the implications of these decisions for tele-access are not widely understood, allowing them to be resolved in ways that can actually exacerbate problems of tele-access.

Box 11.1. Key Features of an Ecology of Games Shaping ICT Policy

- A policy ecology is defined by a set of games, structured by rules and assumptions about how to act in order to achieve a particular set of objectives.
- Policy outcomes are shaped by the interplay of two or more separate, but interdependent, games that often cross traditional policy sectors, such as communication versus industrial policy.
- Not all players are involved in the same game, with most games having a relatively specialized set of players.
- Some players are involved in multiple games, as when politicians seek to influence a variety of related areas.
- Plays within one game can influence the rules, prizes, or moves within one or more other games, such as when the outcome of anti-trust rulings affect the permissible strategies of competing telecommunications companies.
- Plays are not random, as they are constrained by such factors as institutional arrangements and public policy.

Source: Dutton (1992*a*, 1995*b*).

It is possible that tele-access can be more purposively shaped if these interdependencies across policy sectors are understood. However, the size and dynamic complexity of this giant ecology of games shaping tele-access make this improbable. An alternative strategy is to create a more all-encompassing tele-access game that involves players throughout the larger ecology of policy. In many respects, this is exactly what politicians did in creating a grand project tied to visions of an ISH. Therefore, an effective mix of tele-access policies could depend greatly on the continued vitality of such visions.

Stakeholder Analyses

Most discussions of the politics of information and communication policy do not look at the larger ecology of public policy. Instead, they focus on the strengths and interplay of the various stakeholders involved in the communication-policy sector. Some view the political process within this sector, or its outcomes, to be highly skewed towards the interests of transnational business and economic élites (Mosco 1982; D. Schiller 1982; Garnham 1983). Others regard the politics among these stakeholders as being relatively more pluralistic, including frequent divisions between business and economic élites (Krasnow *et al.* 1982; McQuail and Siune 1986). For example, the US Telecommunications Reform Act of 1996 was a heavily lobbied bill before Congress, but this legislation emerged as an outcome of political 'gridlock' among divided stakeholders within the communications industry (Neuman *et al.* 1997).

Studies of the politics of stakeholders capture important dynamics of policy-making, such as in identifying groups that are left out of the policy process. However, they oversimplify the system of action governing the behaviour of individual players, for example, by expecting policy-makers to respond to pressure from interested groups in ways similar to a billiard ball. In focusing on a set of stakeholders in a single sector, fighting to shape the same policy decision, they also often ignore the routine ways in which policy choices and outcomes are influenced by choices made in a number of separate but interdependent policy sectors.

For example, the divestiture of AT&T forced the largest telecommunications company in the world to separate itself from the Regional Bell Operating Companies (RBOCs) and the provision of local telephone services (see Box 11.2). This aimed to encourage AT&T to compete in long-distance telephone services, enter the information-technology business, and move into international markets. Yet this liberalization of telecommunications by furthering competition in long-distance telephone services was a decision by the courts, not by Congress or the President, and was made on the basis of anti-trust rather than telecommunications policy.

Box 11.2. The Divestiture of AT&T: The Role of Anti-Trust

In the USA, efforts in the late 1970s and early 1980s to rewrite the Communications Act of 1934 were overtaken by anti-trust policy, specifically the settlement of the Department of Justice's case against AT&T. In 1949, the Justice Department brought action against AT&T under the rules established by anti-trust law, not by communications policy. In 1956, a consent decree was reached in the form of a Final Judgment that restricted AT&T to providing regulated telecommunications services. A Modified Final Judgment (MFJ) was reached in 1982 that agreed to the divestiture of the Bell Operating Companies from AT&T in order to promote competition through a reduction of cross-subsidies, discriminatory pricing, or other strategies. In return, AT&T was allowed to enter the information-technology arena, which it did. The actions of the key players within the Justice Department and the court appear to have been governed almost exclusively by the goals and rules of anti-trust, rather than by communications.

Source: Dutton (1992*a*).

Stakeholders are also likely to be simultaneously playing in a variety of games within different arenas. What might be irrational in one arena could be driven by the pursuit of objectives in other arenas. For instance, as discussed in Chapter 9, many criticized the Clinton Administration's support for allocating portions of the frequency spectrum to broadcasters to provide HDTV services as 'superhighway robbery' (P. Taylor 1997*b*). The auctioning of the spectrum might have raised up to $70 billion that could have furthered other public objectives, such as public broadcasting. However, the President placed more priority on supporting the microelectronics and computing industry's stakes in digital TV, and gaining free air time from broadcasters for candidates for elected office—two objectives of his administration's digital TV licence plan. Issues as far removed from broadcasting as campaign reform and industrial policy can influence broadcast policy.

The Fragmentation of ICT Policy

To understand the value of the ecology of games, it is important to recognize the fragmentation of the ICT policy-making process. Important ICT policy and regulatory decisions are made in many separate but interdependent policy sectors, and framed in the context of a multiplicity of seemingly unrelated issues.

Fragmentation across Sectors

Communications policy-making is just one sector of ICT policy affecting society, but it is so fragmented across levels and agencies of government that it is difficult to provide even a comprehensive list of the actors directly involved. In the USA, for example, people often believe communications policy is relatively centralized in the Federal Communications Commission (FCC). However, many other actors are involved, including: other independent agencies, such as the Federal Trade Commission (FTC); Congress, along with numerous Congressional committees and Congressional support agencies; many departments of the Executive, such as Defense, and the National Telecommunications and Information Administration (NTIA) of the Department of Commerce; the courts, particularly the DC Circuit Court of Appeals, but also the Federal District Court for the District of Columbia, which presided over the AT&T consent decree; state and local governments, who regulate telecommunications and cable; and many international institutions, such as the Consultative Committee on International Telegraphy and Telephony (CCITT) of the International Telecommunications Union (ITU), which sets standards for wireline communication systems, like international telephone networks.[2]

The USA is only one case. The fragmentation of communication policy has long been recognized in many nations and regions as a fundamental problem of creating a coherent approach to policy (Dutton *et al.* 1987*a*; Locksley 1990). This issue has led some to recommend the establishment of regulatory agencies like the US FCC as a mechanism for integrating communication policy-making. For example, in the UK, the creation of the Office of Telecommunications (Oftel) to regulate a competitive telecommunications industry has been followed by intermittent proposals to combine the regulation of broadcasting and telecommunications within a single agency (Cane 1995, 1997).

At the regional level, issues of fragmentation are also difficult to manage. Within the EU, ICT policy requires some harmonization across nations as well as across sectors (Locksley 1990; Freeman *et al.* 1991). For example, in 1997 the EU had seventeen members. Its executive arm is the European Commission, which is organized into twenty-three separate 'Directorate-Generals' that have many overlapping responsibilities central to communications policy.[3] For instance, the Commissioner for Industrial Affairs has responsibility for information technologies and telecommunications, but so does the Commissioner for Competition. Communications policy is of obvious relevance to the Directorate-General (DG) III, responsible for industry, DG IV for competition, DG V for employment, DG XII for science, research, and development, and DG XIII for telecommunications and information markets. This overlapping responsibility is illustrated by a group of experts on the information society, chaired by Luc Soete, who co-authored a report for DG V that incorporates a wide spectrum of concerns related to employment, ranging from issues concerning the nature of work to the role of ICTs in education, social cohesion, and the quality of life (HLEG 1996).

Tele-access is even broader and more fragmented, also taking in policies affecting issues such as communication, information, trade, industry, defence, and rural development. The ecology of games provides a framework that incorporates these routine interactions across policy sectors. And the concept of tele-access illuminates the connections of issues that cross policy sectors, helping to understand these issues as components of one broader issue—the shaping of tele-access.

Connecting All Policies to Tele-Access

Discussions of ICT policy yield lists of issues, reflecting the many social and economic implications of ICTs. The complexity and diversity of issues are overwhelming, including hot-button issues discussed earlier in this book, such as universal service, privacy, freedom of expression, copyright, standards, employment, cultural sovereignty, social cohesion, and community, to name a few. One illustration of this problem arose when the Clinton Administration established committees to deal with important issues surrounding the information superhighway. It resulted in so many committees and meetings that participants became frustrated and disillusioned with the prospects for working out the details of policy (Dutton *et al.* 1994: 15).

In response to this issue proliferation, some have argued for an increasing focus on competition as a major priority (European Commission 1994; Norman 1995). The pre-eminent position of competition among a proliferation of ICT policy issues has led to initiatives aimed at furthering competition in ICT industries, but has also crippled efforts effectively to broaden the debate beyond the establishment of regulatory structures to facilitate competition. The idea that governments can increasingly restrict their regulatory role to overseeing competition policy and abandon their 'role as arbiter of tariffs and definer of public service and public interest' (Neuman *et al.* 1997: p. xiv) has gained much ground (Baer 1997). However, sceptics of this position, such as Robin Mansell (see Essay 11.1), argue that the debate needs to be broadened beyond competition.[4] Mansell and many of her colleagues maintain that there is no consensus over how competition can be pursued without risking other widely shared values, such as the accessibility, privacy, and security of ICT networks and services (Mansell 1996). This risk is even greater owing to the separation of related issues such as competition and technical standards, even though they have a direct bearing on one another.[5]

A Tele-Access Policy Map

One way to broaden this debate well beyond competition is to simplify the policy map so that policy-makers and the public at large can see the connections that exist across sectors and issues. Tele-access provides one promising perspective from which these connections can be made. Many ICT issues are essentially concerned with closing or opening access for those who provide or use ICT products and services—they regulate tele-access—as illustrated in Table 11.1.

Table 11.1. Tele-Access in Regulation and Public Policy

Focus of policy and regulation	Constraining tele-access	Facilitating tele-access
Provider/producer	Monopoly	Competition
	Spectrum allocation	Increasing bandwidth
	Cross-ownership	No restrictions
	Foreign ownership	Loosening restrictions
	Trade quotas	Free trade
	Secrecy	Open government
	Censorship	Freedom of expression
	Privacy	Surveillance
	Proprietary standards	Open standards
User/consumer	Service to target markets	Universal services
	Private facilities	Public facilities
	Private, proprietary	Public information
	Copyright	Unrestricted use
	Security of records and systems	Techniques for remote and unauthorized hacking
	Screening information	All channel open access

Table 11.1 is not intended to be comprehensive. It is possible to add a variety of issues within this framework. Yet it suggests several general observations:

1. Many prominent ICT policy issues are essentially concerned with shaping access to information, people, services, or technologies (tele-access). For example, through the allocation of spectrum and restrictions on cross-ownership, governments regulate who can provide various ICT products and services. Policies aimed at protecting freedom of expression, such as the first amendment in the USA, and privacy are determining 'who, can say what, about whom?'—regulating access to information.
2. Competition itself can be viewed as an issue of tele-access, as competition policy is essentially regulating the terms on which would-be providers can gain access to facilities, infrastructures, and consumers.
3. Universal service is of obvious relevance to tele-access, but it is only one of a number of critical issues.
4. Overly simplistic policy prescriptions in support of unregulated access raise many anomalies, such as support for surveillance or opposition to copyright, because accepted policies often require restrictions on access. Even policies and regulations aimed at facilitating access, such as competition, also entail restrictions on who gains access to what and under what terms and conditions.

The complexity surrounding issues of tele-access cannot be conveyed in such a summary chart. However, the following sections discuss a few of the many issues in more detail by first looking at policies focused on producers, and then shifting to those focused on users of ICTs and the public at large.

Regulating the Provision and Production of ICTs

Policies and regulations governing the providers of ICT products and services have shifted over time towards reducing restrictions on access—permitting more open entry into the production and provision of ICTs. Examples include:

- increasing the spectrum, through technical innovations and policy initiatives, that has lessened concerns over spectrum scarcity and allowed more channels of communication in broadcasting, telecommunications, cable, and satellite;
- reducing restrictions on media cross-ownership, such as in permitting US telephone companies to enter local cable TV services;
- lowering national barriers to entry, including efforts to lessen restrictions on foreign ownership of media and telecommunications firms in nations such as the UK and USA, and the relaxation of quotas on the importation of foreign programming, although both of these barriers remain formidable;
- moving from proprietary and national standards, such as in analogue TV standards, to standards that permit more heterogenous and proprietary networks to interconnect over systems like the Internet.

This can also be seen in the case of competition policy, and moves towards the protection of freedom of expression over the new media, but it is useful to first discuss developments in a broader sector—industrial policy.

Industrial Policy: Opening and Closing Borders

The role of anti-trust policy in shaping the structure of the US telecommunications industry (see Box 11.1) is only one example of how issues of business and industrial policies are driving the production and utilization of ICTs. A central assumption of proponents of an information economy has been the growing centrality of the ICT industry, but also the importance of using ICTs in all sectors of the economy (Porat 1971; Bell 1974). Cross-national research has supported the major role ICT production and utilization can play in local, national, and regional employment and economic development (Freeman *et al.* 1991; Freeman 1996*a*; Kraemer and Dedrick 1996; Gillespie and Cornford 1996).

However, research has also alerted policy-makers to the tension between two general approaches to harnessing ICTs for economic development. One perspective is to focus on fostering the production of ICTs to build ICT industries, like microelectronics. The other is to emphasize its utilization across the economy. While these can be complementary, the production and utilization of ICTs can call for different policy prescriptions tied to tele-access.

The economic benefits stemming from the production of ICTs is exemplified by the 30 × 10 mile strip of land between San Francisco and San Jose, California—the Silicon Valley—that became a world centre for the microelectronics industry. The success of Apple Computer, Intel, and many other Silicon Valley firms generated initiatives around the world that sought to emulate the synergy created between Stanford University and its surrounding microelectronics firms (Rogers and Larson 1984). One manifestation has been the creation of science parks and incubator programmes aimed at fostering the development of high-tech firms (Charles and Howells 1992: 96–106). Another has been the use of governmental tax and spending initiatives to create a critical mass of multimedia industries within particular locales, such as in the Los Angeles Basin (Scott 1995).

Nevertheless, government and industry have begun to place more emphasis on fostering the use of ICTs as opposed to more single-mindedly pushing production. This can be seen in a shift of emphasis in the Silicon Valley, where a set of social and economic problems began to threaten the vitality of communities and the ability of firms to recruit the best people (Rogers and Larson 1984). Government and industry initiatives within the region to employ ICTs to improve the quality of life spawned a new vision of a 'Smart Valley', which has also been emulated in other nations and regions, such as in Britain's 'Smart Isles' Project.[6]

However, a focus on ICT utilization was demonstrated even earlier by the government of Singapore. The nation's National Computer Board (NCB), which had responsibility for information systems within the government, developed the vision of an 'Intelligent Island' that was accompanied by a variety of large-scale infrastructure projects (Wong 1997). For example, TradeNet was aimed at using ICTs to increase the efficiency of all aspects of international trade handled through the port of Singapore—a centrepiece of this city state's economy. The centralized structure and manageable scale of Singapore permitted the government to implement major ICT innovations in relative short time frames that demonstrated the potential economic benefits of policy focused on use.

In other nations, such as Japan, industrial policy has had more emphasis on the production of ICTs for export, such as in the computer industry, placing less emphasis on the domestic utilization of IT (Flamm 1987; Fransman 1990; Kraemer and Dedrick 1996). Utilization is probably more critical as a policy focus since ICTs are used throughout the economy and production can be improved by greater levels of demand, closer ties between producers and users, and better conceptions of users (Flamm 1987; Kraemer and Dedrick 1996; Woolgar 1996).

British policy in the 1980s identified the utilization of ICTs as a stimulus to the development of greater capacities for production. For example, the vision of an 'entertainment-led' cable policy of the early 1980s led to the lifting of restrictions on the provision of multichannel cable TV systems in the UK (ITAP 1982, 1983). Entertainment proved to be a less powerful and rapid driver behind the development of IT than expected (Dutton and Blumler 1988), but the rate of cable development in the UK began to accelerate in the late 1990s.

Whether or not this entertainment-led policy was successful, the government encouraged foreign investment in British cable and telecommunications industries, and supported greater competition in telecommunications. Also other European nations, such as France and Germany, lifted restrictions on cable and moved to liberalize telecommunications, following a similar industrial logic (Vedel and Dutton 1990). These policies emphasize utilization as a stimulus to production, and show how opening and closing national borders are central to industrial policy related to ICTs.

Competition in Broadcasting and Telecommunications

Empirical evidence and much faith and ideological commitment to competition versus regulation (of a monopoly supplier) have fostered a growing consensus around a new regulatory model that treats competition in communications as a centrepiece of technology and ICT-led industrial policies, as noted earlier. This is a new paradigm among economists focused on regulation (Derthick and Quirk 1985). However, there is also mounting evidence that competition has indeed contributed to reducing prices and increasing technological innovation in telecommunications, which together support the increased utilization of ICTs across sectors of the economy (Baer 1996; Freeman 1996*a*). In communications, competition has been introduced in telecommunications and furthered in other areas, such as broadcasting.

Telecommunications are particularly critical since they have been traditionally operated as a natural monopoly by a public or government-run monopoly, such as many European Postal, Telephone, and Telegraph operators (PTTs) or as a public telephone operator, such as the British Post Office prior to the formation of BT and France Télécom, or as a private monopoly regulated by the public sector, such as in the case of AT&T prior to divestiture. The divestiture of AT&T was a major impetus behind the introduction of competition in long-distance services. However, it took over a decade longer for the US Congress to adopt policies aimed at supporting competition with the Telecommunications Bill of 1996. Even two years after this reform bill, local and regional RBOCs continued to maintain effective monopolies over the provision of local as opposed to long-distance telephone services. Well before the USA and other major industrial nations, the British introduced competition at all levels as a matter of telecommunications policy and then furthered competition in the 'local loop' by licensing cable TV operators to provide telephone services for their customers.

Competition involves a variety of policies on tele-access. For example, Martin Cave (see Essay 11.2) argues that the UK's early decision to permit only two competing telephone companies, BT and Mercury, delayed the full benefits of competition, with the very concept of a duopoly showing how even competition policy restricts at the same time as it opens access. His essay also shows the variety of access issues involved in competition policy, including access to different markets, such as local, long distance, and international, and access to different facilities and infrastructures, such as in interconnecting with the dominant carrier's network and sharing the underground conduits of a cable-service provider.[7]

Other nations of Europe did not emulate the British model, but nevertheless moved towards the liberalization of telecommunications and broadcasting. In 1994 a high-level industry group recommended the full liberalization of telecommunications throughout the EU (Bangemann Group 1994). This recommendation was followed by an Action Plan of the European Commission (1994) that supported competition through such means as relaxing restrictions on media ownership, and creating a more competitive telecommunications sector across nations of the EU (Kubicek *et al.* 1997).[8]

The Action Plan was presented as a means to support the development of an 'information society' and suggested that effective competition in ICT services must be balanced with other social objectives.[9] Increasingly, there is recognition that competition policy also requires continued regulatory and public supervision to avoid abuses by the suppliers of ICT industries, as well as by the users of the technologies in business and the public sector (Dutton 1996*a*: 13–14).

Often the proponents of getting the government out of communications see the consumer's interests to be a natural by-product of free competition (Gates 1995), but meaningful competition is likely to depend on effective regulation (Neuman *et al.* 1997). For example, the US Congress passed the Telecommunications Act of 1996 to promote the dual objectives of competition and deregulation. Yet the legislation was packed with regulations, while the deregulation which was introduced did not encourage greater competition. In the years following the Act, a host of mergers and acquisitions have left ICT industries more consolidated, not more competitive (see Box 11.3).

Box 11.3. Illustrations of Convergence, Fragmentation, and Consolidation

Convergence of media

- Time Warner Inc.'s acquisition of Turner Broadcasting System Corp.
- Microsoft's 1997 acquisition of WebTV Networks Inc., which produces a system for browsing the Web on TV and Comcast Corp., a cable TV company, together with other ventures, like its 1995 partnership with NBC-TV on a new interactive news channel (MSNBC) that aired in 1996 on TV and the Web
- The RBOC US West Inc.'s acquisition of Continental Cablevision Inc.; AT&T's purchase of cable giant Tele-Communications Inc.

Fragmentation of media

- US Pacific Bell selling its cable TV systems in 1997
- US West buying a cable firm in 1996, then splitting in two, separating its cable and phone business the following year

Consolidation of media

- Microsoft's investment in Apple, and Apple's incorporation of Microsoft's Internet Web browser as a primary option on Macintosh computers
- SBC Communications, the Texas-based Bell Holding Company, merging with the Pacific Telesis Group and Ameritech Corporation; the Atlantic Bell-Nynex merger; and the Bell Atlantic-GTE merger
- Worldcom's $ multibillion acquisitions of telephone and data networks, including its 1997 offer to acquire MCI, outbidding BT; Global One, an alliance between Deutsche Telekom, France Télécom, and Sprint of the USA; and the AT&T/BT joint venture
- America Online's acquisition of Compuserve subscriber accounts

The liberalization of telecommunications has been accompanied by the introduction of increased competition in broadcasting. In Europe, this has been approached through policies promoting the development of multichannel cable and satellite services, as well as the introduction of more private, commercial broadcasting, supported by advertising and pay subscriptions. Across Europe, for example, most nations had less than five TV channels in the early 1980s, but dozens by the late 1990s, easily doubling the provision of broadcasting. Terrestrial, cable, and satellite channels are competing for audiences and advertising revenues, alongside public-service broadcasters, such as the BBC, that remain very strong through funding by the broadcast licence fee (Peacock 1986). Around the world, the big digital TV, cable, and satellite packagers promise a 500-channel future.

The close connections between competition and industrial policy can be seen in the creation of policies aimed at promoting the development of the information superhighway. In the 1980s, for example, the UK Cabinet Office report on cable TV (ITAP 1982) recommended a 'two wire' model of information highways into the home. One would be from a public telecommunications operator (PTO) such as BT, and the other from a cable TV operator. One means of encouraging this model was to prevent the PTO—BT—from carrying entertainment services, such as TV.[10] Supporters saw this ban as important to the vitality of new cable companies and other telecommunications companies competing with BT. Critics argued for the elimination of these restrictions on BT and other PTOs (Trade and Industry Select Committee 1994), putting forward a position close to the US NII vision of permitting 'any company to provide any service to any customer'. Here the logic was one of convergence—that any service can be provided over a cable or telecommunications network and therefore the ability to provide entertainment services, such as pay-TV, would create a financial incentive for BT to invest in advanced networks.

The centrality of decisions on tele-access to competition policy can be vividly seen by contrasting efforts to use competition to build ISH in Britain and other nations of Europe, which in an earlier era chose to restrict the market-led development of cable TV as a threat to equity. In the UK and other nations of Europe, where broadcasting was developed as an important public service, access to broadcasting has long been a central aspect of public policy. Terrestrial broadcasting networks were developed nationwide in ways that made public-service broadcasting available to all regions (Negrine 1985). In contrast, the commercial development of local and commercial broadcasting stations in the USA tended to leave many rural and suburban areas with limited access, such as to only one or two of the major commercial networks, while major TV markets, such as New

York and Los Angeles, supported multiple local broadcasting stations in addition to local affiliates of all three broadcasting networks.

In the 1970s, the Labour Party set up a committee to look into broadcast policy shortly after it came into office. In 1977 its report dismissed the prospects for cable-system development in the near future and rejected the 'patchwork system' that it forecasted should the existing cable industry be allowed to evolve under a liberal regulatory regime. It judged the scale of funding required for the systematic development of a nationwide cable grid to be far beyond what could be afforded by the public sector at that time. Referring to cable TV as a 'parasite', living off programming originated by others, the report was sceptical of a cable network developing in Britain before the end of the century (Annan Committee 1977).

Two years after this report, in 1979, the Conservative Party won the general election, ushering Margaret Thatcher into office as Prime Minister. One of her early initiatives was to put forward the idea of an IT-led industrial policy with the establishment of an Information Technology Advisory Panel (ITAP). This panel had a major impact on redirecting the nation's policies in telecommunications, cable, and broadcasting, removing restrictions on cable TV, and introducing competition in telecommunications to encourage the market-led development of telecommunications infrastructures and services (ITAP 1982, 1983).

ITAP saw cable as an information infrastructure that had a strategic role to play in realizing an information-technology-oriented industrial policy for Britain. Cable became an issue relevant to international trade, economic development, and industrial policy—not only cultural policy (Humphreys 1986; McQuail and Siune 1986). Ministers and civil servants from the Department of Trade and Industry and leaders within the telecommunications, electronics, and computing industries joined the main players from the broadcast industry, the Home Office, and cable industry in decisions relevant to the future of broadcasting. In this way, an ecology of policy games played a critical role in the development of cable and telecommunications policy in Britain. The view that cable was a strategic economic resource brought industrial élites into the cable and broadcast policy arena.

Trade Quotas: Technologies without Frontiers

Other reasons for restrictions on the development of cable and satellite TV are also tied to issues of tele-access. One was an effort to protect public-service broadcasters from unrestricted access to foreign broadcasts. For example, this was an element in the early history of broadcasting in the UK (see Box 11.4).

Box 11.4. British Policy on Access to Foreign Broadcasting over Wire

In 1925, small operators in the UK were allowed to use wired systems to relay radio broadcasts to households with inadequate off-air reception. This market was created by delays in the construction of transmission facilities outside the major urban areas. The facilities also provided a vehicle for the production of original programming and the retransmission of foreign broadcasts that posed a threat to the British Broadcasting Corporation. As early as 1930, the BBC convinced the British Post Office to establish a ban on the origination of programming by relay operators and to place restrictions on their use of foreign programming. In the late 1930s, the Post Office began issuing licences to relay operators that required them to carry a predominance of BBC programming. In 1951, the first relay operators began to retransmit television programming in the UK, but these early cable TV operators, like their predecessors, were also limited to the distribution of services of British broadcasters. These restrictions survived two broadcasting inquiries conducted by the government during the early years of cable TV. From the 1960s through the early 1980s, broadcasting authorities sanctioned only small experiments with multichannel cable systems, such as community programming in communities like Milton Keynes (Firnberg and West 1987).

Sources: Hollins (1984: 36); Negrine (1985).

National restrictions on cable TV began to be reduced throughout Europe in the early 1980s. In addition, the view that cable and satellite systems were inherently international in their reach and critical to economic development led the EU to propose regulations that would create a common market in broadcasting. A 1982 resolution of the European Parliament called for the development of a European Affairs Channel and the standardization of rules on such practices as advertising to permit a 'television without frontiers' across Europe. The European Commission supported the establishment of a common market in broadcasting, but many in Europe sought to maintain greater national control (see Box 11.5).

Discussion about pan-European broadcasting was overshadowed by a 1989 'Television without Frontiers' EU directive that required at least 51 per cent of the content broadcast in EU countries to be originated within the Union. However, these were just guidelines to be met whenever practical and were often not met. France sought to strengthen European quotas, making broadcast trade restrictions one of the most controversial topics in international negotiations between the USA and Europe. However, revised versions of this directive retained the 'where practicable' phrase. As in the case of industrial policy generally discussed above, free trade in

Box 11.5. British Arguments Opposing a European Television without Frontiers

A select committee of the British House of Lords was appointed to review EU proposals for a television without frontiers. It raised a number of issues that were in opposition to the proposal and in conflict with dominant policy initiatives of the government. The Cabinet viewed broadcasting and telecommunications as an economic matter as well as a culture concern (ITAP 1982). The committee's concerns included:

- interest in preserving national broadcasting traditions that vary widely across Europe, such as nationally distinct policies towards advertising;
- the belief that TV is a cultural rather than an economic matter, and therefore not within the purview of the Commission;
- the view that national boundaries on TV would discourage foreign 'American' imports, and therefore enhance the development of European content;
- fear that European regulations on copyright, advertising, and other practices would create a new layer of additional regulation; and
- opposition to the establishment of a government channel for European affairs, which was outside the traditions of UK broadcasting policy.

Source: European Communities Committee (1985).

broadcasting is expected to support the development of local production capacity in the longer term (Waterman 1988). Following such economic arguments, Europeans have renewed calls for trans-European information superhighways to facilitate pan-European tele-access (Mackintosh 1986; Dixon 1994).

In the 1990s concerns of Asian nations, such as Iran and China, over unrestricted access to satellite broadcasting and to the Internet both diffused and legitimated the issue of tele-access as a major issue of cultural sovereignty across a wider variety of media. American computer software, like Microsoft Windows, and Japanese game machines, like Nintendo—not only Hollywood movies—became identified with concerns over the cultural sovereignty of nations and regions. And this was a two-way street, at times, with the purchase of Japanese supercomputers by public agencies in the USA touching off debates over threats to American suppliers, such as Cray Research Inc. (Iritani 1996). And the US Congress raised concerns over children obtaining international access to pornography, which in part led to calls for the Communications Decency Act (CDA), discussed in the next section.

Content Controls: Censorship and Freedom of Expression

The new ICTs are said to be impossible, but also inappropriate, to regulate (de Sola Pool 1983*a*). Impossible, because control over them was thought to be inherently distributed and incapable of being centrally controlled or censored. Inappropriate, because computers would be the newspapers of the future (de Sola Pool 1983*a*). In line with this deterministic thesis, many Americans have argued that a strict interpretation of the first amendment, which protects freedom of expression, should be extended to the Internet, and the US courts have strongly supported this view, arguing that the factors justifying regulation of broadcasting are 'not present in cyberspace' (US Supreme Court 1997). Many nations, even those without policies or traditions in line with the first amendment, operate only limited governmental censorship and regulation of the Internet. This can be seen in Malaysia's establishment of a regulatory free zone to encourage companies to locate in its Multimedia Super Corridor, free of concerns over governmental censorship (see Box 6.4).

However, there is extensive worldwide debate over how content should be regulated over the Internet, as well as other electronic media such as satellites that span national boundaries (de Sola Pool 1990). Copyright protection and privacy, for example, entail constraints on content. The dissemination of illegal or offensive material has generated proposals to regulate content on the Internet and Web, such as the Communications Decency Act of 1996 (CDA), which was designed to shield children from harmful material on the Internet. The US Supreme Court (1997) ruled the CDA to be unconstitutional, on the grounds that its provisions would abridge 'freedom of speech'.

As the court's action makes clear, the regulation of tele-access is not equivalent to getting the government out of ICT policy. The first amendment itself regulates tele-access in calling for government to protect the public's right to access what some might deem objectionable or indecent material.

In addition to calls for regulation, ICTs are making content control more feasible, even over the Internet. ICTs like the V-chip can be used by families to screen TV programmes, shifting content controls from the providers and government regulatory agencies to private companies and individuals, who offer content controls as a service to the user. ICTs like Platforms for Internet Content Selection (PICS) make it technically easier for parents and teachers, but also any number of private and public actors, to screen the Internet's content (see Box 11.6). The developers of PICS believe they facilitate user control over access, and provide an alternative to censorship. Their critics contend that PICS make the Internet into a

Box 11.6. Platforms for Internet Content Selection (PICS)

PICS were developed at the WWW Consortium at the Massachusetts Institute of Technology (MIT). They are a set of technical standards for creating and attaching or detaching electronic labels to Web sites that describe their content. Search engines locate sites with particular words, so that they can be viewed, or screened out. In contrast, PICS enable the creator or distributor of a Web page to describe the content as 'educational', 'funny', or 'obscene', for example. A third party, such as an Internet Service Provider, can attach labels on all the items received off the Web. Any user, parent, or teacher can control what can be viewed on the Web by screening out all content labelled as unsuitable for children, for example. Of course, this is a very general-purpose technology, which could be used to screen any type of content. It creates an important gatekeeping role for those who label content, as well as for those who manage gateways into a home, corporate intranet, or national gateway.

Sources: Resnick and Miller (1996); Resnick (1997).

'censorship machine' (Caruso 1997). However, policy will be critical in this and other cases, since the technology enables them to play either role.

Regulating the Use and Consumption of ICTs

Many other ICT policies are focused more on the public's access to ICTs as citizens, consumers, or users. This includes such traditional issues as universal service, applicable to emerging competitive telecommunications regimes as well as to traditional monopolistic regimes, but also a host of other issues. In this area, there has been an opposite tendency—towards constraints on tele-access—such as in the area of copyright.

Intellectual Property Rights: Copyright as a Restriction on Access

Intellectual property rights (IPR), including laws and policies governing patents and copyrights, provide an important example of areas in which restrictions on tele-access are perceived to be critical to the production of ICTs. Efforts to ensure the security of information, such as through encryption, comprise another.

Existing IPRs have been challenged by technological changes that make it easier and cheaper to copy. ICTs also make any content more independent of any particular medium. A book can be printed, or distributed electronically, for example, pushing authors and publishers to update their copyright agreements to cover reproduction, storage, and retrieval in 'any form or by any means'. As Vincent Porter explains in his essay (see Essay 11.3), ICTs have also eroded national boundaries in ways that require greater international harmonization of law and policy in this area.

Copyright is often viewed as non-controversial—a benign restriction on access that benefits the public at large. However, the details of copyright vary across nations and over the years. The adoption of one or another policy can influence the incentive structure of participants by reordering the incentive structure of who benefits and who pays for a given work. Therefore, proposals in Europe and North America have generated debate among authors, print and electronic publishers, librarians, and on-line service providers, all of whose interests will be differentially affected by changes in who has the right to copyright material and use it in a variety of applications and contexts. Much is at stake. The Internet developed on the basis of offering a shared and free distribution of information, but its commercial future will depend increasingly on the ability of providers to restrict access to those who pay for the use of digital works.

Universal Service in Telecommunications

The concept of 'universal service' has also evolved through the years, but has become widely understood to refer to making plain old telephone services (POTS) available to everyone at an affordable price (Blau 1997). In the era of monopolies, telephone operators pursued this objective by averaging costs for the provision of telephone services across all users, so that the operator could subsidize the costs of providing services to rural and other hard-to-reach areas and ensure that low-income households could afford access to some minimal level of basic telephone services.

The introduction of competition in such areas as long-distance telephone services pushes prices towards their actual costs. In combination with technological innovation, this creates major incentives for reducing costs and improving the efficiency of service provision. However, it requires mechanisms other than cost-averaging to support universal service. These could take the form of direct subsidies to users, indirect subsidies to public agencies or service providers for offering a service, or a 'universal service fund', to which all service providers contribute (Dutton *et al.* 1994). For example, the US Telecommunications Act of 1996 calls for these latter two approaches, providing for long-distance telephone companies to contribute a substantial sum to finance access in libraries, schools, and health facilities, in return for reductions in the fees they pay to use the facilities of the local telephone companies.

Concepts of universal service are widely accepted on the grounds of equity and public safety, sometimes referred to as a 'lifeline' service, because the telephone has become viewed as a necessity for a minimum standard of living in advanced industrial societies. It also gains acceptance through an understanding that the value of a two-way telecommunications network is a function of its ubiquity. The more people you can reach from your telephone, the more valuable it is to you. Therefore, universal service is a public good in several respects.

However, controversy surrounds the definition of: what services should be universal; what is affordable; who should be subsidized; and how these services should be paid for. The resources to support universal service raise the costs to all other users. All users benefit from the extended reach of interactive media, and the achievement of greater equity, but increased costs can diminish the utilization of ICTs, and therefore the vitality of the overall economy as well as the ICT industry. This means concerns over equity need to be constantly balanced with efforts to spur economic development and the growth and competitiveness of ICT industries, which itself is expected to be a force for lower prices.

Information Policy: Public Information and Facilities for Access

The promoters of an ISH envision public access to electronic information infrastructures in ways that promise to extend concepts of universal service, such as to e-mail (Anderson *et al.* 1995). Related initiatives to expand public access to electronic information resources include projects supported by the US Telecommunications and Information Infrastructure Assistance Program (TIIAP), pilot projects of the European Commission, and many local, national, and regional experiments (see Chapter 7). For example, the Federal government made the electronic holdings of the National Library of Medicine (NLM)—over 9 million medical articles—available to the public free, over the Internet (Gore 1997). This has involved the NLM in a set of efforts aimed at ensuring access to this information in distressed areas of inner cities and rural communities (Allen and Dillman 1994). A report on the role of the ISH in rural development put it this way: 'The main problem in the future will remain the issue of access: access to infrastructure at reasonable cost in remote areas (a policy issue) and access by individuals to the networks, machines, information, and knowhow of IT' (Bryden *et al.* 1995: 16). It is to the interaction of these access issues at various levels and in various sectors of society that I turn in the final chapter.

Organizing an Ecology of Games: The Role of the Information Superhighway

The concept of tele-access captures the connections across policy issues and sectors that can help provide a foundation for policy debate and actions. This is exactly the role that the ISH played for several years. It provided a vehicle for creating a new giant meta-game that helped organize a very diverse set of policy issues around the development of an ISH. This set of national and global projects promised rewards to those who win, and posed threats to those who lost. The joke in Washington, DC in early discussions of the ISH was: 'If you're not part of the steamroller, you're part of the road' (Baer 1996: 369).

In addition, the vision of an ISH entails all four dimensions of tele-access (see Box 11.7). The ISH should, according to its proponents, provide the infrastructures necessary to connect every business and household to all kinds of electronic services (R. L. Smith 1970, 1972; Gore 1991, 1993). It is expected to democratize access to information, and connect people locally and globally in ways that enhance community (de Sola Pool 1983*a*; Gore 1991).

Box 11.7. The Information Superhighway as a Vision of Tele-Access

The information superhighway proponents—past and present—argue that advances in ICTs will improve access to:

- *information*: democratizing its production and consumption;
- *people*: enhancing community on a local and global scale;
- *services*: ICT networks should be viewed as an electronic highway, bringing all kinds of services to businesses and households; and
- *technology*: more equally distributing the know-how and infrastructure for a networked society and economy.

Sources: de Sola Pool (1983*a*); Dutton *et al.* (1987*a*: 8–9); Gore (1991).

There was leadership, and momentum behind this vision of an ISH. Legislation, directives, committee reports, and projects evolved around the world (Kubicek *et al.* 1997).[11] Most of the technology already existed, and, within the most advanced industrial economies, the resources required to deploy them more widely were within the reach of government and industry. However, discussion of the ISH has become more fragmented once again, as policy debate has become more focused on technological visions such as the Internet, and retreats back into such well-known sectors and issues as copyright, privacy, competition, and universal service.

The tele-access goals of the ISH are more likely to be achieved if players understand the degree to which related policy issues, such as copyright, also entail issues of access. If so, policy-makers can deal with the complexity of regulating tele-access in more effective ways, rather than approaching each communications policy as unique, or simply an issue of deregulation —one of getting the government out of ICT policy. Regulating tele-access is an inherent effect of all ICT policy choices.

Notes

1. This section is based on Dutton (1992*a*, 1995*b*), which provide a more detailed discussion of the ecology of games as a perspective on public policy.
2. A more comprehensive inventory of Federal, state, and local government actors and international institutions shaping communication policy in the USA is provided by Nadel (1991).
3. An overview of the EU and its governance is provided by Leonard (1994).
4. Robin Mansell's work (Mansell 1996) has sought to broaden discussion of competition, and this need was a theme of PICT's synthesis workshops.
5. Technical standards can advantage access of different providers. In addition, the process for setting standards has tended to advantage the largest firms, who have the resources to 'sustain a level of participation in standards committees' that is necessary to influence their direction (Hawkins 1996: 178).
6. Smart Valley Inc. defines itself as a 'non-profit organization committed to creating an electronic community based on an advanced information infrastructure. This mission is being carried out through collaborative infrastructure projects in education, health care, local government, business and the home' [http://www.svi.org/].
7. Nicholas Garnham (1990: 146) identifies four levels of competition: 'in the local loop, in the trunk network, in international circuits and in private leased circuits'. It is also useful to look at data networks, and mobile communications, both analogue and digital (Baer 1996: 355).

8. From January 1998, the EU opened up fixed and mobile telephony to competition, as well as voice and data communications, and infrastructures. Foreign ownership of national telecommunications operators will also be allowed.
9. Ironically, the EU plans for an 'information society' often had a technical focus, while US plans for an 'information infrastructure' often focused on social issues (Kubicek *et al.* 1997: 19–20).
10. This restriction on the PTO was later endorsed by the McDonald report on the liberalization of UK telecommunications (DTI 1988), as well as by a Government White Paper on telecommunications (DTI 1991).
11. The ISH has meant different things internationally (Dutton *et al.* 1996; Kahin and Wilson 1997). For example, the Clinton–Gore Administration had a different mental model of the information highway—one which was based on the technology of the Internet—than did earlier proponents of a trans-European broadband network (Mackintosh 1986). Until the late-1990s, most discussion in Britain was anchored more in the modernization of existing telephone and cable networks than in the next generation of the Internet (DTI 1988). Likewise, entrepreneurs in Germany have placed more emphasis on satellite communications as the highway to households, than has the industry in the USA or Britain (Kleinsteuber 1997).

Essays

11.1. The Interlocking Pieces of the Information Economy

Robin Mansell

An intricate system of networks is needed to underpin an economy in which electronic information is produced and used, exchanged and processed, to a greater degree than ever before. Robin Mansell explores the social and economic implications of technical choices about the configuration of such networks.

11.2. Competition in Telecommunications

Martin Cave

The UK was the first major country to introduce competition in local telecommunications services. Much can, therefore, be learned from its experience, which is analysed here by Martin Cave.

11.3. Intellectual Property Rights

Vincent Porter

The laws of intellectual property, particularly copyright, will influence the structure of the market place in information goods and services. Vincent Porter highlights issues that need to be taken into account in determining the precise nature of these laws.

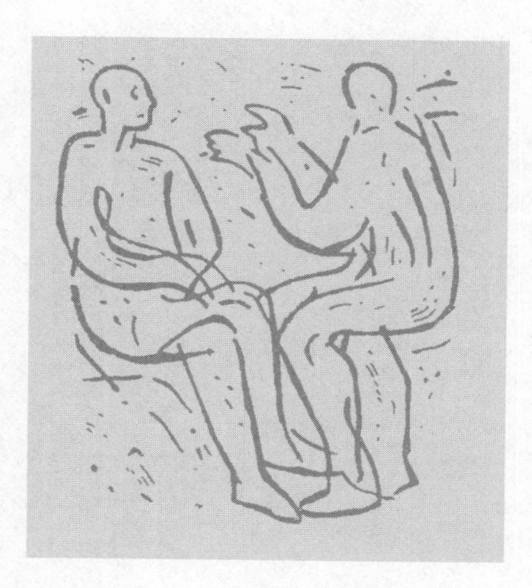

11.1. The Interlocking Pieces of the Information Economy

Robin Mansell

The information economy is like a very complex jigsaw puzzle. When the picture is completed, it promises to bring more prosperity, greater employment opportunities, and an enhanced quality of life. Many observers believe that all these benefits will be achievable once the pieces fall into place for the 'network economy'—the part of the overall picture made up by the underlying system of electronic networks and services (Mansell and Wehn 1998).

This networked information infrastructure is often depicted as a series of technical choices about architectural design, interface protocols, and implementation approaches. However, there are many other dimensions. For example, in the political realm there is agreement on the need for greater competition in the communication industries to dislodge the monopolistic companies and stimulate innovation. In addition, the suppliers building the networks come from a variety of industrial backgrounds—telecommunications, computing, and audiovisual. These industries are technically convergent because of their dependence on digitally encoded information and software and are also beginning to converge at the organizational level in private and public sectors. At the policy and regulatory level, however, they continue to be extremely divergent.

The fact that competitiveness and the capacity to innovate are linked is widely understood and discussed in many public policy and corporate documents which seek to strengthen the links (see e.g. European Commission 1994). Nevertheless, there is a very important piece of the puzzle that continues to be missing: in most parts of the world there is little consensus on policies and regulations that will strengthen innovative capacity and competitiveness as well as promoting the accessibility and affordability of electronic communications networks.

There is an interesting common characteristic in most countries which have addressed this issue. Changes in policy and regulation likely to affect the technical innovation process take place separately from those which address market structure and competition. For instance, changes in policy or regulation that are likely to affect technical choices (such as network architectures and standards) are invariably introduced independently of changes in policy and regulation that will affect the long-term structure of markets and the process of competition (such as with prices and costs).

A re-coupling of these two strands of discussion is essential if information infrastructure policy and regulations are to be effectively coordinated in a way which stimulates competition and innovation in the delivery of, and affordable access to, advanced networks and services. Yet it is not clear that the institutional and policy-making process itself is well attuned to the need to address these issues in a coherent way.

Studies of technical change, innovation, and competitiveness have yielded two particularly useful insights in this context (Mansell 1996). First, market failure and tendencies towards monopolistic or oligopolistic markets are not inherently bad; they are an essential stimulus to innovations. Secondly, the reverse may also be true. Actions aimed at stimulating innovation through co-operation among firms, or by protecting national champions, suffer from all the inefficiencies and lack of responsiveness to customers for which monopolistic markets are notorious.

The production and use of all forms of knowledge are inextricably linked both to the configuration of intricate technical networks and to the industrial and policy networks of institutions which shape the network economy. However, there is often reluctance on the part of private and public actors to sanction the establishment of new policy or regulatory institutions to meet future needs—a delay which could find that a window of opportunity has closed. To achieve greater coherence in policy and regulation in the face of convergent technologies, policy-makers will need to take responsibility for guiding incentives for innovation and competition in ways that are consistent with the social and economic aspirations of the communities that populate the network economy of the twenty-first century.

An effective public policy and regulatory regime must be concerned with more than competition and innovation on the supply side of the network market place. It should also be involved with the user side and the practical ways that electronic information is produced and applied. These social and organizational aspects are evident in the way experimentation with information-related applications is seen in Japan as a way of pulling the services spawned by technical convergence through to commercial realization. This kind of emphasis is less evident in the USA or Europe.

If the network infrastructure is to support the full benefits of an information economy, policies will have to ensure businesses and consumers can signal their preferences for new services in the market place and that affordable access to the networked economy is provided for the majority of citizens. Establishing the information infrastructure and the related network economy therefore cannot be treated purely as a technical process. It must be seen as an integral activity in the piecing-together of the information-society jigsaw.

References

European Commission (1994), *Growth, Competitiveness, Employment: The Challenges and Ways Forward into the 21st Century*. A White Paper (Luxembourg: Office for Official Publications of the European Communities).

Mansell, R. (1996), 'Network Governance', in Mansell and Silverstone (1996), 187–212.

—— and Silverstone, R. (1996) (eds.), *Communication by Design: The Politics of Information and Communication Technologies* (Oxford: Oxford University Press).

—— and Wehn, U. (1988) (eds.), *Knowledge Societies: Information Technology for Sustainable Development* (Oxford: Oxford University Press).

11.2. Competition in Telecommunications

Martin Cave

The UK started to liberalize value-added telecommunications services and equipment supply in the early 1980s. Most significantly, in 1984 it licensed a second operator, Mercury, to compete directly on fixed-link services with the newly privatized incumbent, BT. For the next seven years, competition was restricted to this duopoly.

Mercury rolled out its network quite quickly, but it attained only a small overall market share. However, it made considerable inroads into certain of BT's more lucrative markets, notably large business customers making substantial international calls, especially in the City of London (Cave and Williamson 1996).

In 1991 entry was liberalized into all UK telecommunications markets, except international services. By then, many other countries had caught up with the UK, and some had overtaken it. Meanwhile, there was vigorous resistance to liberalization in some countries. For example, the EU's policy of completely liberalizing its telecommunications markets by 1998 was forced upon many reluctant incumbent monopoly suppliers and some unwilling governments.

In the UK, the full liberalization introduced in 1991 led to an attack on BT and Mercury on all fronts—in local service generally from the cable operators, which supplied over 2½ million lines by 1997; in central business districts from competitive access providers; in long-distance markets from Energis; and in international services from resellers of telecommunications capacity. Although BT continued to dominate in local markets for some time, it soon began to lose significant market share in long-distance and international markets (see Table 11.2).

Table 11.2. UK Telecommunications Market Share, Jan.–Mar. 1997 (%)

	BT	Mercury	Others
Call revenues total 1997	79.0	7.8	13.2
comprising			
local	88.7	2.0	9.3
national	78.4	9.5	12.1
international	58.2	13.6	28.2
Exchange lines total	89.8	1.1	9.1
comprising			
business	90.6	5.4	4.0
residential	89.5	0.0	10.5

Source: Oftel (1997).

The UK has also successfully evolved competitive radio-based wireless services, such as used in cellular mobile phones. By 1997 there were four operators of such networks and over 7½ million subscribers, which is one of the highest rates in Europe for this market. As interconnections between wireless and fixed networks become more widely available, this may lead to a significant threat to BT's monopoly in the provision of access outside cabled areas.

The pattern of regulation in the UK has been critical to these developments. For instance, the UK's Director-General of Telecommunications and the Office of Telecommunications (Oftel) have played crucial roles in the development of competition through price caps on BT and the regulation of interconnection payments.

Price controls have restricted the amount by which BT can raise prices in real terms for a given basket of services. These caps were successively tightened, with cuts of 7.5 per cent required for each of the years 1993 to 1997. From 1997, prices of a smaller basket of services have to fall by 4.5 per cent in each year to 2000. The severity of these controls has led to some concerns that they are a grave constraint on the development of competition. However, any entrant into the sector would have to take into account what BT would charge if subjected to serious competition, so it is not clear that entry would be more attractive in the long run if price-control arrangements were looser.

The amount an operator pays to interconnect with BT's network influences entrants' prospects even more, especially in the early stages of starting up a service when over half an operator's total revenues may have to be passed to BT. Oftel has tended to tip the balance in favour of entrants. First, it granted Mercury relatively favourable interconnection terms in 1985. Then it waived entrants' contributions to BT's access deficit, which is based on the extent to which quarterly rentals and connection charges fail to cover the costs of attaching subscribers to the network. However, this system was considered too unwieldy and Oftel eliminated it.

One possible alternative would be to cease to control many of BT's retail prices. Instead, a price cap could be imposed on the charges BT make for the use of its network by other operators, in the form of interconnection payments, and by its own associated retail business. These reforms were implemented in 1997 and represent a significant withdrawal from intrusive regulation. They make Oftel increasingly like a competition authority policing any abuses of dominant positions, rather than a command and control regulator.

Changing regulatory approaches obviously affect the business strategies of participants in the market. The interaction of incumbents and entrants can usefully be viewed as a race between the incumbent's capacity to adapt to the new environment and the entrant's ability to acquire the assets necessary to survive. One of those assets is assistance from the regulator. But no sensible entrant would rely on that in the medium-to-long term and, possibly, not even in the short term.

Mercury's initial strategy was to follow BT as price leader, maintaining a 15 per cent or so discount which it believed necessary to attract profitable long-distance and international business. During the duopoly period, Mercury made quantitatively small, but profitable, inroads into BT's market share. In other countries, entrants launched a much more vigorous all-out assault on the incumbent's market share, probably to their long-run advantage. Mercury also diversified into other services, such as public call boxes and apparatus supply.

In retrospect, this policy seems to have been misguided. Many of these activities required a scale which Mercury was not able to achieve. After the liberalization of entry in 1991, and the concurrent regulatory change which permitted BT to offer discounts to large customers, Mercury was forced to cut back and refocus its activities. For example, it withdrew from public call box provision in 1995.

Other entrants have based themselves on exploiting an economy of scope or use of a novel technology. Energis, for example, has relied upon access to the National Grid Company's network of electricity transmission pylons. Cable companies take advantage of their ability to use their networks to provide both entertainment and telecommunications services. Ionica uses a radio-based technology with low start-up costs. Strategies like these avoid simple duplication of the BT's technology and are therefore more likely to succeed.

Overall, the UK's experience of competition has been mixed. Benefits have been gained from falling prices, improved BT productivity, and an increasing number of business and residential customers having access to alternative suppliers. On the other hand, the duopoly years seem to have delayed the liberalization process. Other countries seeking gains from liberalization can, therefore, derive some comfort from the UK's experience, as well as learning lessons from it.

References

Cave, M., and Williamson, P. (1996), 'Entry, Competition, and Regulation in UK Telecommunications', *Oxford Review of Economic Policy*, 12/4: 100–21.

Oftel (1997): Office of Telecommunications, *Market Report* (London: Oftel, Information Update, September).

11.3. Intellectual Property Rights

Vincent Porter

Copyright law is increasingly being used to encourage investment in the new information economy being created by ICTs. The technologies themselves are protected by patent rights. The protection of Intellectual Property Rights (IPR) through copyright legislation means that a balance must be found between the interests of authors, investors in various products and services—such as electronic and print publishers—and the public.

The international community relies on a series of national laws, which meet the minimum standards set out in two international conventions—the Berne Convention for the Protection of the Rights of Authors in Literary and Artistic Works and the Rome Convention for the Protection of Performers, Producers of Phonograms and Broadcasting Organizations—which afford reciprocal protection to foreign works and their authors. These have been reinforced by provisions in the General Agreement on Trade in Services of the World Trade Organization, which also protects computer programs as literary works and, unlike the two international conventions, grants rental rights to authors of film, and computer programs, and to record producers. In addition, national legislation must impose criminal penalties as well as civil sanctions for breach of copyright (see Stewart 1989; Porter 1992, 1995).

There are still significant variations in national copyright laws, depending on legal and cultural traditions. For example, the stated aim of US intellectual property laws was to serve the welfare of the public by promoting the progress of science and the useful arts. In most of mainland Europe, however, the underlying principle was to protect the intrinsic human rights of individual authors, which had both an economic and a moral dimension. This meant that the fair-use provisions of common-law countries, such as the USA and the UK, were generally more extensive than those on the mainland of Europe. Although European countries gave a far longer term of protection, they normally only afforded full protection to works by individual authors. They have a lower level of protection for what were termed 'the neighbouring rights' of broadcasters, CD producers, and other business organizations who produced commercial artefacts from works protected by a separate author's right. The harmonization of national laws within the European Union and the European Economic Area has meant that neighbouring rights are protected almost as extensively as author's rights and both have been given the longer term of protection.

The economic protection afforded by copyright laws has been extended to protect computer programs and electronic databases which are not properly in the literary or artistic domain that was the locus of earlier IPR legislation; and in some countries, employers even own the rights of their employees. In contrast, the moral rights of authors have generally been ignored. They were excluded from the protection of the WTO Agreement, and the laws of some countries hardly protect them at all. They are barely protected by ancillary laws in the USA, and in the UK the author has to assert the right to be identified as the author of a work. Furthermore, publishers can require authors to waive their moral rights to prevent any distortion or mutilation of their work.

When the Berne Convention was first signed in 1886, in addition to protecting the rights of authors, it also sought to guarantee freedom of information and to encourage learning. But the increased emphasis which the international community has placed on raising the minimum levels of national protection in order to safeguard the rights of publishers and investors is jeopardizing the freedoms of software engineers, scholars, librarians, and other users. For instance, the term of IPR protection for computer programs and electronic databases has been extended far beyond that necessary to encourage science and the useful arts; and in both the USA and the European Union extensions in the term of protection have also meant that works which were in the public domain have been brought back into a protected market place.

Another series of regulatory developments which threaten the freedom to communicate are the uneven extensions of the protection afforded by national laws to pseudo-literary works, such as compilations. All these are in excess of the mandatory minimum requirements laid down in the Bern Convention.

From an economic perspective, IPR are monopoly rights which, if used improperly, can distort free trade in IP-related goods and services. As IP-related trade moves from the periphery to the centre of advanced economies, it becomes essential to bring the exercise of IPR under the aegis of national and supranational competition laws. In the EU, the competition provisions of the Treaty of Rome already take precedence over the exercise, but not the existence, of IPR, especially when their exercise constitutes an abuse of a dominant position.

The need for the international community to rethink copyright laws and conventions has been intensified by ICT innovations like the potential to convert all media into digital form, and by the establishment of the World Wide Web and other global networks which can relay those works across the world. The ease and speed with which digital information can be transformed and transmitted pose enormous intellectual and administrative challenges in deciding what should be copyrighted and how that protection can be enforced.

The advent of digitization means that the international community will have to draw a clear distinction between protected works and unprotected information. Not all digitized information needs to be protected, although works like films should continue to be protected even in digital form. The international community will therefore need to agree which packets of digitized information are to be protected, and which are not.

If the international community persists with the traditional approach of setting minimum standards of protection for national laws, there is a grave danger that a two-tier system will emerge. In this, an international infrastructure, like the Internet, will operate with virtually no IPR protection. At the same time, a series of national information superhighways will enforce copyright protection based on the laws of that country. Even if international links between national superhighways were established by bilateral treaties, it might still be difficult to decide in which jurisdiction any alleged offence occurred.

In order to establish a truly global information infrastructure, there will have to be both an internationally agreed standard of protection and an internationally recognized regime of enforcement. This could be achieved most simply by establishing a new international jurisdiction for defining and protecting digitally transmitted works. A less satisfactory solution would be to delegate the problem to the businesses using the net. In addition, it may also be necessary to give a non-waivable right of integrity for the digital transmission of works to broadcasters and producers of sound recordings and audiovisual material.

In determining specific IPR laws for digitized information, the full consequences should be carefully considered. For instance, a decision has to be made about whether to protect digital transmissions by reinterpreting traditional rights for communicating a work to the public or by creating a new digitized distribution right. Important policy issues regarding the promotion of science and the arts are at stake in this decision. An extensive reinterpretation of the right to communicate work to the public is likely to entrench the market position of the owners of old rights, which would then have new markets opened to them. The creation of a special digitized distribution right would be a new right, restricted to new works, which would stimulate investment in new digital works.

Many other choices like these will have to be made to ensure an appropriate balance is maintained between the interests of authors and publishers of digitized works, and those who wish to use international information superhighways to access a wealth of knowledge, information, and entertainment.

References

Porter, V. (1992), *Copyright and Information: Limits to the Protection of Literary and Pseudo-Literary Works in the Member States of the European Community* (Luxembourg: Commission of the European Communities, DG IV).

—— (1995), 'Wanted: A New International Law to Project Intellectual Property Rights on the GII', *Intermedia*, 23/4: 31–6.

Stewart, S. M. (1989), *International Copyright and Neighbouring Rights* (2nd edn., London: Butterworths).

12 The Politics of Tele-Access: Social Relations in a Network Society

Technical breakthroughs, electronic gadgets, instant billionaires, and computer nerds interest a broad public. This fascination is reinforced by visionary perspectives on an information society bringing new opportunities for work, expression, and human association and experience (Negroponte 1995; Cochrane 1997; Dyson 1997). The very promise of ICTs also entails a major threat—that most of the world will not be on-line. The fabric of society is at stake.

Charting Uncertain Futures

Many respected futurists who are optimistic about the ICT revolution remain uncertain about its long-term social and economic implications (Drucker 1995; Handy 1996). More pessimistic critics speak in harsh tones of increasing inequality in a society that is distancing itself from reality (Slouka 1995; H. I. Schiller 1996). As the public's euphoria with the Internet calms, a list of hot-button social issues is taking shape once again. Will innovations in ICTs enhance or erode personal privacy; renew democratic processes or stifle debate; narrow or widen gaps between information rich and poor; create or destroy jobs?

Forecasting the future of ICTs is anything but straightforward (see Chapter 4). Nevertheless, individuals make policy on the basis of more or less well-considered images of the future. One nationally coordinated effort in technology forecasting was undertaken by the UK's Office of Science and Technology (OST). This process highlighted areas of widespread agreement, but also uncertainties about the future of ICTs (see Essay 12.1).

Most experts involved in technology foresight studies like this agree that ICTs will become increasingly central across all sectors of the economy, making the economic stakes in the development of ICT networks and services of great importance to firms and nations. This reinforces evidence that local, national, and regional competitiveness will continue to drive investment and innovations in ICTs well into the twenty-first century.

The OST exercise led to general agreement that developments in digital multimedia were leading towards technological convergence, such as in the greater use of computers in telecommunications and broadcasting. However, as Nicholas Garnham (Essay 3.1) pointed out, the fact that business plans and strategies are not driven simply by technology poses major constraints on the convergence of ICT industries. In line with this more open-ended range of possibilities for the communication industry, David Stout (Essay 12.1) suggests that 'technical convergence' could lead instead to a process of 'fusion' that would create new and alternative networks and services.

Stout's essay also warns us that this profusion of ICTs should be a mine, but could indeed become a minefield. Individuals in all settings, whether a household or a global media firm, need a mental map and compass that will guide them in making social and technical choices in the information age.

Tele-Access as a Compass for Social Choice

The ecology of choices shaping tele-access, and the factors constraining these choices, provide a framework for research, policy, and practice (see Fig. 12.1). The chapters of this book have focused on defining the concept of 'tele-access', how it is an outcome of an ecology of social choices, and

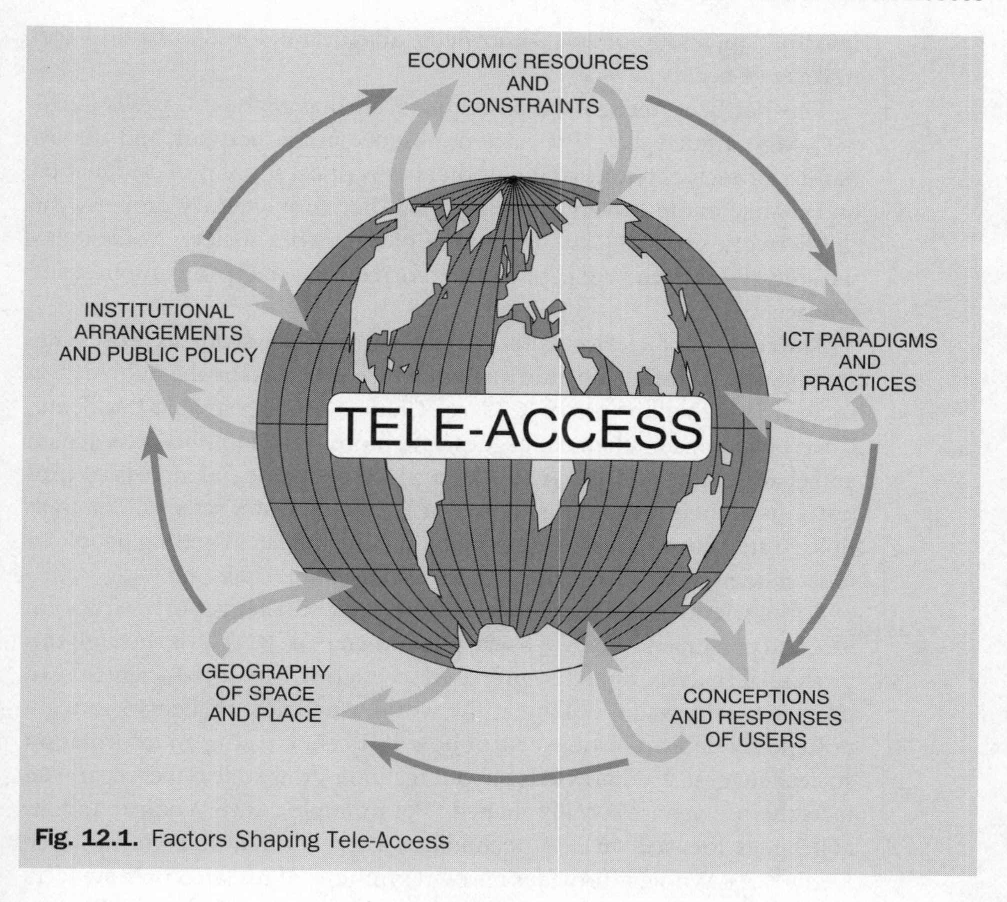

Fig. 12.1. Factors Shaping Tele-Access

how these choices are structured or constrained by a specific set of factors. In addition, this book illustrates the applicability of this framework to a variety of different social and institutional settings, including the household, education, government, and business.

I believe this represents a significant departure from prevailing notions of an 'information society'; and can also inform research based on other more traditional perspectives on the social role of ICTs derived from concepts such as information, influence, or technology (see Chapter 2). This notion of an 'information society' will mislead if taken too literally (see Chapter 1). This is often the case: take the idea that 'information is power', which has become a cliché of the 1990s. But Francis Bacon's (1597) insight was that: 'Knowledge is power.' In fact, information can be very weak.[1] Knowing how and why information is often weak can be critical, but knowledge itself is also not a new resource. It is how we get access to information, and to people—knowledgeable or not, loved or hated—that makes tele-access new.

This has been recognized by scholars, such as Manuel Castells (1996: 469), who argued that: 'Presence or absence in the network and the dynamics of each network *vis-à-vis* others are critical sources of domination and change in our society: a society that, therefore, we may properly call the network society.' Castells's notion of a network society avoids many weaknesses with the concept of an 'information society' by emphasizing tele-access.

The concept of tele-access also avoids an image of individuals processing, storing, and transmitting information, which is evoked by the information society. Studies of the use of ICTs (see Chapter 4) suggest what Luc Soete, Chris Freeman, and their colleagues have more aptly described as a 'screen and chair society' (HLEG 1996: 25) in which people spend increasing proportions of their lives sitting in front of TV or computer screens. The multiple strategies of ICT providers, including the media, in getting people to look at one screen versus another are also new and changing.

Social science poses theoretical and empirical challenges to both optimistic and pessimistic views of social phenomena like ICTs. It is through this process of analysis, driven by practised scepticism, that social scientists can provide new ways of looking at the world. Analytical challenges to technologically deterministic views of how ICTs are creating an information society suggest that knowledge about technology as well as technical artefacts themselves are socially shaped.[2] For example, Steve Woolgar and his colleagues focused on how technology itself is a cultural product (see Essay 12.2). Woolgar provides his own synthesis of research on how ICTs are created by, and sustain, particular sets of social relations—'society made durable'. He also turns these observations back on social scientists, who are producers and users of ICTs, to reflect on how these same theoretical perspectives can be applied to an understanding of how they conduct their work.

The shaping of tele-access focuses on a process that captures the centrality of ICTs to social relations. Yet it also provides a different way of thinking about the social and economic role of ICTs than that afforded by the concept of an information society (see Chapter 2). In addition, the shaping of tele-access provides direction to future research on ICTs.

Directions for Research

The social and technical shaping of tele-access focuses on how individuals, groups, and organizations can better manage tele-access. Tele-access directs attention to understanding technical and social barriers to networks in society, such as income and education (see Chapter 9). This issue generates an extensive agenda of research issues related to technical, geographical, cultural, and physical barriers to access related to such problems as:

- gender inequalities in tele-access, such as those sustained by the culture of the household or the workplace;
- generational differences in attitudes towards technology, and skills related to its use, which might disadvantage elders in the household or workplace;
- physical handicaps or disabilities, which drove some of the pioneering research on ICTs like computer conferencing, but which has been neglected;
- geographical disparities in access, such as in remote regions, or disadvantaged urban areas; and
- minority languages, and cultures, on the Internet and within other electronic communities that are dominated by the English language and middle-class culture.

These are economic as well as social issues in that such barriers can prevent public agencies and firms from gaining access to a trained workforce.

Tele-access is clearly a two-edged sword. Many people wish to live on an electronic cul-de-sac or virtual wilderness trail, rather than next to a superhighway. How do people effectively screen, filter, and protect themselves, when living on an information superhighway? Information overload is not a new issue: libraries have addressed this problem for decades (Biggs 1989). However, tele-marketing, voice mail, and the Internet have made 'overload' more salient for a broader public (Noll 1996). Information has also been linked to some of the more prominent disasters of the past decade (Peltu *et al.* 1996), such as the shooting-down of Iran Air Flight 655 by the *USS Vincennes* in the Persian Gulf (Rochlin 1991).

Between those excluded from ICTs and those overwhelmed by the technologies and lost in information are the multitude of people who routinely use ICTs across many social contexts in ways that shape tele-access. This book has only begun to conceptualize the research issues in these contexts. Research on these and many other important social and institutional settings, from libraries (Agre 1995) to health care (Sackmana 1997) to military strategy and diplomatic affairs (Arquilla and Ronfeldt 1998), might be advanced by research conceived and guided by perspectives on the shaping of tele-access.

In this book I have tried to convey the social role of ICTs to a broader public beyond the social sciences because it:

1. focuses on 'the actual mechanism by which technology leads to social change', which provides added insights about utopian and dystopian perspectives on ICTs (Mesthene 1969); and
2. offers a perspective that can guide policy and practice.

Utopian and Dystopian Perspectives on Tele-Access

The importance of tele-access has long been recognized, even if it has not been explicitly formulated as a focus of enquiry. For example, tele-access is a central theme of some of the most prominent utopian and dystopian views on ICTs. As I argued in discussing policy and regulation (see Chapter 11), one of the most positive of contemporary views on the role of ICTs in shaping tele-access can be found in the proponents of the information superhighway. Proponents of a multimedia digital superhighway believe that the declining costs and increasing capacity and intelligence of converging ICTs will offer the public access to any information, anywhere, and at any time (Gore 1991; Gilder 1994; Negroponte 1995). The architects of the ISH sense a historically unparalleled opportunity for democratizing access to information, reinvigorating community, and extending democracy (see Box 12.1).

Box 12.1. US Goals in Building a Superhighway for the Information Age

Goals of the private-sector Advisory Council to the US government's Information Infrastructure Task Force (IITF) included:

1. advancing 'diverse cultural values, and [a] sense of equity';
2. building 'stronger communities, and a stronger sense of national community';
3. making the ISH 'affordable, easy to use, and accessible from even the most disadvantaged or remote neighbourhood';
4. involving individuals, as well as the private and public sector at all levels in building the ISH; and
5. doing the research and development necessary to maintain 'world leadership' in services, products, and markets created by the ISH.

Source: US Advisory Council (1996*b*: 9).

At the same time, dystopian treatments of technology have also recognized that the networking of communities and nations through electronic highways that link households has major social and political implications for tele-access. One of the best examples can still be found in George Orwell's (1949) classic novel *Nineteen Eighty-Four* (see Chapters 1 and 2). Orwell developed the scenario of a totalitarian society controlled by Big Brother in which Thought Police observed citizens over 'telescreens'—an electronic system for two-way video communications (see Box 12.2).

Box 12.2. An Orwellian Perspective on the Role of ICTs

Not realizing that the Thought Police were eavesdropping on a room he had rented to escape surveillance, the protagonist of *Nineteen Eighty-Four*, Winston Smith, reads the following passage aloud from a book he believed to have been written by revolutionaries, opposed to Big Brother:

> in the past no government had the power to keep its citizens under constant surveillance. The invention of print, however, made it easier to manipulate public opinion, and the film and the radio carried the process further. With the development of television, and the technical advance which make it possible to receive and transmit simultaneously on the same instrument, private life came to an end. Every citizen, or at least every citizen important enough to be worth watching, could be kept for twenty-four hours a day under the eyes of the police and in the sound of official propaganda, with all other channels of communication closed. The possibility of enforcing not only complete obedience to the will of the State, but complete uniformity of opinion on all subjects, now existed for the first time. (Orwell 1949: 206–7)

Winston Smith was overheard through a microphone hidden behind a painting on his wall, and arrested for thought crimes.

The threats conveyed by this classic have been updated by a continuing stream of work on the potential for ICTs, such as two-way cable TV and the Internet, to be used in ways that undermine democratic expression and personal privacy (Flaherty 1989; Gandy 1993; Agre and Rotenberg 1997). As the cover of a popular US news magazine (*Time*, 25 Aug. 1997) declared: 'You have no secrets. At the ATM, on the Internet, even walking down the street, people are watching your every move.'

These utopian and dystopian scenarios highlight the potential social significance of ICT networks in shaping tele-access. Nevertheless, the public has become increasingly comfortable with ICTs. They are not deeply concerned about threats posed to such values as privacy (Dutton and Meadow 1987). The year 1984 has come and gone without Big Brother. As a college professor, I can no longer assume that my students have even read *Nineteen Eighty-Four*, arguably the greatest dystopian novel in the English language. However, many themes of Orwell's dystopian society, such as the politics of language and the rewriting of history, are as relevant today as in 1948, when he wrote this novel. Why such complacency?

One explanation is that more individuals reside in the 'certainty trough' created by the successful application of ICTs across many areas of society (see Essay 8.1). Managers and professionals who simply use ICTs are less concerned about threats to privacy, for example, than those most informed about ICTs, such as computer experts, or those least well informed, such as the poor (Dutton and Meadow 1987).

Another reason is that utopian forecasts are too long term and dramatic to be taken seriously. There has not (yet) been the ICT equivalent of a nuclear disaster, such as the 1986 Chernobyl disaster in the Ukraine, to awaken the public to the risks of an accessible society. But there is also a degree to which utopian and dystopian perspectives do not instruct us on what to do. They tend to be deterministic and therefore pro- or anti-technology. In contrast, the idea of tele-access can be connected both to the day-to-day decisions and to the more open-ended future relating to policy choices.

Social Research: A Guide to Policy and Practice

Like the sciences in general, social and economic research seeks to advance knowledge for its own sake. However, in research programmes like PICT social scientists also have sought to be of value to policy and practice. In the realm of ICTs, social research can help managers and policy-makers make more empirically grounded forecasts of the future of ICTs based on the ways people actually use the technologies. It also enables more realistic assessments of the social implications of ICTs by exploring the actual impacts of ICTs when well implemented. This realism can inform policy and help focus attention on social issues that are tied to patterns of adoption and use, such as concerns over equity across geographical and socio-economic lines. By illuminating the role of social choice in technological design and innovation, and in business, education, government, and all other contexts, social research also shows how it is impossible to disentangle technological from social and organizational innovation (Mansell and Silverstone 1996). For example, social research has repeatedly demonstrated that the nature and impacts of the ICT revolution have been driven by shifts in the paradigms that dominate our thinking about technology and the economy (Freeman 1996*a*, *b*).

More specifically, an understanding of tele-access—as well as the factors shaping tele-access—can reorient the way people think about ICTs and therefore change decisions and societal outcomes in constructive ways. Politicians, managers, industrialists, scientists and engineers, and the public at large make choices that control access to themselves and others. Shaping ICTs is one strategy. Influencing public policy and regulation is another.

Access in a network society will be the outcome of individuals pursuing a diversity of goals and objectives—choosing a TV channel, wiring a classroom, applying for credit, teaching a child to read, training for a job, and negotiating communications policy—in their various roles as viewers, teachers, consumers, parents, students, or public officials. The implications of social and technical choices for tele-access should figure prominently in the equation as a value that individuals consciously considered, but were willing to negotiate and balance with the claims of others. Then the next millennium might well be a qualitatively more democratic and vibrant information society.

One problem with grand, macro-level social theories like the information society is the sense of historical determinism they convey, which lulls individuals to accept their predetermined fate (Popper 1945). They view social and technical change as driven by forces to which individuals simply need to adapt. They can, therefore, be relatively weak guides to policy and practice.

Over a decade after the publication of his book *The Wired Nation* (R. L. Smith 1972), which made one of the most influential cases for viewing ICT infrastructures as an electronic highway, not just a source of more TV, Ralph Lee Smith (with Cole 1987: 126) poked fun at his early work, saying: 'At college . . . a fellow student wrote a paper that was 20 pages long. It was entitled, "The Universe, God, Man and the State." *The Wired Nation* was 80 pages long and had approximately the same content.' Smith sought to convey how little his vision of a wired nation had to do with the day-to-day business issues he faced when he began working within the cable TV industry.

In contrast, an understanding of how access is being regulated by technology and other social factors could make a subtle but profound change in the way we think about every day, as well as once-in-a-lifetime choices influencing the design and use of ICTs. Whether one's choices concern public ICT policies or just getting on the Internet, they can influence an individual's communicative power.

Society on the Line

Shaping tele-access is a process that is relevant to discussion of any information technology, from the printing press to virtual reality. This process is not unique to the electronic, multimedia world we are entering. However, revolutionary change in ICTs has made ICTs an increasingly central element in shaping tele-access. ICTs are becoming more central to a network society (Castells 1996). The choices made over the coming decades in the design and use of ICTs will shape tele-access for future generations.

The Getty Center in Los Angeles opened its museum to the public in 1998. One of its most impressive features is the way in which the architectural design and facilities have purposefully shaped and constrained access to ICTs. State-of-the-art ICTs are readily available to the public in an overview of the center shown on digital video disc (DVD), various 'information' centres, and a centre for the 'digital experience', where visitors can surf the Internet and the Web. However, computers and screens do not invade the art displays, and even the art displays are purposively moderated in order not to overwhelm the viewer with too many paintings or sculptures. In other words, the designers of the Getty regulated access in ways that qualitatively improve the experience of being in this museum.

There are powerful economic and social incentives behind the production and use of ICTs across every level and sector of society and the economy—to provide all channel access to every classroom, household, car, and job. Social outcomes of the ICT revolution will depend on the choices made by individuals, households, but also by regulators, and other policy-makers as they make decisions that shape access in our increasingly multimedia world.

Understanding tele-access is not a silver bullet that will solve all problems related to ICTs. However, actors in all types of social and commercial relationships could benefit by considering how ICTs are used to redistribute communicative power in ways that advantage some actors over others. Advances in ICTs will not inevitably bring more democratic patterns of communication, but they will become a focal point for efforts to shape who gets access to various segments of the public at large. The battle over redistributing communicative power might be called the politics of tele-access, and its outcome could determine important features of our global village.

In September 1997, when over 2 billion people around the world watched the London funeral service for Diana, Princess of Wales, to many it seemed that television had created a sense of community that even the Utopian proponents of a global village could not have dreamt. Here was the Prime Minister of Britain, Tony Blair, reading from I Corinthians 13 that concludes: 'there are three things that last for ever: faith, hope, and love; but the greatest of them all is love.' Like few others, this global media event demonstrated the power of electronic networks to bring people together.

Ironically, over sixty years earlier, another Mr Blair—Eric Blair, who wrote under the pen name of George Orwell (1936)—opened *Keep the Aspidistra Flying*, a book that was highly critical of capitalist society, with an adaptation of I Corinthians 13:

> **Though I speak with the tongues of men and of angels, and have not money, I am become as a sounding brass, or a tinkling cymbal. And though I have the gift of prophecy, and understand all mysteries, and all knowledge; and though I have all faith, so that I could remove mountains, and have not money, I am nothing. . . . And now abideth faith, hope, money, these three; but the greatest of these is money. (Orwell 1936)**

These juxtaposed images of society dramatically illustrate the gap between the visions and fears of the information age. The road to a global village versus an Orwellian society of information inequality is not a choice between these Utopian and Dystopian futures—if only it were so simple—but the outcome of an ecology of interrelated choices about tele-access in all sectors of society. Rather than drift into the twenty-first century thinking our future of access is being determined by technical advances in ICTs or by the strategies of computer and telecommunications firms, you need more consciously to negotiate issues of access—to yourself, to others, to information, to services, and to ICTs.

Notes

1. The weakness of information was one theme of our PICT synthesis workshops, particularly one on IT in organizations, reinforcing some of my own research on the politics of information in decision-making (Dutton and Kraemer 1985).
2. The social shaping of ICTs was another central theme of PICT research, developed in unique ways by MacKenzie (1993), Williams and Edge (1996), and Woolgar and his colleagues (Cooper and Woolgar 1993).

Essays

12.1. ICTs and Technology Foresight

David Stout

The centrality of ICTs to national competitive advantage was highlighted by the UK's Technology Foresight initiative. David Stout draws from his experience with this initiative to explain why it is important that scientists and technologists share their knowledge with industry, policy-makers, and the general public.

12.2. Analytic Scepticism

Steve Woolgar

One common feature of the best social-science research on ICTs is what PICT researcher Steve Woolgar has called 'analytic scepticism,' which he defines as a willingness to challenge commonly accepted views and assumptions of both the proponents and sceptics of technology. His work calls attention to the ways in which educational and other cultural institutions configure the producers and users of ICTs, and uses this insight to reflect on the role of social-science research.

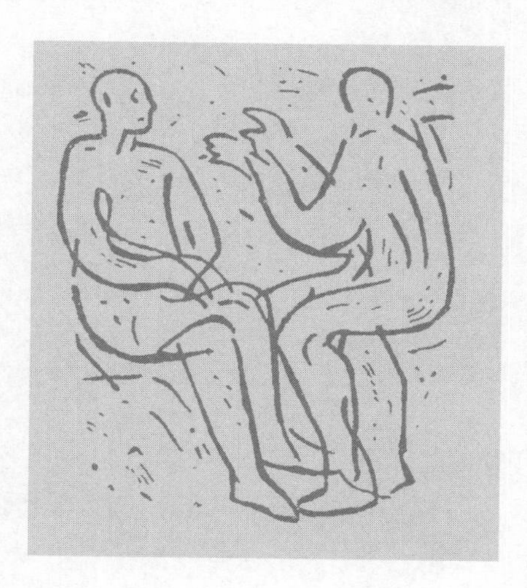

12.1. ICTs and Technology Foresight

David Stout

Some people have informed insights into what technology can make possible; others know what is wanted from technology; and there are those who are experts in delivering what is wanted. Getting these three groups to come together to consider desirable feasible options and the barriers that might prevent them from being achieved is an important element in policy-planning processes at national and regional levels.

For instance, the UK government launched the Technology Foresight programme in April 1994 to bring together these parties. One of its main aims was to create a body of knowledge which could be drawn on by senior managers. In itself, this would require ICT innovations to help provide 'new delivery mechanisms across old boundaries' (Sparrow 1993: 24).

After the first year's work, fifteen expert panels of academics and industrialists had reported on salient probable and possible events that might create major exploitable opportunities to innovate over the subsequent ten to twenty years. They also recommended what research priorities and interactions between academia, industry, and government were needed to create the possible futures. In the two years that have followed, all fifteen panels have taken their messages to the business and academic communities and worked in great depth and across sectors on specific issues. In the IT case, this has included combining with the Finance, Retail, and Distribution, and Leisure and Learning Panels to set up twenty-year scenarios of social change and potential IT application to business and to pinpoint the brakes and accelerators on their adoption in the the UK.

Communications were established as a separate Foresight Panel from IT and Electronics, although they were merged into a single IT, Electronics, and Communications (ITEC) Panel after the initial reports were published (OST 1995*a*; *b*). Other panels focused on non-ICT market sectors, yet information issues permeated almost all panel analyses and new ICTs were regarded as a significant influence on the performance of each sector. Novel combinations of information were identified as creating fresh competencies and market openings. The management and diffusion of these innovations were seen as the source of future national competitive advantage.

The means to generate, interpret, transmit, and act on knowledge was the substance of all outputs from the non-ICT sectors. This was evident from the distillation of the fifteen sector reports prepared by the Technology Foresight Steering Group (1995), which identified software design, multimedia, networked systems, automatic sensors, total business process re-engineering, and genetic engineering as being among the many ICT-related priorities that had been highlighted.

For instance, the Health and Life Sciences Panel (OST 1995*c*) was absorbed by the applications of biotechnology, which is itself about the decoding, modification, and use of genetic information. Biotechnology is more than a single industry. It is also a powerhouse of techniques that can be applied to human wants as various as health, energy provision, new natural materials, and waste management.

My own panel was Manufacturing, Production, and Business Processes. Using Delphi forecasting techniques, eighty-two future key events were pinpointed. Of these, twenty-eight were ICT-related (OST 1995*d*). These included: self-organizing production facilities; integrated ICT-based training capabilities on all equipment; remote monitoring of all plant; interactive customer access to product information at time of purchase; and integrated IT systems to support diffuse cooperation on individual projects. In the Leisure and Learning Panel (OST 1995*e*), practically every issue highlighted by the Delphi survey referred to the information superhighway in one way or another.

It is useful to think of innovations as happening in one of three ways (McClelland 1994: 21). Most innovation is incremental and largely driven by competition between independent economic agents in oligopoly markets. Then, occasionally, there are major scientific breakthroughs which draw a train of applications behind them for decades. This indicates that all innovations in the breakthrough category relevant to Foresight's twenty-year look-ahead had occurred already.

The third kind of innovation results from the combination of hitherto independent and seemingly separate technologies, such as the fusion of different scientific disciplines into biotechnology and the convergence of computing, communications, consumer electronics, and many other technologies into ICT.

A study by the IT and Electronics Sector Panel (1994) found that ICTs are also intrinsic to all other innovations across the economy. At the same time, there is a convergence within the ICT sector caused by the collapsing of previously diversely delivered forms of voice, text, and image communication—including telephone, e-mail, fax, and video—into one broadband facility. Such convergence has two consequences for future national competitive advantage in terms of costs and the potential for interaction.

When access to knowledge is both expensive and time-consuming, it is more likely to be dispensed with. This can lead to decisions being taken too late, which loses innovative advantage. However, the World Wide Web, CD-ROMs, and other ICT innovations make information access cheap and quick. In addition, the most important contribution by ICT to innovation may be to create new possibilities through the fusion of hitherto separated bases of knowledge by enabling the interaction of people from different corporate and academic institutions in any location. The advantages of specialization might then be sacrificed for the sake of shared incentives and the smooth management of information flows.

The falling costs of knowledge transactions between entities is stripping away the traditional advantage of large enterprises in being able to reconcile the benefits of scale and specialization. Deintegration, contracting-out, and combining special skills from independent locations are becoming more advantageous. New content-based information businesses are appearing in close association with the particular skills, technologies, and markets of individual client sectors in areas such as health, education, entertainment, finance, and retailing.

The ubiquity of global markets resolves the paradox of collaborating while yet competing. Global competition has become the spur to alliances and the clustering of niche firms around mother-ship companies and centres of scientific excellence. Research, production, and service and control units may then be continents apart, linked by contracts rather than common ownership. Alliances between industry leaders can also secure early advantage through the spread of common standards or interfacing platforms.

The first phase of an exercise like Foresight has identified key generic technologies, market priorities, and the steps to be taken by industry, academia, research sponsors, and public policy-makers. In order to create the envisaged future, economy-wide communication between and within sectors needs to take place and current ICT strengths to be mobilized. Informational challenges must also be met in all sectors—for instance, in shifting the bias of management training and reward in all companies away from traditional skills—such as accountancy and monitoring—towards new and part-intuitive talents—for example, in the efficient sifting, filtering, and application of vast amounts of technical, competitor, and market data.

Preparations are well in hand for the next round ('Foresight 2000') starting in 1999 and learning from the first exercise. What has emerged from the first round of Foresight is the pervasiveness of information management across every area of innovative opportunity. The challenge for IT is to ensure that what might be a minefield actually becomes a mine.

References

IT and Electronics Sector Panel (1994), *The Technology Demographics Roadmap* (TFP.ITE (11) Version 6; London: Office of Science and Technology).

McClelland, J. F. (1994), *Technology Foresight 4: An Information Technology View* (London: Office of Science and Technology).

OST (1995*a*); Office of Science and Technology, *Technology Foresight Panel 6: Progress through Partnership—Communications* (London: HMSO).

—— (1995*b*), *Technology Foresight Panel 8: Progress through Partnership—IT and Electronics* (London: HMSO).

—— (1995*c*), *Technology Foresight Panel 4: Progress through Partnership—Health and Life Sciences* (London: HMSO).

—— (1995*d*), *Technology Foresight Panel 9: Progress through Partnership—Manufacturing, Production and Business Processes* (London: HMSO).

—— (1995*e*), *Technology Foresight Panel 14: Progress through Partnership—Leisure and Learning* (London: HMSO).

Sparrow, O. (1993), *The Context of Foresight* (London: HMSO).

Stout, D. K. (1995), 'Technology Foresight: A View from the Front', *Business Strategy Review*, 6/4: 1–16.

Technology Foresight Steering Group (1995), *Progress through Partnership: The Report of the Technology Foresight Steering Group* (London: HMSO).

12.2. Analytic Scepticism

Steve Woolgar

One of the most striking features about new technology is that it almost always engenders extraordinary degrees of both optimism and pessimism. Perhaps this has always been the case. It seems even more so with information and communication technologies (ICTs): electronic calculators will make schoolchildren lazy or they will enable teachers to take the drudgery out of mathematics; video games will enable interactive learning or they will induce epilepsy; the very latest Windows software release is an over-hyped rip-off or it is an essential extension of quality document-production techniques; the explosive growth of the use of the Internet will promote unfettered democratic communication systems or it will encourage the global spread of race hatred. This capacity to engender such polar opposite expectations probably also explains why putative consumers and users of ICTs are frequently disappointed.

Clearly, an important task for the social analysis of technology is to explain the peculiarity of this phenomenon. As part of our efforts to understand the creation, diffusion, and use of ICTs, social scientists have to figure out what manner of social relations are being forged (or reinforced) between producers and consumers, such that different people will at different times hold widely differing expectations of ICT products. At the same time it is vital that we, as analysts and observers of ICT, are cognizant of our own attitudes towards, and expectations of, new technology. Our efforts to acquire a sophisticated understanding of the complex social processes involved will be severely compromised to the extent that we ourselves, whether explicitly or implicitly, buy into a particular version of 'what the technology can do'.

'Analytic scepticism' denotes the attempt to resist an unreflexive subscription to the claims and achievements of science and technology. It signals a perspective on ICTs which is committed to question taken-for-granted assumptions, particularly as these relate to 'technical' matters.

Analytic scepticism finds expression in the pursuit of the central contention that 'culture' is a key factor affecting the design, production, and use of technologies. Scientific knowledge and technological systems are to be understood as social and cultural artefacts, rather than as simply the linear or logical development from existing knowledge and technologies. The design, capability, nature, use, and effects of new technology are the upshot of myriad social and cultural influences, not the result of purely technical design decisions. These influences are embedded in technology and in its use: technology is society made durable.

Of course, 'culture' affords a variety of interpretations relating to different aspects of the social, economic, and organizational circumstances of production and use. Thus, it is important to ask questions about a wide range of aspects of science and technology, including human factors, organizational style, the representation of technical capacity, market dynamics, practitioners' attitudes and beliefs, company politics, and designers' preconceptions about users. A central aim of this approach is to assess which are the more valuable senses of 'culture'.

One particular focus is those aspects of beliefs, language, and practice which display and reaffirm features of the relationship between the social and the technical, designers and users, producers and consumers, insiders and outsiders. In this interpretation, culture includes reference to the construction, maintenance, and negotiation of significant social boundaries. Thus, for example, in focusing upon computerization and software, this line of research examines the ways in which relations between producers and consumers bear upon the design and use of information technology. We need to understand how designers' preconceptions about users affect development; the social processes involved in attempts to discern users' 'requirements'; the variation in users' attitudes towards IT products as a function of proximity to the site of production; the extent to which these relations are mediated by organizational and other social structural boundaries; and the consequences of these relations for uptake and use. An understanding of the influence of these aspects of culture is of particular significance in the context of the increasing globalization of the products of information and communication technology.

The analytic scepticism which informs this kind of research is captured in the description of the approach as 'ethnographic': the combination of theoretical questioning and detailed empirical study. Under this broad rubric, the research deploys a variety of techniques from participant observation to interviews, the use of documentary evidence, and of audio and video recording, and analysis of interaction. However, although this kind of research adopts much of the anthropologist's methodological paraphernalia, the main import of analytic scepticism is more conceptual than methodological. The point is not so much to 'tell it like it really is' in, say, software development practice—useful though this may be in disabusing idealized versions of technological development. Rather, the main point is that by deliberately maintaining a studied distance with respect to the claims of the natives, be they producers or users, about their products, we can generate a profoundly different way of thinking about the social basis of technology.

'Analytic scepticism' provides the basis for a research policy which enables interdisciplinary cooperation between disciplines such as sociology, anthropology, computer science, and history. In links with computer science, for example, this research investigates the social and cultural dimensions of human–computer interaction (HCI), and of document and information management, with a view to informing the processes of design and development. Significantly, analytic scepticism involves a commitment to engage with ICT natives. This is the kind of ethnography which finds there is much to be learned by feeding back findings to industrialists and ICT practitioners. It is research done with natives, not upon them.

In focusing upon aspects of culture denoted by 'producer–consumer relations' research informed by analytic scepticism also addresses a more general problem: what is to count as effective communication across social organizational boundaries? This is especially important at a time when skills are highly differentiated—between science and non-science, between technology and the arts, between academics and non-academics, between public and private sectors—and at a time of increasing demands for accountability and utility across disciplinary and organizational divides.

Analytic scepticism is central to efforts to address fundamental problems of social science. It denotes an effort, in the study of technology, both to draw upon, and to contribute to, major themes in social constructivism, deconstruction, discourse analysis, textual analysis, literary theory, feminist, and historical scholarship. Through an empirical focus on science and technology—arguably the most pervasive and significant forms of cultural artefact in our society—we can seek new approaches to long-standing problems of societal organization.

Appendix

The Programme on Information and Communication Technologies (PICT)

Britain's Economic and Social Research Council (ESRC) launched PICT in 1985, in the wake of major UK policy initiatives in information technology, broadcasting, cable, and telecommunications. PICT's mission was to conduct high-quality academic research in the UK on the long-term social and economic implications of advances in information and communication technology, which would inform policy and practice.

PICT evolved into a ten-year programme involving over sixty researchers at six university research centres across Britain, as well as numerous other researchers in the UK and abroad. It was sustained until 1995 by two phases of research grants from the ESRC, a public agency established by Royal Charter in 1965 to support research and training in UK higher-education and research institutes. This funding permitted the research to be independent of commercial interests and the agenda of particular government agencies.

Researchers generated additional support from a wide variety of public and private organizations for activities organized by the PICT National Office—which directed programme-wide activities—and the six research centres. The independence of the research was reinforced by the fact that sponsorship was distributed across a large number of agencies and firms.

Work at the PICT centres covered a diverse range of research projects and workshops, conferences, forums, lectures, and publications. Innovative approaches to dissemination were an essential part of achieving PICT's objective of bringing the results of its research to the attention of policy-makers, IT and telecommunications practitioners, government, and the public at large.

Researchers placed an emphasis on case studies and qualitative analyses in four main areas, which also characterize social research on ICTs in general (Dutton 1996*a*). These are:

1. **production:** the wide range of social, cultural, and political processes that shape innovations in products, services, and industries;
2. **utilization:** the ways in which ICTs are used in firms, management, and work to reinforce or transform the structure, geography, and processes of organizations;
3. **consumption:** the many ways households, citizens, consumers, and the general public consume and actively adapt ICTs to fit into their everyday lives; and
4. **governance:** the criteria and processes by which public policy and regulation balances competing values and interests.

PICT research draws many connections across these areas, such as the degree to which users are also involved in production as they reinvent and reconfigure technologies in the workplace and the household.

The PICT Centres

PICT developed a decentralized, federal network of social scientists which could span the UK and build bridges to colleagues in other countries. In many respects, the six university research centres that made up the larger network began and remained quite distinct. Each brought a different mix of multidisciplinary skills, took a unique approach to its work, and was funded at different levels to conduct research in separate but interrelated areas of enquiry. Nevertheless, ongoing communication and coordination across the centres was maintained through the National Office, the PICT Committee, the PICT Management Group, and a variety of programme-wide activities.

The six centres were at:

- **Brunel University, West London.** The Centre for Research into Innovation, Culture and Technology (CRICT) explored the social and cultural dimensions of science and technology, with particular reference to ICTs. Its ethnographic research on technology and, in particular, software development provides insights into the ways in which the beliefs, language, and practices of producers and users—such as developers' preconceptions of users—affect the way information technology is designed and used.
- **University of Edinburgh.** The Research Centre for Social Sciences (RCSS) investigated the social shaping of ICTs—how technical choices in the design, development, and use of advanced computing and telecommunications systems are influenced by the players involved and their particular social and economic setting. Edinburgh researchers have pursued this line of enquiry with respect to the generation of enabling technologies, such as microprocessors; the development of high-performance computer hardware and software; and the application of ICTs, particularly within industry.
- **University of Manchester Institute of Science and Technology (UMIST), Manchester School of Management.** The Centre for Research on Organizations, Management and Technical Change (CROMTEC) studied the role of information technology in business and management practices, with particular emphasis on the ways in which an organization's informal social arrangements and formal structures change as they become more committed to the use of new ICTs.
- **University of Newcastle.** The Centre for Urban and Regional Development Studies (CURDS) has injected a geographical concern for space and place into debates about the emerging information economy. The work of CURDS focuses on how the use of ICT influences what tasks are carried out where—within and

between organizations—and on the changing supply and demand of advanced communications services in cities, regions, and rural areas.

- **University of Sussex.** The Centre for Information and Communication Technologies (CICT) in the University's Science Policy Research Unit (SPRU) carried out research on international and regional policy and regulatory issues, innovation and technical change in the communication field, and the production and consumption of new products and services for consumers and businesses. This research considered telecommunications policy, standardization practices, software development, and consumer ICTs from social and economic perspectives.
- **University of Westminster.** The Centre for Communication and Information Studies (CCIS) focused on the regulation of broadcasting and telecommunications and the development of the information and communication industries. Initiated through PICT funding, CCIS generated support outside PICT that sustained the Centre's work and contributed to Westminster's high profile among those UK universities specializing in media and communication studies.

Structure of the Programme

PICT was directed and coordinated by the PICT Committee, appointed by the ESRC's Research Programmes Board. The Director of PICT made recommendations to the PICT Committee and took responsibility for ensuring that the mission and objectives of the programme were met by the combined efforts of a federal network of PICT centres. The research centres made recommendations to the Director and PICT Committee through the PICT Management Group, consisting of the heads of PICT centres and the Director of PICT. Through the life of the programme, there were four directors: William Melody (1985–8), Nicky Gardner (1988–92), John Goddard (1992–3), and William Dutton (1993–5).

Over sixty research projects and assignments were conducted during the course of Phase I (1986–90) and Phase II (1991–5) of PICT. All six centres received major research assignments in the first phase. During the second, the ESRC concentrated its funding on research at Edinburgh, Newcastle, Sussex, and UMIST. However, all six centres continued to function as a network through funding from other sources, the continued role of the PICT Management Group, a variety of joint activities, such as conferences and policy forums, and the development of PICT-wide strategies to synthesize and disseminate the research.

Disseminating PICT Results

The programme sought to use its research to inform and stimulate policy debate through several complementary activities, which included (see PICT 1995 for full details):

1. **The Charles Read Lectures.** These ran annually from 1988 to 1995 and were sponsored by BT from 1991. They commemorated Charles Read, who was highly influential in the establishment of PICT when he was an Adviser to Prime Minister Margaret Thatcher's Cabinet Office in the early 1980s. The series included lectures by William Melody, John Goddard, Roger Silverstone, Rod Coombs, Robin Mansell, Alan C. Kay, and Dr Martin Bangemann.
2. **Conferences.** These events brought together researchers from within and outside PICT centres to discuss findings relevant to the programme's goals. PICT's final conference on the social and economic implications of ICTs was held at the Queen Elizabeth II Conference Centre, Westminster, London, 1995.
3. **Policy Research Paper (PRP) Series.** A total of thirty-three PRPs were published, providing a summary of results from PICT research in a concise and readable format accessible to policy-makers, practitioners, and researchers.
4. **Policy Research Forums.** PICT launched a series of Policy Research Forums in its last two years. Their aim was to establish a neutral meeting ground for constructive dialogues across academic, practitioner, and policy communities. Special PRPs were produced to summarize discussions at the Forums and related background reports and position statements from participants (Dutton *et al.* 1994*a*; Dutton *et al.* 1994*b*; Dutton *et al.* 1995). All three forum reports were revised and updated for Dutton (1996*a*).
5. **Occasional lectures.** During the last two years of the programme, eleven lectures open to a broadly based invited PICT mailing list were given by leading experts from the UK, North America, and East Asia.
6. **Books and other publications.** Researchers in the PICT network have published a large number of books and research papers. Many are referenced in this book and Dutton (1996*a*).
7. **Synthesis workshops.** In PICT's final two years, specialists from around the world in the fields covered by the programme were invited to workshops to assist PICT researchers in developing a synthesis of their work during the programme. Each workshop concentrated on a major substantive area (see Box A.1). The workshops were designed as closed meetings, with six to ten PICT researchers giving brief presentations in areas of their particular expertise, seeking to identify new insights and perspectives gained from the research. Various UK and international scholars from outside PICT also attended to help us identify original contributions and to cross-cut themes of most significance to research, policy, and practice.
8. **Synthesis books.** This book and a companion volume, *Information and Communication Technologies—Visions and Realities* (Dutton 1996*a*),

Box A.1. Topics of the PICT Synthesis Workshops

1. *Cross-Cutting Themes* (Alexander House, Sussex, March 1993): the PICT Management Group launch of the synthesis.
2. *The Public at Large: Consumers, Citizens, Audiences, Users, Households* (Brunel University, December 1993), discussing home telematics and the information society, the household and ICTs, and conceptions of the user.
3. *IT Development and Innovation* (Brunel University, December 1993), with sessions on technical knowledge is socially shaped, IT development and transfer—R&D labs and commercial developer organizations, and development in organizations.
4. *IT in Organizations* (Brunel University, December 1993), focusing on social shaping and construction of IT in organizations, gender and technology relationships, and restructuring organizations.
5. *(Tele)communication Industries and Public Policy* (Policy Studies Institute, February 1994), discussing telecommunications, broadcasting, and development.
6. *The Use of Social Science Research* (Brunel University, September 1994), which focused on the role of social-science research in policy and practice (Skolnick 1995).

provide an overall synthesis and summary of key research findings from PICT. Syntheses of the work at particular PICT centres include books by Robins (1992), MacKenzie (1996), Mansell and Silverstone (1996), and Bloomfield *et al.* (1997).

Glossary

Advanced Research Projects Agency US Department of Defense agency responsible for initial sponsorship of the ARPANET, which evolved into the Internet

analogue information represented as a continuously changing physical quantity, such as the amplitude of a sound wave

artificial intelligence techniques to enable computer-based systems to respond in ways similar to human beings

Asynchronous Transfer Mode network communication technique capable of switching high-bandwidth multimedia communications, including video

automatic teller machine device for delivering cash and carrying out other transactions for authorized customers using a bank debit or credit card

bit (bi)nary digi(t) used in a mathematical system that recognizes only two states, typically represented as '0' and '1'

bandwidth indication of amount of information a telecommunications channel can carry (analogous to a measure of the bits per second rate of a digital channel)

broadband telecommunications medium, such as coaxial cable, or optical fibre, which can cope with the large volumes of data required for multimedia applications

bulletin board system a computer system that allows users of an electronic network to leave messages that can be read by many other users

business process re-engineering approach to radically restructuring organizations by optimizing the processes needed to meet specific goals, which normally entails reductions in staff and changes in existing departmental boundaries

CCITT international standards-making body representing telecommunications operators, suppliers, and other interested parties

cellular radio a wireless telephone service which divides the areas covered into small cells to assist in managing the network efficiently

circuit switching way of linking systems and devices on a network by directly connecting transmission circuits, as with traditional telephone exchanges

coaxial cable transmission medium used for cable networks, with a bandwidth narrower than optical fibre but broader than ordinary copper wire pairs

common carrier telecommunications network supplier which carries communications from others

computer conferencing asynchronous group discussion based on the exchange of electronic messages on a computer network

computer numerically controlled machine software-controlled machine tools

convergence coming together of all information and communication forms into common underlying approaches based on digital techniques

cross-ownership where one company owns more than one major media outlet, such as television, cable, and newspapers in the same geographical area

cross-subsidy use of revenues from a profitable activity for less profitable ones—for instance, to support telecommunications services to remote areas

cultural imperialism strong influence by one country over other nations through a domination of electronic media and computer software production and distribution

cyberspace term indicating the electronically simulated space or virtual territory created by networked information flows, such as an electronic forum

dial-up access connecting to a network over the telephone line by dialling a number, rather than being directly connected to it, such as over a local area network

digital information represented by strings of 1's and 0's, such as the bits (0/1 or on/off) used by digital information and communications technologies, such as the PC

digital compression techniques which enable large volumes of digital information to be represented and sent using fewer bits

digital democracy applications of computer-based networks, such as the Internet and the Web, to support democratic processes, such as discussion, polling, or voting

direct satellite broadcasting transmission of television or radio programmes directly from a satellite to an antenna connected to viewers' TV sets

distance learning use of electronic networks to deliver educational services to the household, workplace, or other location different from that of the instructor

download to retrieve an electronic document, software, or other file over an electronic network that is stored on another computer

electronic data interchange ability to exchange electronic documents and other information automatically, such as orders and invoices, between organizations

electronic democracy applications of electronic networks, such as two-way cable TV, to support democratic processes, such as discussion, or voting (digital democracy)

electronic funds transfer using electronic networks for money-based transactions

electronic mail network service allowing typed messages to be sent between people using personal electronic mailboxes to store messages until they are read

electronic service delivery using electronic networks to provide customers and clients directly with a variety of information and transactional services

ethnography social-science method for observing and analysing group behaviour through direct observation and participation in its activities

expert system artificial-intelligence technique for developing software incorporating human expertise on particular subjects

file transfer protocol standard for exchanging computer files across the Internet

Fordism rigid, routinized assembly-line work processes, based on Taylorism, which Henry Ford introduced in the early twentieth century to build cars

graphical user interface use of icons and pointer devices to simplify users' interaction with a computer, as in Microsoft Windows and Apple Macintosh systems

hacking accessing a computer-based system without appropriate authorization, which can be unlawful

high-definition television television pictures with a high resolution, such as using more scanning lines, to display sharper images than traditional lower-resolution images; this requires greater bandwidth, or higher speed networks

High Performance Computing and Communication an initiative of the US Office of Science and Technology Policy to develop advanced information infrastructures

information and communication technologies all the kinds of electronic systems used for broadcasting, telecommunications, and computer-mediated communications

information economy an economy in which the processing and transmission of information is a central activity

information infrastructure provision of underlying network capabilities to support a variety of services based on computing and telecommunications capabilities

information service provider organization, group, or individual who creates and packages information content carried by electronic networks

information society refers to the increasing centrality of information to all forms of social and economic activity

information superhighway term coined by US Vice-President Al Gore for an advanced telecommunications infrastructure accessible to all

information technology computer-based techniques for storing, processing, managing, and transmitting information, including telecommunications and related management science techniques

Integrated Digital Services Network service using digital techniques throughout the network for transmitting voice, data, and video services over the same infrastructure

interactive television cable network service allowing TV sets to be used for two-way communication between the home and cable operator for services such as shopping

Internet an international 'network of networks' that uses a set of standards to permit the interconnection of millions of computers, enabling such services as electronic mail and remote access to information, such as on the Web

intranet a private 'network of networks' using ICTs to regulate access in and out of a defined network organization

local area network computer-based network which directly links computers within a limited geographical area, such as a particular room, building, or campus

leased line link from a telecommunications operator dedicated to a particular customer for the payment of a regular fee

liberalization opening up of public telecommunications supply to competition

magnetic stripe method of storing digital information, as on most bank and credit cards

modem MOdulator/DEModulator which converts between analogue and digital techniques to allow computers to be connected via non-digital networks

Mosaic an early browser for the World Wide Web, employing a simple graphical user interface, which has influenced the design of many other user interfaces

multimedia integration of text, video, and audio capabilities in computer and telecommunications systems

narrowband telecommunication channel which can handle only relatively small volumes of data, such as wire pairs used for e-mail or voice-only telephony

narrowcasting targeting communication media to specific segments of the audience

National Research and Education Network component of the High Performance Computing and Communication initiative of the US Office of Science and Technology Policy

numerically controlled machine machine tool controlled by a paper-tape loop

on-line Once used to denote direct interaction with a computer-based system via a telecommunications link; increasingly refers to any use of digital networks

open system the aim of enabling any computer or telecommunications system to interconnect and be compatible with any other

Open Systems Interconnection model for open-system compatibility developed by the International Standards Organization and CCITT

operating system software, such as Microsoft Windows, Unix, and IBM MVS, which manages the computer's basic functions so the system can be exploited by users

optical fibres broadband telecommunications links using light to transmit information

packet switching method for coding and transmitting digital information as small packets of information rather than a single continuous stream and then reassembling them at their final destination

plain old telephone service Basic voice-only telephony services

post-Fordism new forms of work organization which move away from the automated mass-production line of Fordism

private automatic branch exchange system located on a user's premises which links phones inside the organization and connects them to the public network

privatization opening-up of the public telecommunications supply industry to private ownership

protocol detailed definition of the procedures and rules required to transmit information across a telecommunications link

Public Switched Telephone Network telecommunications network available to the public

public telecommunications operator supplier offering telecommunications infrastructure capabilities to individuals and companies

Regional Bell Operating Company seven regional telecommunications companies created in the USA after the break-up of AT&T in 1984, which began to consolidate after the 1996 Telecommunication Act

semiotics study of the underlying meanings of symbols and metaphors

smartcard credit-card-sized device employing a microprocessor for storing and processing information

social shaping of technology a perspective of research which acknowledges the role of social and economic forces influencing the design and implications of technology

spammer sender of junk e-mail

tariff rebalancing shifting the basis of telecommunications charges to reflect the direct costs of each service, without allowing for cross-subsidies

Taylorism way of organizing work which emphasizes routinization as a means of optimizing productivity, originally developed by engineer F. W. Taylor in the late nineteenth century but also often employed in modern computer-based automation

telebanking interactive networked service allowing transactions with banks to be undertaken from the home, or office

telecommuting using telecommunications to perform work at home or at a work centre that would otherwise involve commuting physically to a more distant place of work

teleconferencing meeting involving people in different locations communicating simultaneously, or asynchronously, through electronic media, such as video

telematics information and communications networks and their applications

teleport site-specific telecommunications infrastructure, such as a land-station link to a satellite, associated with related land and building development

teleshopping ability to order goods and services from home directly through an interactive network, such as cable TV, the telephone, or the Internet

teletext broadcast service over a television signal which allows users to call up a wide variety of information on a TV screen, such as movie listings, news, and weather

television licence required in many countries for the use of a TV set, as in the UK, where the public service BBC is funded from annual TV licence fees

telework use of an electronic network to enable individuals to work from home, a decentralized work centre, a car, or the customer's premises

Transmission Control Protocol/Internet Protocol set of interconnection standards used for the Internet

universal service provision of a minimum set of telecommunications services to all households

Usenet International collection of electronic discussion groups on a multiplicity of topics, accessible through the Internet

V-chip an electronic device that can be installed in a TV set to block out 'objectionable' material, which is detected by a rating that must be encoded in the television signal

value-added network an enhanced service built on the basic telecommunications network, such as e-mail and telebanking

video conferencing teleconferencing involving video communication

video on demand interactive network service which allows customers to view a video whenever they wish

videotex computer-based network service which delivers textual and graphical information, typically as pages of information stored on remote computers

virtual organization operation involving many individuals, groups, and firms in different locations using electronic networks to act as if they were a single organization at one site

virtual reality computer-based visualization of a total environment that gives the user a perception of being within the environment rather than viewing it on a screen

voice mail the ability to store spoken messages on a network for subsequent retrieval by the recipient

Web see World Wide Web

Webmaster manager of a Web site or home page

wide area network telecommunications network that extends beyond individual buildings or campuses

World Wide Web hypertext system which allows information sites distributed on networked computers to be accessed through a computer employing a Web browser

References

Abramson, J. B., Arterton, F. C., and Orren, G. R. (1988), *The Electronic Commonwealth* (New York: Basic Books).

Ackoff, R. (1969), 'Management Misinformation Systems', *Management Science*, 14/4: 147–57.

ACLU v. *Reno* (1996), *American Civil Liberties Union* v. *Reno*, No. 96–963 (ED Pa., 11 June 1996) (1996 WL 3118).

Adams, S. (1996), *The Dilbert Principle: A Cubicle's-Eye View of Bosses, Meetings, Management Fads & Other Workplace Afflictions* (New York: HarperCollins Publishers).

Agre, P. E. (1995), 'Institutional Circuitry: Thinking about the Forms and Uses of Information', *Information Technology and Libraries*, 14/4: 225–30.

—— and Rotenberg, M. (1997) (eds.), *Technology and Privacy: The New Landscape* (Cambridge, Mass.: MIT Press).

Allen, J. C., and Dillman, D. A. (1994), *Against All Odds: Rural Community in the Information Age* (Boulder, Col.: Westview Press).

Anderson, R. H., Bikson, T. K., Law, S. A., Mitchell, B. M. (1995), *Universal Access to E-Mail* (Santa Monica, Calif.: Rand).

Annan Committee (1977), The Home Office, *Report of the Committee on the Future of Broadcasting* (Cmnd. 6753, London: HMSO).

Armstrong, R. (1988), *The Next Hurrah: The Communications Revolution in American Politics* (New York: William Morrow).

Arquilla, J., and Ronfeldt, D. (1998), 'Doctrinal and Strategic Dimensions of Information-Age Conflict', *Information, Communication, and Society*, 1/2: 121–43.

Asper, G., Dutton, W., Nascimento, M. E. M., de Lourdes Teodoro, M. (1996), 'Perspectivas de Implementação da Technologia de Reuniões Electrõnicas na Câmara de Pesquisa e Pós-Graduação da Universaidade de Brasília, no Contexto da Difusão de Inovações', paper delivered at the National Meeting of Graduate Schools of Administration, Angra Bos Reis, Rio de Janero, Brazil, 22–5 Sept.

Arlen, G. (1991), *SeniorNet Services: Toward a New Electronic Environment for Seniors* (Forum Report, No. 15); (Queenstown, Md.: Aspen Institute).

Attali, J., and Stourdze, Y. (1977), 'The Birth of the Telephone and Economic Crisis: The Slow Death of Monologue in French Society', in de Sola Pool (1977), 97–111.

Attewell, P. (1987*a*), 'Big Brother and the Sweatshop: Computer Surveillance in the Automated Office', *Sociological Theory*, 5: 87–100.

—— (1987*b*), 'The Deskilling Controversy', *Work and Occupations*, 14/3: 323–46.

—— and Rule, J. (1984), 'Computing and Organizations: What We Know and What We Don't Know', *Communications of the ACM*, 27/12: 1184–92.

Aumente, J. (1987), *New Electronic Pathways: Videotex, Teletext, and Online Databases* (London: Sage Publications).

Bachrach, P. (1967), *The Theory of Democratic Elitism: A Critique* (Boston: Little, Brown, and Company).

—— and Baratz, M. S. (1970), *Power and Poverty: Theory and Practice* (London: Oxford University Press).

Baer, W. S. (1996), 'Telecommunication Infrastructure Competition: The Costs of Delay', in Dutton (1996*a*), 353–70.

—— (1997), 'Will the Global Information Infrastructure Need Transnational (or Any) Governance?', in Kahin and Wilson (1997), 532–52.

—— Botein, M., Johnson, L. L., Pilnick, C., Price, M., and Yin, R. K. (1974), *Cable Television: Franchising Considerations* (New York: Crane, Russak & Company).

Bagdikian, B. H. (1997), *The Media Monopoly* (5th edn., Boston: Beacon Press).

Baldwin, T. F., McVoy, D. S., and Steinfield, C. (1996), *Convergence: Integrating Media, Information & Communication* (Thousand Oaks, Calif.: Sage Publications).

Bangemann Group (1994): High Level Group on the Information Society, *Europe and the Global Information Society: Recommendations to the European Council* (Brussels: European Commission).

Banker, R. D., Kauffman, R. J., and Mahmood, M. A. (1993) (eds.), *Strategic Information Technology Management* (Harrisburg: Idea Group Publishing).

Barber, B. R. (1984), *Strong Democracy: Participatory Politics in a New Age* (Los Angeles: University of California Press).

Barbrook, R., and Cameron, A. (1996), 'The Californian Ideology', paper presented at the 9th Colloquium on Communication and Culture, Piran, Slovenià, 10–14 Apr.

Barnett, S., and Curry, A. (1994), *The Battle for the BBC* (London: Aurum Press).

Battle, J. (1995), 'Deeping Democracy—The Challenge of the New Technology', speech to PICT's International Conference on the Social and Economic Implications of Information and Communication Technologies, Queen Elizabeth II Conference Centre, Westminster, London,12 May.

Bauer, R. A., de Sola Pool, I., and Dexter, L. A. (1972), *American Business and Public Policy: The Politics of Foreign Trade* (2nd edn., Chicago: Aldine Atherton, Inc.; first edn., 1963).

Becker, L. B. (1987), 'A Decade of Research on Interactive Cable', in Dutton *et al.* (1987*a*), 102–23.

—— and Schoenbach, K. (1989) (eds.), *Audience Responses to Media Diversification: Coping with Plenty* (Hillsdale, NJ: Lawrence Erlbaum Associates).

Becker, T. (1981), 'Teledemocracy', *Futurist* (Dec.), 6–9.

—— and Scarce, R. (1986), 'Teledemocracy Emergent: State of the American Art and Science', in Dervin and Voigt (1986), 263–87.

Bekkers, V., Koops, B.-J., and Nouwt, S. (1996), *Emerging Electronic Pathways: New Challenges for Politics and Law* (London: Kluwer Law International).

Bell, D. (1974), *The Coming of Post-Industrial Society: A Venture in Social Forecasting* (London: Heinemann; originally published, New York: Basic Books, 1973).

—— (1980), 'The Social Framework of the Information Society', in Forester (1980), 500–49.

Bellamy, C., and Taylor, J. A. (1998), *Governing in the Information Age* (Buckingham: Open University Press).

Bellotti, V. (1997), 'Design for Privacy in Multimedia Computing and Communications Environments', in Agre and Rotenberg (1997), 63–98.

Beniger, J. N. (1986), *The Control Revolution* (Cambridge, Mass.: Harvard University Press).

Berelson, B. R., Lazarsfeld, P. F., and McPhee, W. N. (1954), *Voting: A Study of Opinion Formation in a Presidential Election* (Chicago: University of Chicago Press).

Berleur, J., Clement, A., Sizer, R., and Whitehouse, D. (1990) (eds.), *The Information Society: Evolving Landscapes* (New York: Springer-Verlag).

—— Beardon, C., and Laufer, R. (1993) (eds.), *Facing the Challenge of Risk and Vulnerability in an Information Society* (Amsterdam: North-Holland).

Bessant, J. (1984), 'Information Technology and Employment', *Prometheus*, 2/2: 176–89.

Beynon, J., and MacKay, H. (1992) (eds.), *Technological Literacy and Education* (Falmer, UK: Falmer Press).

Biggs, M. (1989), 'Information Overload and Information Seekers: What We Know About Them, What to Do About Them', *Reference Librarian*, 25–6: 411–29.

Biocca, F., and Levy, M. R. (1995) (eds.), *Communication in the Age of Virtual Reality* (Hillsdale, NJ: Lawrence Erlbaum Associates).

Blau, A. (1997), 'A High Wire Act in a Highly Wired World: Universal Service and the Telecommunications Act of 1996', in Kubicek *et al.* (1997), 247–63.

Bloomfield, B. P., Coombs, R., Knights, D., and Littler, D. (1997) (eds.), *Information Technology and Organizations: Strategies, Networks, and Integration* (Oxford: Oxford University Press).

Blumler, J. G. (1989), 'The Role of Public Policy in the New Television Marketplace', *Benton Foundation Project on Communications and Information Policy Options* (Washington: Benton Foundation).

—— (1992) (ed.), *Television and the Public Interest: Vulnerable Values in West European Broadcasting* (London: Sage Publications).

—— and Katz, E. (1974) (eds.), *The Uses of Mass Communications: Current Perspectives on Gratifications Research* (Beverly Hills, Calif.: Sage Publications).

—— and Nossiter, T. J. (1991) (eds.), *Broadcasting Finance in Transition: A Comparative Handbook* (Oxford: Oxford University Press).

—— McLeod, J. M., and Rosengren, K. E. (1992) (eds.), *Comparatively Speaking: Communication and Culture Across Space and Time* (Newbury Park, Calif.: Sage Publications).

Bortz, P. I. (1985), 'Being Realistic about Videotex', in Greenberger (1985), 113–24.

Brand, S. (1987), *The Media Lab* (New York: Penguin Books).

Braverman, H. (1974), *Labor and Monopoly Capital: The Degradation of Work in the Twentieth Century* (New York: Monthly Review Press).

Brownstein, C. N. (1978), 'Interactive Cable TV and Social Services', *Journal of Communications*, 28/2: 142–7.

Bruyn, S. T. (1966), *The Human Perspective in Sociology: The Methodology of Participant Observation* (Englewood Cliffs, NJ: Prentice-Hall, Inc.).

Bryden, J., Fuller, T., and Rennie, F. (1995), 'Implications of the Information Highway for Rural Development and Education', Report of the Arkleton Trust Seminar, Douneside, Aberdeenshire, Scotland, Feb.

Bureau of the Census (1996), *Statistical Abstract of the United States* (116th edn., Washington: US Department of Commerce).

Burstein, D., and Kline, D. (1995), *Road Warriors: Dreams and Nightmares along the Information Highway* (New York: Dutton).

Campbell, A., Converse, P. E., Miller, W. E., and Stokes, D. E. (1964), *The American Voter: An Abridgement* (New York: John Wiley & Sons, Inc.).

Campbell, A. K., and Bahl, R. W. (1976) (eds.), *State and Local Government: The Political Economy of Reform* (New York: Free Press).

—— and Birkhead, G. S. (1976), 'Municipal Reform Revisited: The 1970s Compared with the 1920s', in Campbell and Bahl (1976), 1–15.

Cane, A. (1995), 'Regulate or Face a Nightmare', *Financial Times*, 14 Aug., 10.

—— (1997), 'Telecoms Regulator Declines Second Term', *Financial Times*, 24 Sept.

Carey, J. W. (1989), *Communication as Culture: Essays on Media and Society* (Boston: Unwin Hyman).

Caruso, D. (1997), 'Technology: Digital Commerce', *New York Times*, 15 Dec.

Carvel, J. (1997), 'British Pupils Lead World in Computer Access', *Guardian*, 4 Sept.

Castells, M. (1989), The Informational City (Oxford: Basil Blackwell).

—— (1996), *The Rise of the Network Society: The Information Age: Economy, Society and Culture*, i (Oxford: Blackwell Publishers).

Cawson, A., Haddon, L., and Miles, I. (1995), *The Shape of Things to Consume* (Aldershot, UK: Avebury).

CCTA (1994), Government Centre for Information Systems (formerly Central Computing and Telecommunications Agency), Office of Public Service and Science, *Information Superhighways: Opportunities for Public Sector Applications in the UK: A Government Consultative Report* (London: CCTA).

Central Statistical Office (1996), *Annual Abstract of Statistics* (London: HMSO).

Charan, R. (1991), 'How Networks Reshape Organizations—for Results', *Harvard Business Review*, 69/5: 104–15.

Charles, D., and Howells, J. (1992), *Technology Transfer in Europe: Public and Private Networks* (London: Belhaven Press).

Chesbrough, H. W., and Teece, D. J. (1996), 'When Is Virtual Virtuous?', *Harvard Business Review* (Jan.–Feb.), 65–73.

Child, J. (1987), 'Managerial Strategies, New Technology, and the Labour Process', in Finnegan *et al.* (1987), 76–97.

Chomsky, N. C. (1984), 'The Manufacture of Consent', in Peck (1987), 121–36.

Cimons, M. (1996), 'Gore Calls for More Efforts to Link Schools to Internet', *Los Angeles Times*, 30 June, A18.

Clark, Rt Hon. D. (1997), Chancellor of the Duchy of Lancaster, Speech at the House of Commons, 26 June.

Clegg, C., Axtell, C., Damodaran, L., Farbey, B., Hull, R., Lloyd-Jones, R., Nicholls, J., Sell, R., Tomlinson, C., Ainger, A., and Stewart, T. (1996), 'The Performance of Information Technology and the Role of Human and Organizational Factors' (Sheffield, UK: ESRC Centre for Organization and Innovation, Institute of Work Psychology, University of Sheffield).

Cleveland, H. (1985), 'Twilight of Hierarchy', *Public Administration Review* (Jan.–Feb.), 185–95.

Cochrane, P. (1995), 'The Information Wave', in Emmott (1995), 17–33.

—— (1997), *Tips for Time Travellers* (London: Orion Business Books).

Collins, R. (1989), 'The Language of Advantage: Satellite Television in Western Europe', *Media, Culture, and Society*, 11/3: 351–71.

—— (1992), *Satellite Television in Western Europe* (rev. edn., London: John Libbey).

—— Garnham, N., and Locksley, G. (1988), *The Economics of Television* (London: Sage Publications).

Collins-Jarvis, L. (1992), 'Gender Representation in an Electronic City Hall', paper presented at the International Communication Association Meeting, Miami, May.

Comstock, G., Chaffee, S., Katzman, N., McCombs, M., and Roberts, D. (1978), *Television and Human Behavior* (New York: Columbia University Press).

Coombs, R., and Hull, R. (1996), 'The Politics of IT Strategy and Development in Organizations', in Dutton (1996*a*), 159–76.

Cooper, G., and Woolgar, S. (1993), *Software is Society Made Malleable: The Importance of Conceptions of Audience in Software and Research Practice* (PICT Policy Research Paper, No. 25, Uxbridge, UK: Brunel University).

Crane, R. (1979), *The Politics of International Standards: France and the Color TV War* (Norwood, NJ: Ablex Publishing Company).

Crichton, B. (1993), 'Large Employer Attitudes to New Working Practices Using Telematics', unpublished paper (Cambridge, UK: The Home Office Partnership).

Crichton, M. (1994), *Disclosure* (London: Random House, Arrow).

Cringely, R. X. (1992), *Accidental Empires: How the Boys of Silicon Valley Make their Millions, Battle Foreign Competition, and Still Can't Get a Date* (London: Penguin Books).

Croft, R. (1994), 'Computers and Classroom Culture: A Grounded Theory of the Problems and Practicalities', paper presented at the PICT International Student Conference on 'The Study of Information and Communication Technology: Doctoral Research', Policy Studies Institute, London, 28–30 Mar.

Cuban, L. (1997), 'Unless Teachers Get Involved, Wiring Schools Just Enriches Computer Makers', *Los Angeles Times*, Metropolitan Section, 10 Aug., 1–3.

Culnan, M. J., and Markus, M. L. (1987), 'Information Technologies', in Jablin *et al.* (1987), 420–43.

Curcuru, V. (1997), 'Animation's Global Shift', unpublished paper (Los Angeles, Calif.: Annenberg School for Communication, University of Southern California).

Curran, J., and Seaton, J. (1991), *Power without Responsibility: The Press and Broadcasting in Britain* (4th edn., London: Routledge).

Daft, R. L., and Lengel, R. H. (1986), 'Organizational Information Requirements, Media Richness and Structural Design', *Management Science*, 32: 554–71.

Danziger, J. N. (1986), 'Computing and the Political World', *Social Science Computer Review*, 2/4: 183–200.

—— and Kraemer, K. L. (1986), *People and Computers: The Impacts of Computing on End Users in Organizations* (New York: Columbia University Press).

—— Dutton, W. H., Kling, R., and Kraemer, K. L. (1982), *Computers and Politics* (New York: Columbia University Press).

Davidge, C. (1987), 'America's Talk-Back Television Experiment: QUBE', in Dutton *et al.* (1987*a*), 75–101.

Dempsey, M. (1993), 'Corporate Computing Seeks Safe Pair of Hands', *Financial Times*, 20 July, 9.

Denning, D. E., and Lin, H. S. (1994) (eds.), National Research Council, *Rights and Responsibilities of Participants in Networked Communities* (Washington: National Academy Press).

Denning, P. J. (1996), 'The University's Next Challenges', *Communications of the ACM*, 39/5: 27–31.

Denton, N. (1997), 'Popular Growth Fuels Online Services', *Financial Times*, 4 Sept., 4.

—— (1998), 'mainstream.com', *Financial Times*, 4 Jan.

Derthick, M., and Quirk, P. J. (1985), *The Politics of Deregulation* (Washington: Brookings Institution).

Dertouzos, M. L., Lester, R. K., Solow, R. M. (1989), *Made in America: Regaining the Competitive Edge* (Cambridge, Mass.: MIT Press).

Dervin, B., and Voigt, M. J. (1986) (eds.), *Progress in Communication Sciences* (Belmont, NJ: Ablex Publishing Corporation).

DeSanctis, G., and Fulk, J. (1998) (eds.), *Shaping Organizational Form: Communication, Connection, and Community* (Thousand Oaks, Calif.: Sage).

de Sola Pool, I. (1977) (ed.), *The Social Impact of the Telephone* (Cambridge, Mass.: MIT Press).

—— (1983*a*), *Technologies of Freedom* (Cambridge, Mass.: Harvard University Press, Belknap Press).

—— (1983*b*), *Forecasting the Telephone* (Norwood, NJ: Ablex Publishing Corporation).

—— (1990), *Technologies without Boundaries: On Telecommunications in a Global Age* ed. E. M. Noam (Cambridge, Mass.: Harvard University Press).

Dholakia, R. R., Mundorf, N., and Dholakia, N. (1996) (eds.), *New Infotainment Technologies in the Home: Demand-Side Perspectives* (Mahwah, NJ: Lawrence Erlbaum Associates).

Dickson, E. M. (1974), *The Video Telephone: Impact of an New Era in Telecommunications* (New York: Praeger).

Digital Domain, Inc. (1996), *Barbie Fashion Designer CD-ROM* (El Segundo, Calif.: Mattel, Inc.).

Dixon, H. (1994), 'Super-highways sans frontières', *Financial Times*, 21 Feb.

Docter, S. (1997), 'The First Amendment and the Shaping of Communication Technology' unpublished Ph.D. dissertation (Los Angeles, Calif.: Annenberg School for Communication, University of Southern California).

—— and Dutton, W. H. (1998), 'The First Amendment On-Line: Santa Monica's Public Electronic Network', in Tsagarousianou *et al.* (1998), 125–51.

Downs, A. (1967), 'A Realistic Look at the Payoffs from Urban Data Systems', *Public Administration Review*, 27/3: 304–40.

—— (1985), 'Living with Advanced Telecommunications', *Society*, 23/1: 26–34.

Droege, P. (1997*a*), 'Tomorrow's Metropolis—Virtualization Takes Command', in Droege (1997*b*), 1–17.

—— (1997*b*) (ed.), *Intelligent Environments: Spatial Aspects of the Information Revolution* (Amsterdam: Elsevier).

Drucker, P. F. (1993), *Post-Capitalist Society* (Oxford: Butterworth-Heinemann Ltd.).

—— (1995), *Managing in a Time of Great Change* (New York: Truman Talley Books/Dutton).

DTI (1988): Department of Trade and Industry, Communications Steering Group, *The Infrastructure for Tomorrow: Communications Steering Group Report* (London: HMSO).

—— (1991), *Competition and Choice: Telecommunications Policy for the 1990s* (Cm. 1461; London: HMSO).

Dunkle, D. E., King, J. L., Kraemer, K. L., and Danziger, J. N. (1994), 'Women, Men, and Information Technology: A Gender-Based Comparison of the Impacts of Computing Experienced by White Collar Workers', in Gattiker (1994), 31–63.

Dutton, W. H. (1990), 'The Political Implications of Information Technology', in Berleur *et al.* (1990), 173–95.

—— (1992*a*), 'The Ecology of Games Shaping Communication Policy', *Communication Theory*, 2/4: 303–28.

—— (1992*b*), 'The Social Impact of Emerging Telephone Services', *Telecommunications Policy*, 16/5: 377–87.

—— (1992*c*), 'Political Science Research on Teledemocracy', *Social Science Computer Review*, 10/4 (Winter), 505–22.

—— (1993), 'Electronic Service Delivery and the Inner City: The Risk of Benign Neglect', in Berleur *et al.* (1993), 209–28.

—— (1994), 'Lessons from Public and Nonprofit Services', in Williams and Pavlick (1994), 105–37.

—— (1995*a*), 'On the Line: Regulating Access to the Information Society' presentation to the UK's Parliamentary Information Technology Committee (PITCOM), Westminster, 22 Mar.

—— (1995*b*), 'The Ecology of Games and its Enemies', *Communication Theory*, 5/4: 379–92.

—— (1995*c*), 'Driving into the Future of Communications? Check the Rear View Mirror', in Emmott (1995), 79–102.

—— (1996*a*) (ed.), with Malcolm Peltu, *Information and Communication Technologies: Visions and Realities* (Oxford: Oxford University Press).
—— (1996*b*), 'Network Rules of Order: Regulating Speech in Public Electronic Fora', *Media, Culture and Society*, 18/2: 269–90.
—— (1997*a*), 'Multimedia Visions and Realities', in Kubicek *et al.* (1997), 133–55.
—— (1997*b*) (ed.), 'Virtual Innovations in Organizations', unpublished research report to Fujitsu Research Institute for Advanced Information Systems & Economics, Tokyo, Japan, Mar. (Los Angeles, Calif.: Annenberg School for Communication, University of Southern California).
—— (1997*c*), 'Reinventing Democracy', in Droege (1997*b*), 152–9.
—— (1998), 'The Virtual Organization: Tele-Access in Business and Industry', in DeSanctis and Fulk (1998).
—— and Anderson, R. E. (1989) (eds.), 'Symposium on Computer Literacy: Implications for the Social Sciences', *Social Science Computer Review*, 7/1: 1–45.
—— and Blumler, J. G. (1988), 'The Faltering Development of Cable Television in Britain', *International Political Science Review*, 9/4: 279–303.
—— and Guthrie, K. (1991), 'An Ecology of Games: The Political Construction of Santa Monica's Public Electronic Network', *Informatization and the Public Sector*, 1/4: 1–24.
—— and Kraemer, K. L. (1985), *Modeling as Negotiating: The Political Dynamics of Computer Models in the Policy Process* (Norwood, NJ: Ablex Publishing Corporation).
—— and Meadow, R. G. (1987), 'A Tolerance for Surveillance: American Public Opinion Concerning Privacy and Civil Liberties', in Levitan (1987), 147–70.
—— and Vedel, T. (1992), 'Dynamics of Cable Television in the U.S., Britain, and France', in Blumler *et al.* (1992), 70–93.
—— Blumler, J. G., and Kraemer, K. L. (1987*a*) (eds.), *Wired Cities: Shaping the Future of Communications* (New York: G. K. Hall, Macmillan).
—— Rogers, E. M., and Jun, S. (1987*b*), 'Diffusion and Social Impacts of Personal Computers,' *Communication Research*, 14/2 (Apr.), 219–50.
—— Sweet, P. L., and Rogers, E. M. (1989), 'Socioeconomic Status and the Early Diffusion of Personal Computing in the United States', *Social Science Computer Review*, 7/3: 259–71.
—— Guthrie, K., O'Connell, J., and Wyer, J. (1991), 'State and Local Government Innovations in Electronic Services', unpublished report for the US Office of Technology Assessment, 12 Dec.
—— Wyer, J., and O'Connell, J. (1993), 'The Governmental Impacts of Information Technology: A Case Study of Santa Monica's Public Electronic Network', in Banker *et al.* (1993), 265–96.
—— Blumler, J. G., Garnham, N., Mansell, R., Cornford, J., and Peltu, M. (1994*a*), *The Information Superhighway: Britain's Response* (PICT Policy Research Paper, No. 29; Uxbridge, UK: Brunel University).
—— Taylor, J., Bellamy, C., and Peltu, M. (1994*b*), *Electronic Service Delivery: Themes and Issues in the Public Sector* (PICT Policy Research Paper, No. 28; Uxbridge, UK: Brunel University).

—— MacKenzie, D., Shapiro, S., and Peltu, M. (1995), *Computer Power and Human Limits: Learning from IT and Telecommunications Disasters* (PICT Policy Research Paper, No. 33; Uxbridge, UK: Brunel University).

—— Blumler, J., Garnham, N., Mansell, R., Cornford, J., and Peltu, M. (1996), 'The Politics of Information and Communication Policy: The Information Superhighway', in Dutton (1996*a*), 387–405.

—— Russell, S., and Portman, C. E. (1997) (eds.), *The Programme on Information and Communication Technologies: A Collection of Research and Publications* (Swindon, UK: Economic and Social Research Council).

Dyson, E. (1997), *Release 2.0: A Design for Living in the Digital Age* (New York: Broadway Books).

Earl, M. J. (1996) (ed.), *Information Management: The Organizational Dimension* (Oxford: Oxford University Press).

The Economist (1995), 'Democracy and Technology: e-lectioneering', 17 June, 21–3.

Eisenstadt, E. L. (1968), 'Some Conjectures about the Impact of Printing on Western Society and Thought: A Preliminary Report', *Journal of Modern History* (Mar.), 1–56.

—— (1980), 'The Emergence of Print Culture in the West', *Journal of Communications*, 30/1: 99–106.

Ellul, J. (1964), *The Technological Society*, trans. from the French by John Wilkinson (New York: Vintage Books).

Eliot, T. S. (1969), *The Complete Poems and Plays of T. S. Eliot* (London: Faber & Faber).

Ellison, N., and Dutton, W. H. (1997), 'Fujitsu Business Communication Systems', in Dutton (1997*b*), 23–35.

Elton, M. C. J. (1980), 'Educational and other Two-way Cable Television Services in the United States', in Witte (1980), 142–55.

—— (1985), 'Visual Communication Systems: Trials and Experiences', *Proceedings of the IEEE*, 73/4: 700–5.

—— (1991), 'Integrated Broadband Networks: Assessing the Demand for New Services', unpublished paper presented at the Berkeley Round Table on the Industrial Economy (New York: Columbia University).

—— (1992), 'The US Debate on Integrated Broadband Networks', *Media, Culture and Society*, 14: 369–95.

—— and Carey, J. (1984), 'Teletext for Public Information: Laboratory and Field Studies', in J. Johnston (1984), 23–41.

Emmott, S. J. (1995) (ed.), *Information Superhighways: Multimedia Users and Futures* (New York: Academic Press).

EURICOM (1995): The European Institute for Communication and Culture, 'Access to the Media', *The Public: Javnost*, special issue, 2/4: 1–98.

European Audiovisual Observatory (1997), *Statistical Yearbook 1997* (Brussels: Council of Europe).

European Commission (1994), *Europe's Way to the Information Society: An Action Plan*, COM (94) 347 (Brussels: European Commission).

European Communities Committee (1985), Select Committee on the European Communities, House of Lords, *Television without Frontiers* (London: HMSO).

Evans, C. (1979), *The Micro Millennium* (New York: Washington Square Press).

Faulkner, W., and Senker, J. (1995), *Knowledge Frontiers: Public Sector Research and Industrial Innovation in Biotechnology, Engineering Ceramics, and Parallel Computing* (Oxford: Clarendon Press).

Fincham, R., Fleck, J., Procter, R., Scarbrough, H., Tierney, M., and Williams, R. (1994), *Expertise and Innovation: Information Technology Strategies in the Financial Services Sector* (Oxford: Clarendon Press).

Finnegan, R., Salaman, G., and Thompson, K. (1987) (eds.), *Information Technology: Social Issues* (London: Hodder & Stoughton and the Open University).

Firnberg, D., and West, D. (1987), 'Milton Keynes: Creating an Information Technology Environment', in Dutton *et al.* (1987*a*), 392–408.

Fishman, C. (1997), 'Inside the 1–800 Factory', *Los Angeles Times Magazine*, 3 Aug., 14–29.

Flaherty, D. H. (1985), *Protecting Privacy in Two-Way Electronic Services* (White Plains, NY: Knowledge Industry Publications, Inc.).

—— (1989), *Protecting Privacy in Surveillance Societies* (Chapel Hill, NC: University of North Carolina Press).

Flamm, K. (1987), *Targeting the Computer: Government Support and International Competition* (Washington: Brookings Institution).

—— (1998), 'Don't Let Gates Railroad the Antitrust Laws', *Los Angeles Times*, 16 Jan.

Forester, T. (1980) (ed.), *The Microelectronics Revolution* (Oxford: Basil Blackwell).

—— (1987), *High-Tech Society: The Story of the Information Technology Revolution* (Cambridge, Mass.: MIT Press).

—— (1989*a*), 'The Myth of the Electronic Cottage', in Forester (1989*b*), 213–27.

—— (1989*b*) (ed.), *Computers in Human Context: Information Technology, Productivity, and People* (Cambridge, Mass.: MIT Press).

Foulger, D. A. (1990), 'Medium as Process', unpublished dissertation (Philadelphia, Pa.: Temple University).

Fountain, J. E., Kaboolian, L., and Kelman, S. (1992), 'Services to the Citizen: The Use of 800 Numbers in Government', paper presented at the Association for Public Policy and Management, Denver, Colorado, 29–31 Oct. (Cambridge, Mass.: Harvard University).

Fransman, M. (1990), *The Market and Beyond: Cooperation and Competition in Information Technology Development in the Japanese System* (Cambridge: Cambridge University Press).

Freeman, C. (1987), 'The Case for Technological Determinism', in Finnegan *et al.* (1987), 5–18.

Freeman, C. (1994), 'The Diffusion of ICT in the World Economy in the 1990s', in Mansell (1994), 8–41.

—— (1996*a*), 'The Two-Edged Nature of Technical Change: Employment and Unemployment', in Dutton (1996*a*), 19–36.

—— (1996*b*), 'The Factory of the Future and the Productivity Paradox', in Dutton (1996*a*), 123–41.

—— Sharp, M., and Walker, W. (1991) (eds.), *Technology and the Future of Europe: Global Competition and the Environment in the 1990s* (London: Pinter Publications).

Fulk, J., and DeSanctis (1995), 'Electronic Communication and Changing Organizational Forms', *Organization Science*, 6/4: 337–49.

—— and Dutton, W. H. (1984), 'Videoconferencing as an Organizational Information System: Assessing the Role of Electronic Meetings', *Systems, Objectives and Solutions*, 4: 105–18.

—— and Steinfield, C. (1990) (eds.), *Organizations and Communication Technology* (Newbury Park, Calif.: Sage Publications).

—— Schmitz, J., and Steinfield, C. (1990), 'A Social Influence Model of Technology Use', in Fulk and Steinfield (1990), 117–40.

Gandy, O. H., Jr. (1989), 'The Surveillance Society: Information Technology and Bureaucratic Social Control', *Journal of Communication*, 39/3: 61–76.

—— (1993), *The Panoptic Sort: A Political Economy of Personal Information* (Boulder, Colo.: Westview Press).

GAO (1997): General Accounting Office, *Internet and Electronic Dial-Up Bulletin Boards* (Washington: GAO Report GGD-97-86, 16 June).

Garnham, N. (1983), 'Public Service versus the Market', *Screen*, 24: 6–27.

—— (1990), *Capitalism and Communication: Global Culture and the Economics of Information* (London: Sage Publications).

—— (1994*a*), 'What Ever Happened to the Information Society?', in Mansell (1994), 42–51.

—— (1994*b*), presentation to '(Tele)communication Industries and Public Policy: A PICT Synthesis Workshop', Policy Studies Institute, London, 24 Feb.

Garson, G. D. (1996), 'The Political Economy of Online Education', *Social Science Computer Review*, 14/4: 394–409.

Gates, B. with Myhrvoid, N., and Rinearson, P. (1995), *The Road Ahead* with CD-Rom (London: Viking).

Gattiker, U. E. (1994) (ed.), *Women and Technology* (Berlin: Walter de Gruyter).

Gell, M., and Cochrane, P. (1996), 'Learning and Education in an Information Society', in Dutton (1996*a*), 249–63.

Gérin, F., and de Tavernost, N. (1987), 'Biarritz and the Future of Videocommunications', in Dutton *et al.* (1987*a*), 237–54.

Gerstenzang, J. (1997), '70 Nations Agree to Open Up their Telephone Markets', *Los Angeles Times*, 16 Feb., A16.

Gilbreth, F. B., Jr., and Carey, E. G. (1948), *Cheaper by the Dozen* (Cutchogue, NY: Buccaneer Books).

Gilder, G. (1994), *Life After Television: The Coming Transformation of Media and American Life* (rev. edn., New York: W. W. Norton; originally published 1990).

Gillespie, A., and Cornford, J. (1996), 'Telecommunication Infrastructures and Regional Development', in Dutton (1996*a*), 335–51.

Gingrich, N., and Gingrich, M. (1981), 'Post Industrial Politics', *The Futurist* (Dec.), 29–32.

Goddard, J. (1994), 'ICTs, Space, and Place', in Mansell (1994), 274–85.

—— and Cornford, J. (1994), 'Superhighway Britain: Eliminating the Divide between the Haves and Have-Nots', *Parliamentary Brief* (May–June), 48–50.

—— and Richardson, R. (1996), 'Why Geography Will Still Matter: What Jobs Go Where?', in Dutton (1996*a*), 197–214.

Gold, M. (1997), 'The Paycheck Road to Better Health', *Los Angeles Times*, Business Section, 8 Aug., 1, 5.

Goldmark, P. C. (1972), 'Communication and the Community', in *Scientific American* (1972), 103–8.

Gonzalez, I. M. (1986), 'Utilization Patterns of Hi-OVIS Interactive Broadcasting System in Higashi Ikoma, Japan', unpublished Ph.D. Dissertation (Los Angeles, Calif.: Graduate School, University of Southern California).

Gonzalez, J. M. (1997), 'Grant to Extend Hours of 10 County Computer Learning Centers', *Los Angeles Times*, 27 Mar., B6.

Gore, A. (1991), 'Infrastructure for the Global Village', *Scientific American*, 265 (Sept.), 108–11.

—— (1993), 'National Information Infrastructure Initiatives' [http:///www.hpcc.gov/white-house/gore.nii.html], June.

—— (1994), The White House, Office of the Vice-President, *Re-Engineering through Information Technology*, pt. I (Washington: Government Printing Office, 27 May).

—— (1997), *MEDLINE Gives American's Immediate Access to Useful Information*, news release (Washington: The White House, Office of the Vice-President, 26 June).

—— and Brown, R. (1995), *Global Information Infrastructure: Agenda for Cooperation* (Washington: Government Printing Office).

Gotlieb, C. C., and Borodin, A. (1973), *Social Issues in Computing* (New York: Academic Press).

Graham, S., and Marvin, S. (1996), *Telecommunications and the City: Electronic Spaces, Urban Places* (London: Routledge).

Greenberger, M. (1983), *Caught Unawares: The Energy Decade in Retrospect* (Cambridge, Mass.: Ballinger Publishing Company).

—— (1985) (ed.), *Electronic Publishing Plus* (White Plains, NY: Knowledge Industries Publications).

—— (1992) (ed.), *Multimedia in Review: Technologies for the 21st Century* (Santa Monica, Calif.: The Voyager Company and Council for Technology and the Individual).

Greenfield, P. M. (1984), *Mind and Media: The Effects of Television, Video Games, and Computers* (Cambridge, Mass.: Harvard University Press).

Greider, W. (1981), 'The Education of David Stockman', *Atlantic Monthly*, 248/6: 27–54.

Grint, K. (1992), 'Sniffers, Lurkers, Actor Networkers: Computer Mediated Communications as a Technical Fix', in Beynon and MacKay (1992), 148–70.

Grossman, L. K. (1995), *The Electronic Republic: Reshaping Democracy in the Information Age* (New York: Viking).

GTE (1996): GTE Telephone Operations, *Caller ID*, 7–96:CA:F#90 (San Angelo, Tex.: GTE Telephone Operations).

Guthrie, K. K. (1991), 'The Politics of Citizen Access Technology: The Development of Communications and Information Utilities in Four Cities', unpublished Ph.D. Dissertation (Los Angeles, Calif.: Annenberg School for Communication, University of Southern California).

—— and Dutton, W. H. (1992), 'The Politics of Citizen Access Technology', *Policy Studies Journal*, 20/4: 574–97.

—— Schmitz, J., Ryu, D., Harris, J., Rogers, E., and Dutton, W. (1990), 'Communication Technology and Democratic Participation: PENners in Santa Monica', paper presented at the Association for Computer Machinery (ACM) Conference on Computers and the Quality of Life, Washington, Sept.

GVU (1995): Graphic, Visualization & Usability Center, 'GVU's 3rd WWW User Survey' [http://www.cc.gatech.edu/gvu/user_surveys/survey-04-1995], Apr.

Haddon, L., and Skinner, D. (1991), 'The Enigma of the Micro: Lessons from the British Home Computer Boom', *Social Science Computer Review*, 9: 435–49.

Haeckel, S. H., and Nolan, R. L. (1993), 'Managing by Wire', *Harvard Business Review* (Sept.–Oct.), 122–32.

Hafner, K., and Lyon, M. (1996), *Where Wizards Stay Up Late: The Origins of the Internet* (New York: Simon & Schuster).

Hale, M. L. (1997), 'California Cities and the World Wide Web', unpublished master's thesis (Los Angeles, Calif.: The Graduate School, University of Southern California).

Hamelink, C. J. (1988), *The Technology Gamble* (Norwood, NJ: Ablex Publishing Company).

Hammer, M., and Champy, J. (1993), *Reengineering the Corporation: A Manifesto for Business Revolution* (London: Nicholas Brealey Publishing; repr. 1994).

Handy, C. (1993), *Understanding Organizations* (4th edn., London: Penguin Books).

—— (1995), *Beyond Certainty: The Changing Worlds of Organizations* (Boston: Harvard Business School Press).

—— (1996), *Beyond Certainty: The Changing World of Organizations* (Boston: Harvard Business School Press).

—— (1997), *The Hungry Spirit: Beyond Capitalism—A Quest for Purpose in the Modern World* (New York: Hutchinson).

Hawkins, R. (1996), 'Standards for Communication Technologies: Negotiating Institutional Biases in Network Design', in Mansell and Silverstone (1996), 157–86.

Hayward, E. G. (1996), 'A Classified Guide to the Frederick Winslow Taylor Collection' [http://www.lib.stevens-tech.edu/oldwebpage/special/taylor/hayward/], July.

Healy, T. (1968), 'Transportation or Communications, Some Broad Considerations', *IEEE Transactions on Communication Technology*, 16/2: 195–8.

Heap, N., Thomas, R., Einon, G., Mason, R., and Mackay, H. (1995) (eds.), *Information Technology and Society* (London: Sage Publications).

Hecht, B. (1997), 'Net Loss', *New Republic*, 17 Feb., 15–18.

Heckscher, C., and Donnelon, A. (1994) (eds.), *The Post-Bureaucratic Organization: New Perspectives on Organizational Change* (Thousand Oaks, Calif.: Sage Publications).

Heeter, C. (1995), 'Communication Reseach on Consumer VR', in Biocca and Levy (1995), 191–218.

Hepworth, M. E. (1989), *The Geography of the Information Economy* (London: Belhaven).

Hiltz, S. R., and Turoff, M. (1978), *The Network Nation: Human Communication via Computer* (Reading, Mass.: Addison-Wesley Publishing Company, Inc.).

HLEG (1996): High Level Expert Group on the Information Society, *Building the European Information Society for Us All: First Reflections of the High Level Group of Experts* (Brussels: European Commission, DGV, Interim Report).

HOK Facilities Consulting (1994): Hellmuth, Obata & Kassabaum Inc., 'AO: The Second Wave', *The HOK Consulting Report*, 4: 3.

Hollingshead, A. B. (1996), 'Information Suppression and Status Persistence in Group Decision Making: The Effects of Communication Media', *Human Communications Research*, 23/2: 193–219.

Hollins, T. (1984), *Beyond Broadcasting: Into the Cable Age* (London: BFI Publishing, British Film Institute).

Hooper, R. (1985), 'Lessons from Overseas: The British Experience', in Greenberger (1985), 181–200.

Hopkins, N. (1995), 'Doing Business with Government Electronically', paper presented at the PICT International Conference on the Social and Economic Implications of Information and Communication Technologies, Queen Elizabeth II Conference Centre, Westminster, 10–12 May).

Humphreys, P. (1986), 'Legitimating the Communications Revolution', *West European Politics*, 9/4: 163–94.

Innis, H. (1950), *Empire and Communications* (Oxford: Oxford University Press; rev. edn., Toronto: University of Toronto Press, 1972).

Iritani, E. (1996), 'Supercomputer Bid Touches Off Political Flap', *Los Angeles Times*, 17 Nov.

ITAP (1982): Information Technology Advisory Panel to the Cabinet Office, *Report on Cable Systems* (London: HMSO).

—— (1983), *Making a Business of Information* (London: HMSO).

Ito, Y., and Oishi, Y. (1987), 'Social Impacts of the New Utopias', in Dutton *et al.* (1987*a*), 201–20.

ITU (1994): International Telecommunications Union, *World Telecommunication Development Report* (Geneva: ITU).

Jablin, F. M. (1987), 'Formal Organization Structure', in Jablin *et al.* (1987), 389–419.

—— and Putnam, L. L. (forthcoming) (eds.), *New Handbook of Organizational Communication* (Newbury Park, Calif.: Sage Publications).

—— —— Roberts, K. H., Porter, L. W. (1987) (eds.), *Handbook of Organizational Communication: An Interdisciplinary Perspective* (Newbury Park, Calif.: Sage Publications).

Johansen, R. (1984), *Teleconferencing and Beyond: Communications in the Office of the Future* (New York: McGraw Hill).

—— (1988), *Groupware: Computer Support for Business Teams* (New York: Free Press).

—— and Swigart, R. (1994), *Upsizing the Individual in the Downsized Organization* (Boston: Addison-Wesley Publishing Company).

Johansson, H. J., McHugh, P., Pendlebury, A. J., Wheeler, W. A., Jr. (1993), *Business Process Reengineering: Breakpoint Strategies for Market Dominance* (New York: John Wiley and Sons).

Johnson, L. B. (1971), *The Vantage Point* (New York: Holt, Rinehart, and Winston).

Johnston, J. (1984) (ed.), *Evaluating the New Information Technologies* (San Francisco, Calif.: Jossey-Bass).

Jones, S. G. (1997), *Virtual Culture: Identity and Communication in Cybersociety* (Thousand Oaks, Calif.: Sage Publications).

Jupiter Communications (1996), 'Jupiter Study: Total Online Households Worldwide Will Rise to 66.6 Million by Year 2000' [http://www.jup.com/jupiter/release/nov96/market.shtml], Nov.

Kahan, H. (1984), 'How Americans React to Communications Technology: Technological Craps', paper presented at the wired cities forum, Annenberg Schools of Communications, Washington.

Kahin, B., and Wilson, E. (1997) (eds.), *National Information Infrastructure Initiatives: Vision and Policy Design* (Cambridge, Mass.: Harvard University Press).

Kahlenberg, R. (1997), 'TV Smarts', *Los Angeles Times*, 2 Oct. 1997.

Katz, E. (1996), 'Mass Media and Participatory Democracy: Manifest Functions, Latent Dysfunctions', Presentation at the Annenberg School for Communication, USC, Los Angeles, Calif., 21 Nov.

—— and Lazersfeld, P. F. (1955), *Personal Influence: The Part Played by People in the Flow of Mass Communication* (New York: Free Press).

Katz, J. E. (1990), 'Caller-ID, Privacy, and Social Processes', *Telecommunications Policy* (Oct.), 372–411.

—— and Graveman, R. F. (1991), 'Privacy Issues of a National Research and Education Network', *Telematics and Informatics*, 8/1–2: 71–120.

—— with Aspden, P. (1997), 'Barriers to and Motivations for Using the Internet: Results of a National Opinion Survey', *Internet Research Journal: Technology, Policy, & Applications*, 7/3: 170–88.

—— —— (forthcoming), 'Internet Dropouts: The Invisible Group', *Telecommunications Policy*.

Kawahata, M. (1987), 'Hi-OVIS', in Dutton *et al.* (1987*a*), 179–200.

Kay, A. C. (1991), 'Computers, Networks, and Education', *Scientific American*, 265/3 (Sept.), 100–7.

—— (1994), 'The Best Way to Predict the Future is to Invent it', PICT's 1994 Charles Read Lecture, British Academy of Film and Television Arts, London, 26 May.

Kehoe, L. (1997), 'Intel Tries to Net TV', *Financial Times*, 10 Dec.

Keller, S. (1977), 'The Community Telephone', in de Sola Pool (1977), 281–98.

Kenny, A. (1994), 'Digitised Beowulf Takes Library into the Future', *Times Higher Education Supplement*, 13 May, p. iv.

Kiesler, S., Siegel, J., and McGuire, T. W. (1984), 'Social Psychological Aspects of Computer-Mediated Communication', *American Psychologist*, 39/10: 1123–34.

King, J. L. (1997), unpublished presentation to the Japan American Institute for Management Sciences, Honolulu, Aug.

Kleinsteuber, H. J. (1997), 'Crippled Digitalization: Superhighways or One-Way Streets?', in Kubicek *et al.* (1997), 79–96.

Kling, R., and Iacono, S. (1990), 'Computerization Movements and the Mobilization of Support for Computing', in Berleur *et al.* (1990), 62–83.

Knights, D., and Willmott, H. (1988) (eds.), *New Technology and the Labour Process* (London: Macmillan Press Ltd.).

Kraemer, K. L. (1969), 'The Evolution of Information Systems for Urban Administration', *Public Administration Review*, 29/4: 389–402.

—— and Dedrick, J. (1996), 'IT and Economic Development: International Competitiveness', in Dutton (1996*a*), 319–33.

—— Dutton, W. H., and Northrop, A. (1981), *The Management of Information Systems* (New York: Columbia University Press).

Krasnow, E. G., Longley, L. D., and Terry, H. A. (1982), *The Politics of Broadcast Regulation* (3rd edn., New York: St Martin's Press).

Kraut, R., Scherlis, W., Mukhopadhyay, T., Manning, J., and Kiester, S. (1996), 'The Home Net Field Trial of Residential Internet Services', *Communications of the ACM*, 39 (Dec.), 55–63.

Kubicek, H., Dutton, W., and Williams, R. (1997) (eds.), *The Social Shaping of the Information Superhighway: European and American Roads to the Information Society* (Frankfurt: Campus Verlag).

La Porte, T. R. (1991) (ed.), *Social Responses to Large Technical Systems* (Netherlands: Kluwer Academic Publishers).

Lamberton, D. (1997), 'The Knowledge-based Economy: A Sisyphus Model', *Prometheus*, 15/1: 73–81.

Lasswell, H. D. (1971), 'The Structure and Function of Communication in Society', in Schramm and Roberts (1971), 84–99.

Laudon, K. L. (1974), *Computers and Bureaucratic Reform: The Political Functions of Urban Information Systems* (New York: John Wiley).

—— (1977), *Communications Technology and Democratic Participation* (New York: Praeger).

—— (1986), *Dossier Society: Value Choices in the Design of National Information Systems* (New York: Columbia University Press).

Laufer, R. (1990), 'The Question of the Legitimacy of the Computer', in Berleur *et al.* (1990), 31–61.

Leavitt, H. J., and Whisler, T. L. (1958), 'Management in the 1980's', *Harvard Business Review*, 36/6: 41–8.

Leonard, D. (1994), *Guide to the European Union* (London: Penguin Books).

Levitan, K. B. (1987) (ed.), *Government Infostructures* (Westport, Conn.: Greenwood Press).

Lippmann, W. (1922), *Public Opinion* (New York: Free Press; repr. 1965).

Lips, M., and Frissen, P. (1997), 'Wiring Government: Integrated Public Service Delivery through ICT', *National Programma Informatietechnologie en Recht (ITeR)*, 8: 67–164.

Locksley, G. (1990) (ed.), *The Single European Market and the Information and Communication Technologies* (London: Belhaven Press).

Long, N. E. (1958), 'The Local Community as an Ecology of Games', *American Journal of Sociology*, 64: 251–61.

Lucky, R. W. (1989), *Silicon Dreams* (New York: St Martin's Press).

Lytel, D. (1997), 'Media Regimes and Political Communication', presentation to the Annenberg School for Communication, University of Southern California, Apr.

McChesney, R. W. (1997), *Corporate Media and the Threat to Democracy* (New York: Seven Stories Press).

McDermott, J. (1969), 'Technology: The Opiate of the Intellectuals', repr. in Teich (1981), 130–63.

McGilly, K., Kawahata, M., and Dutton, W. H. (1990), 'Lessons from the Fibre-to-the-Home Trails' (Los Angeles, Calif.: Annenberg School for Communication, University of Southern California).

McGrath, J. E., and Hollingshead, A. B. (1994), *Groups Interacting with Technology* (Thousand Oaks, Calif.: Sage Publications).

Machlup, F. (1962), *The Production and Distribution of Knowledge in the United States* (Princeton: Princeton University Press; repr. 1972).

Mackintosh, I. (1986), *Sunrise Europe* (Oxford: Basil Blackwell).

MacKenzie, D. (1993), *Inventing Accuracy: A Historical Sociology of Nuclear Missile Guidance* (Cambridge, Mass.: Cambridge University Press).

—— (1996), *Knowing Machines* (Cambridge, Mass.: MIT Press).

—— and Wajcman, J. (1985) (eds.), *The Social Shaping of Technology: How a Refrigerator Got its Hum* (Milton Keynes: Open University Press).

McLain, L. (1994), 'OU Sees Net Gain for Real', *The Times Higher Education Supplement*, 16 Sept.

McLaughlin, M. L., Osborne, K. K., and Ellison, N. B. (1997), 'Virtual Community in a Telepresence Environment', in Jones (1997), 146–168.

McLuhan, M. (1964), *Understanding Media: The Extensions of Man* (London: Routledge; repr. 1994).

McQuail, D. (1987), 'Research on New Communication Technologies: Barren Terrain or Promising Arena?', in Dutton *et al.* (1987*a*), 431–45.

—— (1994), *Mass Communication Theory: An Introduction* (3rd edn., London: Sage Publications).

—— and Siune, K. (1986) (eds.), *New Media Politics: Comparative Perspectives in Western Europe* (London: Sage Publications).

Mahoney, S., DeMartino, N., and Stengel, R. (1980), *Keeping PACE with the New Television* (New York: VNU Books International).

Mankin, D., Cohen, S. G., and Bikson, T. K. (1996), *Teams and Technology: Fulfilling the Promise of the New Organization* (Boston: Harvard Business School Press).

Mansell, R. (1993), *The New Telecommunications: A Political Economy of Network Evolution* (London: Sage Publications).

—— (1994) (ed.), *Management of Information and Communication Technologies: Emerging Patterns of Control* (London: Aslib).

—— (1996), 'Innovation in Telecommunication Regulation: Realizing National Policy Goals in a Global Marketplace', in Dutton (1996*a*), 371–86.

—— and Silverstone, R. (1996) (eds.), *Communication by Design: The Politics of Information and Communication Technologies* (Oxford: Oxford University Press).

Martin, J. (1977), *Future Developments in Telecommunications* (2nd edn., Englewood Cliffs, NJ: Prentice-Hall, Inc.).

Massey, D., Quintas, P., and Wield, D. (1992), *High Tech Fantasies: Science Parks in Society, Science, and Space* (London: Routledge).

Masuda, Y. (1980), *The Information Society as Post-Industrial Society* (Tokyo: Institute for the Information Society; first English language publication, Bethesda, Md.: World Future Society, 1981).

Meadow, R. (1985) (ed.), *New Communication Technologies in Politics* (Washington: The Washington Program, Annenberg Schools of Communications).

Mediamark Research Inc. (1996), *Multimedia Audiences* (New York: Mediamark Research Inc.).

Melody, W. H. (1996), 'The Strategic Value of Policy Research in the Information Economy', in Dutton (1996*a*), 303–17.

Memmott, F., III (1963), 'The Substitutability of Communications for Transportation', *Traffic Engineering*, 33 / 5: 20–5.

Mesthene, E. G. (1969), 'Some General Implications of the Research of the Harvard University Program on Technology and Society', *Technology and Culture* (Oct.), repr. as 'The Role of Technology in Society', in Teich (1981), 99–129.

Milbrath, L. W. (1963), *The Washington Lobbyists* (Chicago: Rand McNally & Co.).

—— (1996), 'Psychological, Cultural, and Informational Barriers to Sustainability', *Journal of Social Issues*, 51 / 4: 101–20.

Miles, I. (1988), *Home Informatics: Information Technology and the Transformation of Everyday Life* (London: Pinter Publishers).

—— (1993), 'Services in the New Industrial Economy', *Futures* (July–Aug.), 653–72.

—— and Robins, K. (1992), 'Making Sense of Information', in Robins (1992), 1–26.

—— Rush, H., Turner, K., and Bessant, J. (1988), *Information Horizons: The Long-Term Social Implications of New Information Technologies* (Aldershot, UK: Edward Elgar).

—— with Brady, T., Davies, A., Haddon, L., Matthews, M., Rush, H., and Wyatt, S. (1990), *Mapping and Measuring the Information Economy* (Library and Information Research Report, 77; Boston Spa, UK: British Library).

Miller, G. (1997), 'Intel Backs Down on Digital TV', *Los Angeles Times*, 6 Dec.

Molina, A. H. (1989), *The Social Basis of the Microelectronics Revolution* (Edinburgh: Edinburgh University Press).

—— (1990), 'Transputers and Transputer-Based Parallel Computers: Socio-Technical Constituencies and the Build-Up of British–European Capabilities in Information Technologies', *Research Policy*, 19: 309–33.

Monck, C. S. P., Porter, R. B., Quintas, P., Storey, D. J., with Wynarczyk, P. (1988), *Science Parks and the Growth of High Technology Firms* (London: Routledge).

Mosco, V. (1982), *Pushbutton Fantasies: Critical Perspectives on Videotext and Information Technologies* (Norwood, NJ: Ablex Publishing Corporation).

—— (1989), *The Pay-Per Society: Computers and Communication in the Information Age* (Norwood, NJ: Ablex Publishing Corporation).

—— and Wasco, J. (1988) (eds.), *The Political Economy of Information* (Madison, Wis.: University of Wisconsin Press).

Moss, M. L. (1981) (ed.), *Telecommunications and Productivity* (New York: Addison-Wesley).

—— (1987), 'Telecommunications and the Economic Development of Cities', in Dutton *et al.* (1987*a*), 139–51.

Mossberg, W. S. (1996), 'Going On-Line is Still Too Difficult to Lure A Mass Audience', *Wall Street Journal*, 22 Feb.

Mulgan, G. J. (1991), *Communication and Control: Networks and the New Economies of Communication* (Cambridge: Polity Press).

—— (1994) (ed.), 'Liberation Technology?', *DEMOS Quarterly*, 4: 1–34.

Murakami, T., Nishiwaki, T. *et al.* (1991), *Strategy for Creation* (Cambridge, UK: Woodhead Publishing Limited).

Murata, T. (1987), 'Competition for Shaping the New Utopias', in Dutton *et al.* (1987*a*), 165–78.

Murray, F., and Willmott, H. (1997), 'Putting IT in its Place: Towards Flexible Integration in the Network Age?', in Bloomfield *et al.* (1997), 160–80.

Nadel, M. S. (1991), 'US Communications Policymaking: Who and Where', *Hastings Communications and Entertainment Law Journal*, 13/2: 273–323.

NCES (1993): National Center for Education Statistics, *Adult Literacy in America* (Washington: NCES).

Negrine, R. M. (1985) (ed.), *Cable Television and the Future of Broadcasting* (London: Croom Helm).

Negroponte, N. (1995), *Being Digital* (London: Hodder & Stoughton).

Nelson, T. (1974), *Dream Machines* (Chicago, Ill.: Hugo's Book Service).

Network Wizards (1997), *Internet Domain Survey* [http://www.nw.com], Aug.

Neuberger, C., Tonnebacher, J., Biebl, M., and Duck, A. (1997), 'Die deutschen Tageszeitungen im World Wide Web: Redaktionem, Nutzer, Angebote', *Media Perspecktiven*, 12: 652–62.

Neuman, R. (1985), 'The Media Habit', in Greenberger (1985), 5–12.

Neuman, W. R., McKnight, L., Solomon, R. J. (1997), *The Gordian Knot: Political Gridlock on the Information Highway* (Cambridge, Mass.: MIT Press).

Nielson (1997), *CommerceNet/Nielson Internet Demographics Recontact Study: March/April 1996: Executive Summary* [http://www.nielsonmedia.com/commercenet/exec.html], Aug.

Nilles, J. M. (1992), *Telework and Business Strategy: Leading the Information Age*, (JALA Technical Reports, 3/1; Los Angeles, Calif.: JALA Associates).

—— (1994), *Making Telecommuting Happen* (New York: Van Nostrand Reinhold).

—— Carlson, F. R., Jr., Gray, P., and Hanneman, G. J. (1976), *The Telecommunications –Transportation Tradeoff: Options for Tomorrow* (New York: John Wiley & Sons).

Noam, E. (1991), *Television in Europe* (Oxford: Oxford University Press).

—— (1995), 'Electronics and the Dim Future of the University', *Science*, 270/13: 247–9.

Noble, D. F. (1998), 'Digital Diploma Mills: The Automation of Higher Education', *First Monday*, 3: 1 [http://www.firstmonday.dk/issues/issue3_1/noble/index.html], Jan.

Nohria, N., and Berkley, J. D. (1994), 'The Virtual Organization: Bureaucracy, Technology, and the Implosion of Control', in Heckscher and Donnelon (1994), 108–28.

Noll, A. M. (1972), 'Man-Machine Tactile Communication', *SID Journal*, 1/2 (July–Aug.), 5–11.

—— (1976), 'Teleportation through Communication', *IEEE Transactions on Systems, Man, and Cynbernetics* (Nov.), 753–6.

—— (1978), 'The Effect of Communications Medium on the Fundamental Frequency of Human Speech', *Communication Quarterly*, 26/2 (Spring), 51–6.

—— (1985), 'Videotex: Anatomy of a Failure', *Information and Management*, 9: 99–109.

—— (1986), 'Teleconferencing Target Market', *Information Management Review*, 2/2: 65–73.

—— (1991), *Introduction to Telephones and Telephone Systems* (2nd edn., Boston, Mass.: Artech House).

—— (1992), 'Anatomy of a Failure: Picturephone Revisited', *Telecommunications Policy*, 16/4: 307–16.

—— (1994), 'A Study of Long-Distance Rates', *Telecommunications Policy*, 18: 355–62.

—— (1996), 'The Hazards of Cyber Overload', *Telecommunications*, 30/3: 44.

—— (1997), *Highway of Dreams: A Critical View Along the Information Superhighway* (Mahwah, NJ: Lawrence Erlbaum Assoc.).

—— (1998), 'The Digital Mystique: A Review of Digital Technology and its Application to Television', *Prometheus*, 16: 145–53.

—— and Mays, J. (1971), 'Computer Literacy: An Education-Technology Initiative', unpublished draft of staff report to Edward E. David, Jr., Science Adviser to President Richard Nixon (Washington: Office of Science and Technology).

Nonaka, I., and Takeuchi, H. (1995), *The Knowledge Creating Company: How Japanese Companies Create the Dynamics of Innovation* (Oxford: Oxford University Press).

Nora, S., and Minc, A. (1981), *The Computerization of Society: A Report to the President of France* (Cambridge, Mass.: MIT Press), originally published in 1978 as *L'Informatisation de la société* (Paris: La Documentation Française).

Norman, A. R. D. (1995), 'The Political Agenda for IT', *Information Technology & Public Policy*, 13/2: 115–20.

NRENAISSANCE Committee (1994): Computer Science and Telecommunications Board, National Research Council, *Realizing the Information Future: The Internet and Beyond* (Washington: National Academy Press).

NTIA (1996): National Telecommunications and Information Administration, *Lessons Learned from the Telecommunications and Information Infrastructure Assistance Program* (Washington: NTIA, Dept. of Commerce).

Oettinger, A. G., Berman, P. J., and Read, W. H. (1977), *High and Low Politics: Information Resources for the 80s* (Cambridge, Mass.: Ballinger Publishing Company).

ONS (1996): Office of National Statistics, *Family Spending: A Report on the 1995–96 Family Expenditure Survey* (London: HMSO).

—— (1997), *Social Trends, 1997 Edition* (London: HMSO).

OPCS (1993): Office of Population Censuses and Surveys, *General Household Survey 1993* (London: HMSO).

Oppenheimer, T. (1997), 'The Computer Delusion', *Atlantic Monthly*, July, 45–62.

Orwell, G. (1936), *Keep the Aspidistra Flying* (London; paperback edn., Harmondsworth: Penguin Books 1962, repr. 1972).

—— (1949), *Nineteen Eighty-Four* (New York: Harcourt Brace & Company, Inc.; Cutchogue, NY: Buccaneer Books).

OST (1995): Office of Science and Technology, *Technology Foresight Panel: Progress through Partnership—Transportation* (London: HMSO).

OTA (1985): Office of Technology Assessment, US Congress, *Automation of America's Offices* (Washington: US Government Printing Office).

—— (1993), *Making Government Work: Electronic Service Delivery of Federal Services*, OTA-TCT-578 (Washington: US Government Printing Office).

Papert, S. (1980), *Mind-Storms: Children, Computers, and Powerful Ideas* (New York: Basic Books).

Parkes, C. (1997), 'Media Growth Outstrips US Rise in GDP', *Financial Times*, 4 Nov.

Peacock, A. (1986), *Report of the Committee on Financing the BBC* (London: HMSO).

Peck, J. (1987) (ed.), *The Chomsky Reader: Noam Chomsky* (New York: Pantheon Books).

Peltu, M., McKenzie, D., Shapiro, S., and Dutton, W. H. (1996), 'Computer Power and Human Limits', in Dutton (1996*a*), 177–96.

Penzias, A. (1995), *Harmony: Business, Technology, and Life after Paperwork* (New York: HarperCollins Publishers).

Perez, C. (1983), 'Structural Change and the Assimilation of New Technologies in the Economic and Social System', *Futures*, 15: 357–75.

Perkins, D. N. (1990), 'Person Plus: A Distributed View of Thinking and Learning', paper delivered to the Annual Meeting of the American Educational Association, Boston, Mass., 16–20 Apr.

Pew Research Center (1996*a*), *One-in-Ten Voters Online for Campaign '96*, news release (Washington: The Pew Research Center for the People & The Press).
—— (1996*b*), 'TV News Viewership Declines', unpublished news release of 13 May (Washington: The Pew Research Center for the People & The Press).
—— (1997), 'Fewer Favor Media Scrutiny of Political Leaders', unpublished news release of 21 Mar. (Washington: The Pew Research Center for the People & The Press).
PICT (1995): Programme on Information and Communication Technology, *PICT: A Profile of Research and Publications 1995* (Swindon, UK: Economic and Social Research Council).
Pierce, J. R., and Noll, A. M. (1990), *Signals: The Science of Telecommunications* (New York: Scientific American Library).
Popper, K. R. (1945), *The Open Society and its Enemies, ii* (London: Routledge).
Porat, M. U. (1976), 'The Information Economy', unpublished Ph.D. dissertation (Stanford University).
—— (1977), *The Information Economy: Definition and Measurement* (Washington: Office of Telecommunications, US Department of Commerce).
Porter, V., and Hasselbach, S. (1991), *Pluralism, Politics and the Marketplace: The Regulation of German Broadcasting* (London: Routledge).
Quintas, P. (1996), 'Software by Design', in Mansell and Silverstone (1996), 75–102.
Quittner, J. (1997), Invasion of Privacy, *Time*, 27 Aug., 28–35.
Raab, C., Bellamy, C., Taylor, J., Dutton, W. H., and Peltu, M. (1996), 'The Information Polity: Electronic Democracy, Privacy, and Surveillance', in Dutton (1996*a*), 283–99.
Ranney, A. (1983), *Channels of Power: The Impact of Television on American Politics* (New York: Basic Books).
Ratnesar, R. (1997), 'A Bandwidth Bonanza', *Time*, 1 Sep. 60.
Rawsthorn, A. (1997), 'TV Revolution Will Fuel Sales', *Financial Times*, 4 Feb.
Resnick, P. (1997), 'Filtering Information on the Internet', *Scientific American* (Mar.), 106–8.
—— and Miller, J. (1996), 'PICS: Internet Access Controls without Censorship', *Communications of the ACM*, 39/10: 87–93.
Rheingold, H. (1994), *The Virtual Community: Finding Connection in a Computerized World* (London: Secker & Warburg).
Rice, R. E. (1984), 'Mediated Group Communication', in Rice and Associates (1984), 129–54.
—— and Associates (1984), *The New Media: Communication, Research, and Technology* (Beverly Hills, Calif.: Sage Publications).
—— and Gattiker, U. (forthcoming), 'New Media and Organizational Structuring: The Sublimation of Boundaries in Meaning and Relations', in Jablin and Putnam (forthcoming).
Richardson, R. (1994), 'Back Officing Front Office Functions—Organizational and Locational Implications of New Telemediated Services', in Mansell (1994), 309–35.

Robins, K. (1992) (ed.), *Understanding Information: Business, Technology and Geography* (London: Belhaven).

—— and Webster, F. (1989*a*), *The Technical Fix: Education, Computers, and Industry* (London: Macmillan).

—— —— (1989*b*), 'Computer Literacy: The Employment Myth', *Social Science Computer Review*, 7/1: 7–26.

Rochlin, G. I. (1991), 'Iran Air Flight 655 and the USS *Vincennes*: Complex, Large-Scale Military Systems and the Failure of Control', in La Porte (1991), 99–125.

Rogers, E. M. (1986), *Communication Technology: The New Media in Society* (New York: Free Press).

—— and Larson, J. K. (1984), *Silicon Valley Fever: Growth of High-Technology Culture* (New York: Basic Books, Inc.).

Ronfeldt, D. (1992), 'Cyberocracy is Coming', *Information Society*, 8: 243–96.

Rosengren, K. E. (1997), 'Different Sides of the Same Coin: Access and Gate-keeping', *Nordicom*, 18/2: 3–12.

Sackman, H. (1997), *Biomedical Information Technology: Global Social Responsibilities for the Democratic Age* (San Diego, Calif.: Academic Press).

—— and Boehm, B. (1972) (eds.), *Planning Community Information Utilities* (Montvale, NJ: AFIPS Press).

—— and Nie, N. (1970) (eds.), *The Information Utility and Social Choice* (Montvale, NJ: AFIPS Press).

Sampler, J. (1996), 'Exploring the Relationship between Information Technology and Organizational Structure', in Earl (1996), 5–22.

Sapolsky, H. M., Crane, R. J., Neuman, W. R., and Noam, E. M. (1992) (eds.), *The Telecommunications Revolution* (London: Routledge).

Sawhney, H. (1996), 'Information Superhighway: Metaphors as Midwives', *Media, Culture and Society*, 18/2: 291–314.

Schalken, C., and Tops, P. (1995), 'Democracy and Virtual Communities: An Empirical Exploration of the Amsterdam Digital City', in Van de Donk *et al.* (1995), 143–54.

Schattschneider, E. E. (1960), *The Semi-Sovereign People: A Realist's View of Democracy in America* (New York: Holt, Rinehart, & Winston).

Schiller, D. (1982), *Telematics and Government* (Norwood, NJ: Ablex Publishing Corporation).

—— (1988), 'How to Think About Information', in Mosco and Wasko (1988), 27–43.

Schiller, H. I. (1981), *Who Knows: Information in the Age of the Fortune 500* (Norwood, NJ: Ablex Publishing Corporation).

—— (1989), *Culture Inc.: The Corporate Takeover of Public Expression* (Oxford: Oxford University Press).

—— (1996), *Information Inequality: The Deeping Social Crisis in America* (New York: Routledge).

Schorr, H., and Stolfo, S. J. (1997), 'Towards the Digital Government of the 21st Century', A Report from the Workshop on Research and Development

Opportunities in Federal Information Services, 24 June [http://www.isi.edu/nsf/final.html], June.

Schramm, W. (1977), *Big Media, Little Media: Tools and Technologies for Instruction* (Beverly Hills, Calif.: Sage Publications).

—— and Roberts, D. F. (1971) (eds.), *The Process and Effects of Mass Communication* (rev. edn., Chicago: University of Chicago Press).

Schroeder, R. (1995), 'Virtual Reality in the Real World', in Heap *et al.* (1995), 387–99.

—— (1996), *Possible Worlds: The Social Dynamic of Virtual Reality Technology* (Boulder, Colo: Westview Press).

Scientific American (1972), Communication (San Francisco, CA: W. H. Freeman).

Scott, A. J. (1995), *From Silicon Valley to Hollywood: Growth and Development of the Multimedia Industry in California* (Working Paper Series; Los Angeles, Calif.: The Lewis Center for Regional Policy Studies, University of California, Los Angeles).

Scott Morton, M. S. (1991) (ed.), *The Corporation of the 1990s: Information Technology and Organizational Transformation* (Oxford: Oxford University Press).

Segal, B. (1995), 'A Short History of Internet Protocols at CERN' (CERN: PDP-NS) [http://wwwcn.cern.ch/pdp/ns/ben/TCPHIST.html], Apr.

Shannon, C. E., and Weaver, W. (1949), *The Mathematical Theory of Communication* (Urbana, Ill.: University of Illinois Press; repr. 1964).

Shaw, D. (1997), 'Internet Gold Rush Hasn't Panned Out Yet for Most', *Los Angeles Times*, 19 June, A1, A20.

Shaw, D. L., and McCombs, M. E. (1977) (eds.), *The Emergence of American Political Issues* (St Paul, Min.: West Publishing Company).

Shields, R. (1996) (ed.), *Cultures of Internet: Virtual Spaces, Real Histories, Living Bodies* (Thousand Oaks, Calif.: Sage Publications).

Shiver, J., Jr. (1997*a*), 'FCC's Digital Channel Allotments May Leave 2 Million Out of Picture', *Los Angeles Times*, 23 Apr.

—— (1997*b*), 'Broadcasters Slam Rumored Pick to Head Digital TV Panel', *Los Angeles Times*, 24 Apr.

Shogren, E. (1997), 'Gore Finds Brain Trust in Silicon Valley Group', *Los Angeles Times*, 25 Aug.

Short, J., Williams, E., and Christie, B. (1976), *The Social Psychology of Telecommunications* (London: John Wiley & Sons).

Silverstein, S. (1997), 'Telecenters in Southland Didn't Make It', *Los Angeles Times*, 5 June, D2.

Silverstone, R. (1991), *Beneath the Bottomline* (PICT Policy Research Paper, No. 17; Uxbridge, UK: Brunel University).

—— (1996), 'Future Imperfect: Information and Communication Technologies in Everyday Life', in Dutton (1996*a*), 217–31.

—— and Hirsch, E. (1992) (eds.), *Consuming Technologies: Media and Information in Domestic Spaces* (London: Routledge).

Sirbu, M. (1992), 'The Struggle for Control within the Telecommunication Networks', in Sapolsky *et al.* (1992), 140–8.

Siune, K., and Truetzschler, W. (1992) (eds.), *Dynamics of Media Politics: Broadcast and Electronic Media in Western Europe* (Newbury Park, Calif.: Sage Publications).

Skolnick, B. P. (1995), *The Uses of Social Science Research* (SPRU Working Paper; Brighton, Sussex: SPRU).

Slouka, M. (1995), *War of the Worlds: Cyberspace and the High-Tech Assault on Reality* (New York: Basic Books).

Smith, A. (1980), *Goodbye Guttenberg: The Newspaper Revolution of the 1980s* (Oxford: Oxford University Press).

Smith, M., and Kollock, P. (forthcoming) (eds.) *Communities in Cyberspace* (London: Routledge).

Smith, R. L. (1970), 'The Wired Nation', *Nation*, 18 May.

—— (1972), *The Wired Nation: Cable TV: The Electronic Communications Highway* (New York: Harper & Row).

—— with Cole, B. (1987), 'The American Way of Wiring a Nation', in Dutton *et al.* (1987*a*), 124–30.

Snoddy, R. (1997), 'Time Warner to End Interactive Service', *Financial Times*, 2 May.

Softkey International (1997), *Design Center 3-D™* (El Dorado Hills, Calif.: The Learning Company).

Sparkes, V. (1985), 'Cable Television in the United States: A Story of Continuing Growth and Change', in Negrine (1985), 15–46.

Sproull, L., and Kiesler, S. (1991), *Connections: New Ways of Working in the Networked Organization* (Cambridge, Mass.: MIT Press).

Stoll, C. (1995), *Silicon Snake Oil* (New York: Macmillan).

Streeter, T. (1996), *Selling the Air: A Critique of the Policy of Commercial Broadcasting in the United States* (Chicago: University of Chicago Press).

Tapscott, D. (1995), *The Digital Economy: Promise and Peril in the Age of Networked Intelligence* (New York: McGraw-Hill).

—— (1997), *Growing up Digital: The Rise of the Net Generation* (New York: McGraw-Hill).

Taylor, F. W. (1911), *Principles of Scientific Management* (New York: Harper).

—— (1916), 'Government Efficiency', *Bulletin of the Taylor Society*, 2: 7–13.

Taylor, J., Bellamy, C., Raab, C., Dutton, W. H., and Peltu, M. (1996), 'Innovation in Public Service Delivery', in Dutton (1996*a*): 265–82.

Taylor, P. (1997*a*), 'Survey Predicts Rise in Use of Intranets', *Financial Times*, 2 May.

—— (1997*b*), 'Superhighway Robbery: American Broadcasters v. the Public Good', *New Republic*, 5 May, 20–2.

Tehranian, M. (1996), 'The End of University?', *Information Society*, 12: 441–7.

Teich, A. H. (1981) (ed.), *Technology and Man's Future* (3rd edn., New York: St Martin's Press).

Temin, P. (1987), *The Fall of the Bell System: A Study in Prices and Politics* (Cambridge: Cambridge University Press).

Thomas, R. (1995), 'Access and Inequality', in Heap *et al.* (1995), 90–9.

Thompson, W. B. (1991) (ed.), *Controlling Technology: Contemporary Issues* (Buffalo, NY: Prometheus Books).

Times Mirror (1990), Times Mirror Center for the People & the Press, 'The American Media: Who Reads, Who Watches, Who Listens, Who Cares', unpublished report (Washington: Times Mirror Center for the People & the Press).

—— (1994), 'Technology in the American Household', unpublished report (Washington.: Times Mirror Center for the People & the Press, 24 May).

—— (1995), *Technology in the American Household: Americans Going Online . . . Explosive Growth, Uncertain Destinations* (Washington: Times Mirror Center for the People & the Press).

Toffler, A. (1970), *Future Shock* (New York: Random House).

Traber, M. (1986) (ed.), *The Myth of the Information Revolution* (London: Sage Publications).

Trade and Industry Select Committee (1994), House of Commons, *Optical Fibre Networks* (London: HMSO).

Tsagarousianou, R., Tambini, D., and Bryan, C. (1998) (eds.), *Cyberdemocracy: Technology, Cities, and Civic Networks* (London: Routledge).

Turkle, S. (1984), *The Second Self: Computers and the Human Spirit* (New York: Simon & Schuster).

Uhlig, R. P., Farber, D. J., and Bair, J. H. (1979), *The Office of the Future: Communication and Computers* (Amsterdam: North-Holland).

Uncapher, W. (1996), 'Premonitions of Digital Montana', unpublished paper presented to the Annual Meeting of the American Anthropology Association, San Francisco, California.

Uncapher, W. (forthcoming), 'Electronic Homesteading on the Rural Frontier: Big Sky Telegraph and its Community', in Smith and Kollock (forthcoming).

US Advisory Council (1996*a*): United States Advisory Council on the National Information Infrastructure, Common Grounds, *KickStart Initiative: Connecting America's Communities to the Information Superhighway* (Washington: NTIA).

—— (1996*b*), US Advisory Council on the National Information Infrastructure, Common Grounds, *A Nation of Opportunity: Realizing the Promise of the Information Highway* (Washington: US Department of Commerce, NTIA).

US Supreme Court (1997), *Reno, Attorney General of the United States, et al.* v. *American Civil Liberties Union et al.*, Appeal from the United States District Court for the Eastern District of Pennsylvania, No. 96–511, Argued 19 Mar., decided 26 June.

Valente, T. W. (1995), *Network Models of the Diffusion of Innovations* (Cresskill, NJ: Hampton Press, Inc.).

Valle, J. (1982), *The Network Revolution: Confessions of a Computer Scientist* (Berkeley, Calif.: And/Or Press, Inc.).

Van de Donk, W., Snellen, I., and Tops, P. (1995) (eds.), *Orwell in Athens: A Perspective on Information and Democracy* (Amsterdam: IOS Press).

Vedel, T., and Dutton, W. H. (1990), 'New Media Politics: Shaping Cable Television Policy in France', *Media, Culture and Society*, 12: 491–524.

Vitalari, N., and Venkatesh, A. (1987), 'In-Home Computing and Information Services: A Twenty Year Analysis of the Technology and its Impacts,' *Telecommunications Policy*, 11/1: 65–81.

Wallace, A. (1996), 'State May Launch Own Online College', *Los Angeles Times*, 10 July.

Waterman, D. (1988), 'World Television Trade', *Telecommunications Policy* (June), 141–51.

Waterson, P. E., Clegg, C. W., Bolden, R., Pepper, K., War, P. B., and Wall, T. D. (1997), 'The Use and Effectiveness of Modern Manufacturing Practices in the United Kingdom' (Sheffield, UK: ESRC Centre for Organization and Innovation, Institute of Work Psychology, University of Scheffield).

Weber, A. M. (1993), 'What's So New About the New Economy?', *Harvard Business Review* (Jan.–Feb.), 24–42.

Webster, F., and Robins, K. (1986), *Information Technology: A Luddite Analysis* (Norwood, NJ: Ablex Publishing Company).

Webster, J. (1996*a*), 'Revolution in the Office? Implications for Women's Paid Work', in Dutton (1996*a*), 143–58.

—— (1996*b*), *Shaping Women's Work: Gender, Employment, and Information Technology* (London: Longman).

Weick, K. E. (1987), 'Theorizing about Organizational Communication', in Jablin, *et al.* (1987), 97–122.

Weicklein, J. (1979), *Electronic Nightmare: The New Communications and Freedom* (New York: Viking Press).

Weinberg, A. M. (1966), 'Can Technology Replace Social Engineering?', repr. in Teich (1981), 29–39.

Weiser, M. (1991), The Computer for the 21st Century, *Scientific American*, 265 (Sept.), 66–75.

Weizenbaum, J. (1976), *Computer Power and Human Reason: From Judgement to Calculation* (San Francisco, Calif.: W. H. Freeman & Company).

Wells, D. J. (1996) (ed.), 'Virtual Enterprise', a special issue of *IEEE Engineering Management Review*, 24/2: 2–103.

Wessel, M. R. (1974), *Freedom's Edge: The Computer Threat to Society* (Reading, Mass.: Addison-Wesley).

Westin, A. F. (1997), personal correspondence to the author and Reference Point's Board of Directors, 16 Apr.

—— and Baker, M. A. (1972), *Databanks in a Free Society: Computers, Record-Keeping, and Privacy* (New York: Quadrangle).

Weston, M. (1994), 'Rights and Wrongs', *Times Higher Education Supplement*, 13 May, vii.

Whisler, T. L. (1970), *The Impact of the Computer on Organizations* (New York: Holt, Rinehart, & Winston).

Whitman, M. V. N. (1994), 'Flexible Markets, Flexible Firms', *American Enterprise* (May–June), 27–37.

Wilhelm, A. (1997), 'A Resource Model of Computer-Mediated Political Life', *Policy Studies Journal*, 25(4), 519–34.

Williams, F. (1982), *The Communications Revolution* (Beverly Hills, Calif.: Sage Publications).

—— and Pavlick, J. V. (1994) (eds.), *The People's Right to Know: Media, Democracy, and the Information Highway* (Hillsdale, NJ: Lawrence Erlbaum Assoc.).

Williams, R., and Edge, D. (1996), 'The Social Shaping of Technology', *Research Policy*, 25: 865–99.

Wilson, M., and Dutton, W. H. (1997a), 'General Telephone and Electronics and iLAN', in Dutton (1997*b*), 36–53.

—— —— (1997*b*), 'Telecenters: Learning from the Telecommuting Generation', unpublished report for Fujitsu Research Institute (Los Angeles, Calif.: Annenberg School for Communication, University of Southern California).

Winner, L. (1977), *Autonomous Technology: Technics-out-of-Control as a Theme in Political Thought* (Cambridge, Mass.: MIT Press).

—— (1986), *The Whale and the Reactor: A Search for Limits in an Age of High Technology*. (Chicago: University of Chicago Press).

Winston, B. (1989), 'The Illusion of Revolution', in Forester (1989*b*), 71–81.

Witte, E. (1980) (ed.), *Human Aspects of Telecommunication* (Berlin: Springer Verlag).

Womack, J. P., Jones, D. T., and Roos, D. (1990), *The Machine that Changed the World* (New York: Maxwell MacMillan International).

Wong, P.-K. (1997), 'Implementing the NII Vision: Singapore's Experience and Future Challenges', in Kahin and Wilson (1997), 24–60.

Woo, E. (1996), 'Task Force Backs Spending $11 Billion on Technology for Schools', *Los Angeles Times*, 11 July.

—— (1997), 'Classroom News Program Derided by a Pair of Studies', *Los Angeles Times*, 23 Jan., 1.

Woolgar, S. (1996), 'Technologies as Cultural Artefacts', in Dutton (1996*a*), 87–102.

—— (1997), 'A New Theory of Innovation?', *3M Innovation Lecture* (Uxbridge, UK: Brunel University).

Wurtzel, A. H., and Turner, C. (1977), 'Latent Functions of the Telephone: What Missing the Extension Means', in de Sola Pool (1977), 246–61.

Yates, N. (1997), 'Millions Visit Mars—on the Internet', *Los Angeles Times*, 14 July, A16.

Zuboff, S. (1988), *In the Age of the Smart Machine: The Future of Work and Power* (New York: Basic Books).

Index

Index

Index

Index

Index